168
Advances in Biochemical Engineering/Biotechnology

Aims and Scope

This book series reviews current trends in modern biotechnology and biochemical engineering. Its aim is to cover all aspects of these interdisciplinary disciplines, where knowledge, methods and expertise are required from chemistry, biochemistry, microbiology, molecular biology, chemical engineering and computer science.

Volumes are organized topically and provide a comprehensive discussion of developments in the field over the past 3–5 years. The series also discusses new discoveries and applications. Special volumes are dedicated to selected topics which focus on new biotechnological products and new processes for their synthesis and purification.

In general, volumes are edited by well-known guest editors. The series editor and publisher will, however, always be pleased to receive suggestions and supplementary information. Manuscripts are accepted in English.

In references, Advances in Biochemical Engineering/Biotechnology is abbreviated as *Adv. Biochem. Engin./Biotechnol.* and cited as a journal.

More information about this series at http://www.springer.com/series/10

Toshiyuki Itoh • Yoon-Mo Koo
Editors

Application of Ionic Liquids in Biotechnology

With contributions by

D. C. V. Belchior · C.-W. Cho · I. F. Duarte · M. G. Freire · K. Fujita · M. Goto · T. Itoh · Y.-M. Koo · S. H. Lee · N. L. Mai · M. Moniruzzaman · K. K. Oh · H. Ohno · S. Park · T. P. T. Pham · R. D. Rogers · J. L. Shamshina · M.-H. Song · T. Usuki · Z. Yang · M. Yoshizawa-Fujita · Y.-S. Yun · O. Zavgorodnya

Springer

Editors
Toshiyuki Itoh
Center for Research on Green Sustainable Chemistry, Graduate School of Engineering
Tottori University
Tottori, Japan

Yoon-Mo Koo
Department of Biological Engineering, College of Engineering
Inha University
Incheon, Korea (Republic of)

ISSN 0724-6145 ISSN 1616-8542 (electronic)
Advances in Biochemical Engineering/Biotechnology
ISBN 978-3-030-23080-7 ISBN 978-3-030-23081-4 (eBook)
https://doi.org/10.1007/978-3-030-23081-4

This Springer imprint is published by the registered company Springer Nature Switzerland AG.
The registered company address is: Gewerbestrasse 11, 6330 Cham, Switzerland

Preface

A Short Glance into the Past

Ionic liquids (ILs) are the less volatile, less flammable, less toxic so-called "room temperature molten salts" encompassing a unique solubility for organic and inorganic materials. The history of research on ILs is very interesting. The first room temperature molten salt, ethylammonium nitrate, was reported in 1904, more than 100 years ago! [1]. However, no attention was given toward similar salts for the next 75 years. In 1979, Robinson et al. investigated the electrochemical properties and spectroscopic behavior of a number of aromatic hydrocarbons in the low-temperature molten salt system of $AlCl_3$-*n*-butylpyridinium chloride [2]. In 1986, Wilkes and his coworkers demonstrated the first organic reaction, the Friedel–Crafts reaction, using the same salt as both the solvent and the catalyst [3]. Since these ILs were moisture-sensitive compounds, these reports were unable to trigger any further IL chemistry. In 1992, the air and moisture stable IL, 1-ethyl-3-methylimidazolium tetrafluoroborate ([Bmim][BF_4]), was synthesized by Wilkes [4], and Dupont prepared the moisture stable hydrophobic IL 1-butyl-3-methylimidazolium hexafluorophosphate ([Bmim][PF_6]) in 1996 [5]. This was followed by a milestone paper presented by Seddon and coworkers in 1999 [6]; the authors demonstrated the Heck reaction using an IL and clearly indicated the merit of ILs as solvents for organic reactions. In the same year, Welton presented the first review of ILs [7] wherein he defined the compounds that should be categorized as ILs. These papers triggered the eruption of IL chemistry even in the field of biotechnology. The first three examples of biotransformation in ILs were reported in 2000 [8–10]; since then, ILs have been implicated in numerous applications in the field of bioscience and biotechnology, and the number of papers has reached more than 7,000 in the last two decades [Number of publications in the field of ILs from 1975–2018 (searched by "Web of Science", on March 27, 2018). 84,796 references are listed for the topic of "ionic liquid*" in 99 fields of the database of Web of Science Core Collections of THOMSON REUTERS, Ltd.]. The science of ILs is considered as one of the highlights in modern chemistry, with 86,000 papers published by this year due to the wide application of ILs in various fields [Number of publications in the field of ILs from 1975–2018 (searched by "Web of Science", on March 27, 2018). 84,796 references are listed for the topic of "ionic liquid∗" in 99 fields of the database of Web of Science Core Collections of THOMSON REUTERS, Ltd.].

Today and Future

The usages of ILs in biotechnology are not limited to their application as solvents/cosolvents for enzyme or whole-cell catalyzed reactions but have extended to various fields such as solvents/agents for extraction, isolation, and processing of natural originated compounds (e.g., biopolymers, active pharmaceutical ingredients, protein). In addition, to overcome the concerns regarding the environment and toxicity of using IL in biological processes, many studies investigated the toxicity mechanism and focused on the development of a new generation of ILs derived totally from nature, which are inexpensive, biocompatible, and biodegradable with low toxicity. These new ILs have the potential as alternatives for fossil originated ILs in many biological applications as well as being used as the active pharmaceutical ingredients in biomedical applications [11]. Based on the expectations that ILs may produce a breakthrough in the fields of biotechnology, this book presents the current state of the art in environmentally benign biotechnology using IL engineering, such as application in the reaction media for biotransformation, stabilization and activation of biocatalysts, bioseparation processes, pretreatment of biomass for enzymatic reactions, synthesis of biopolymer-based hydrogels, isolation of medicinal compounds from bioresources, environmental concerns of ILs in biotechnical applications, and the development of ILs and their derivatives from nature. Compared to conventional solvents, the unique properties of ILs have unlocked a wide range for their applications in biotechnology. This book contains a number of trailblazing reviews that describe the power of ILs in biotechnology and discuss the fields where investigations need to be focused for utilizing ILs in biotechnology. We believe this book will aid you in acquiring ideas for future research in the field of biotechnology.

Tottori, Japan — Toshiyuki Itoh
Incheon, Korea (Republic of) — Yoon-Mo Koo

References

1. Walden P (1914) Ueber die Molekulargrösse und elektriseche Leitfähigkeit eunger geschmoltezenen Salze. Bull Acad Imper Sci (St. Petersburg), 405–422
2. Robinson J, Osteryoung RA (1979) An electrochemical and spectroscopic study of some aromatic hydrocarbons in the room temperature molten salt system aluminum chloride-n-butylpyridinium chloride. J Am Chem Soc 101:323–327

3. Boon JB, Levisky JA, Pflug L, Wilkes JS (1986) Friedel–Crafts reactions in ambient-temperature molten salts. J Org Chem 51:480–483
4. Wilkes JS, Zaworotko M (1992) Air and water stable 1-ethyl-3-methylimidazolium based ionic liquids. J Chem Soc Chem Commun, 965–966
5. Suarez PAZ, Dullius JEL, Sandra E, De Souza RF, Dupont J (1996) The use of new ionic liquids in two-phase catalytic hydrogenation reaction by rhodium complexes. Polyhedron 15:1217–1219
6. Carmichael AJ, Earle MJ, Holbrey JD, McCormac PB, Seddon KR (1999) The Heck reaction in ionic liquids: a multiphasic catalyst system. Org Lett 1:997–1000
7. Welton T (1999) Room-temperature ionic liquids. Solvents for synthesis and catalysis. Chem Rev 99:2071–2084
8. Cull SG, Holbre JD, Vargas-Mora V, Seddon KR, Lye GJ (2000) Room-temperature ionic liquids as replacements for organic solvents in multiphase bioprocess operations. Biotechnol Bioeng 69:227–237
9. Erbeldinger M, Mesiano AJ, Russell AJ (2000) Enzymatic catalysis of formation of Z-aspartame in ionic liquid – an alternative to enzymatic catalysis in organic solvents. Biotechnol Prog 16:1129–1131
10. Lau RM, van Rantwijk F, Seddon KR, Sheldon RA (2000) Lipase-catalyzed reactions in ionic liquids. Org Lett 2:4189–4191
11. Egorova KS, Gordeev EG, Ananikov VP (2017) Biological activity of ionic liquids and their application in pharmaceutics and medicine. Chem Rev 117:7132–7189

Contents

Adv Biochem Eng Biotechnol (2019) 168: 1–30
DOI: 10.1007/10_2018_66

Published online: 10 May 2018

Ionic Liquids in Bioseparation Processes

Diana C. V. Belchior, Iola F. Duarte, and Mara G. Freire

Contents

Abstract Bioseparation processes are a relevant part of modern biotechnology, particularly regarding the development of efficient and biocompatible methods for the separation and purification of added-value biologically active compounds. In this field, ionic liquids (ILs) have been proposed, either in liquid–liquid extractions, in which non-water miscible ILs or aqueous biphasic systems (ABS) formed by ILs can be used, or in solid–liquid extractions, in which they are covalently attached to create supported IL phases (SILPs). Aprotic ILs possess unique properties, such as non-volatility and designability, which are valuable in their use in bioseparation processes. In this chapter, we summarize and discuss bioseparation processes based on ILs, including both liquid–liquid and solid–liquid extractions, applied to amino acids and proteins. The most recent and remarkable advances in this area are emphasized, and improvements brought by the use of ILs properly discussed. New insights and envisaged directions with IL-based bioseparation processes are suggested.

D. C. V. Belchior, I. F. Duarte, and M. G. Freire (✉)
CICECO-Aveiro Institute of Materials, Chemistry Department, University of Aveiro, Aveiro, Portugal
e-mail: maragfreire@ua.pt

Graphical Abstract

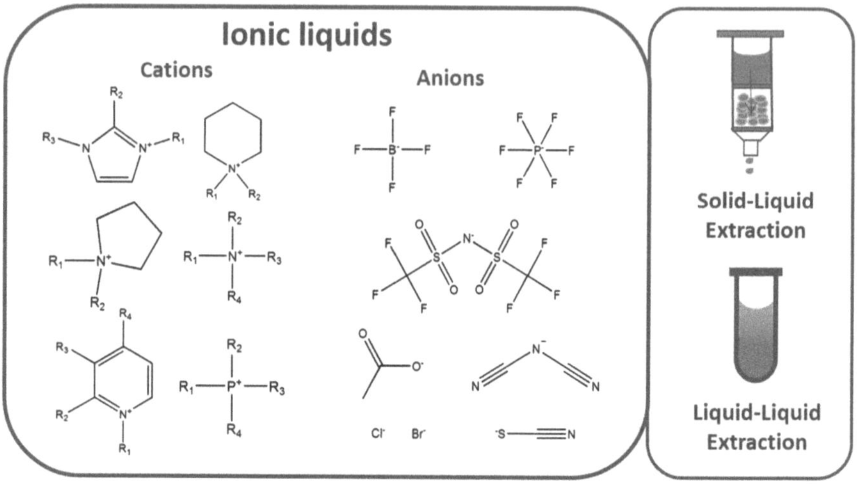

Keywords Amino acids, Ionic liquid, Liquid–liquid extraction, Proteins, Solid–liquid extraction

1 Introduction

In bioengineering and biotechnological processes, the high cost associated with downstream processing, aimed at the purification and recovery of target products, is one of the major issues limiting the widespread use of many bio-based products. Bioseparations are an important part of modern biotechnology, particularly regarding the development of new and biocompatible methods for the separation and purification of added-value amino acids, proteins, and enzymes. Biocompatibility is, however, a crucial feature in the design of these separation platforms, mainly when dealing with biologically active products [1].

Separation methods exploit the differences in physicochemical properties of target compounds and contaminants present in the original media [2]. Separation and purification stages in biotechnological processes usually require numerous steps, associated with high energy and chemicals consumption, and typically represent 20–60% of the final cost of the product; this value may even reach 90% for particular products [1]. In addition to the well-established chromatographic approaches, the separation of bio-based products with liquid–liquid systems is typically carried out using volatile organic solvents because of their immiscibility with aqueous media [3]. However, the most commonly used organic solvents (e.g., toluene, ethyl acetate, hexane) [3] present some disadvantages, such as high volatility and toxicity, and the possibility of denaturing the proteins and enzymes to be recovered [1]. Aiming at replacing the use of volatile organic solvents as extraction media, non-water soluble ionic liquids (ILs) and IL-based aqueous biphasic systems (ABS) have been investigated [4]. ILs belong to the molten salts group; they are composed of inorganic or organic anions and relatively large and asymmetric organic cations, which do not easily form an ordered crystal and therefore may be liquid at or near room temperature [5]. In addition to their negligible vapor pressure, non-flammability, and high chemical stability [5], ILs present excellent solvation qualities [5] and have thus proved to be effective in liquid–liquid separation methods [4]. In addition to these features, which are valuable in the development of sustainable liquid–liquid separation processes, ILs are considered as designer solvents because of the large number of cation–anion combinations [5]. Therefore, specific ILs can be designed to act as enhanced extraction solvents for a wide variety of bio-based compounds.

Within the field of liquid–liquid separation techniques, ABS have been reported as benign alternatives because of the possibility of combining two non-volatile components in water-rich media [6]. ABS consist of two aqueous-rich phases that coexist in equilibrium after dissolution, at given concentrations, of pairs of appropriate solutes, such as two polymers or a polymer and a salt [6]. The aqueous two-phase extraction of biomolecules was first proposed by Albertsson [6] in the 1950s, and has been widely studied over the past decades. This technique has attracted attention because it is able to combine several processing stages, such as clarification, concentration, and primary purification, in one step [6]. ABS are considered a gentler and more biocompatible alternatives than other extraction techniques as they are mainly composed of water. Moreover, ABS are amenable to be scaled-up, and allow continuous operation and process integration. Iqbal et al. [7] recently reviewed the potential and recent advances of ABS as separation platforms.

ABS have been investigated for the separation of a wide range of biologically active substances, such as amino acids, proteins, including enzymes and antibodies, and nucleic acids [8]. Most of the polymers used as phase-forming components of traditional ABS have a stabilizing effect over the proteins tertiary structure [9], thus contributing to the enhanced performance of these systems. More traditional ABS are composed of two polymers or a polymer and a salt [6]. However, this type of systems is limited in terms of the range of polarities offered by the coexisting phases, thus making selective and enhanced separations difficult to achieve in a single step [10]. To overcome this drawback, several approaches have been proposed, for example, the functionalization of polymers [11, 12], the use of additives [13], designing hybrid processes [14], among others. In 2003, ILs were proposed as alternative phase-forming components of ABS, being able to replace conventional polymers and to form ABS combined with inorganic salts [15]. Their large selection of chemical structures allows the design of ILs for specific tasks or applications, which can be extended to ABS. In addition to the possibility of forming ABS by combining ILs with salts, polymers, amino acids, and carbohydrates [16], it has been demonstrated that their separation performance can be tuned via the IL chemical structure [10]. The number of estimated ILs is around one million, suggesting that almost any desirable property may be obtained with a particular IL [5]. Even so, most reported IL-based ABS are formed by imidazolium-based fluids, usually combined with the non-water stable $[BF_4]^-$ anion [4]. This selection has however started to be used less, and some recent works have been focused on the use of more attractive ILs to create ABS [4].

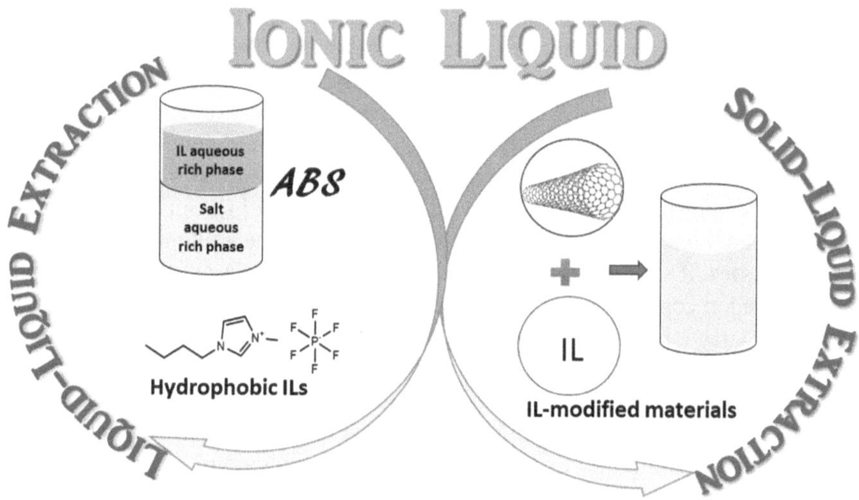

Fig. 1 Overview of the use of ILs in liquid–liquid extractions, employing hydrophobic ILs and ABS, and solid–liquid extractions based on IL-modified materials

Still taking advantage of the ILs unique properties, supported ionic liquid phase (SILP) materials have been proposed as a more recent concept in separation processes, where ILs are immobilized (covalently attached) in a solid phase, which could be used as new chromatographic columns or in solid-phase extraction (SPE) approaches [17]. Materials such as silica, polymers, magnetic nanoparticles, and carbon nanotubes have been successfully modified by ILs [17]. Overall, ILs have been reported as enhanced extraction solvents, both neat and in aqueous solutions, being able to increase the stability of added-value biomolecules, such as proteins, enzymes, and nucleic acids. SILPs also have been successfully used in the separation of amino acids and proteins. Based on the high performance of ILs in bioseparation processes, we present and discuss herein recent advances achieved by the use of ILs in liquid–liquid extractions, employing hydrophobic ILs and ABS, and solid–liquid extractions based on the use of IL-modified materials for the separation of amino acids and proteins, including enzymes and biopharmaceuticals (Fig. 1).

2 ILs in Liquid–Liquid Bioseparation Processes

At the beginning of the twenty-first century, with the appearance of air and water-stable ILs, and more recently with the synthesis and characterization of bio-based ILs, research on novel ILs and their potential applications increased significantly. The major reasons behind this trend are the ILs unique properties, which can be tailored, including their ability to solvate a large array of compounds, coupled with the need to find "greener" solvents to be used in separation processes. The research carried out to date in liquid–liquid extractions from aqueous media using water-immiscible ILs and water-soluble ILs by the creation of ABS for the separation of amino acids and proteins is reviewed and discussed below.

2.1 Water-Immiscible ILs in Liquid–Liquid Extractions

Table 1 lists the ILs investigated and discussed in this section, including their description, abbreviations, and chemical structures. IL-based liquid–liquid approaches employing hydrophobic ILs to separate amino acids were pioneered by Carda-Broch et al. [18] using a crown ether (dibenzo-18-crown-6)-modified IL, followed by a report on a similar modified IL [19], namely dicyclohexano-18-crown-6-based, to extract the amino acids tryptophan, glycine, alanine, leucine, arginine, valine, and lysine. Both groups of researchers showed that the partitioning behavior of amino acids is pH-dependent, and that ILs enable partition coefficients two orders of magnitude higher than those achieved without the respective crown ether modifications [18, 19].

In addition to the functionalized ILs described above, more conventional imidazolium-based ILs ($[C_4C_1im][PF_6]$, $[C_6C_1im][PF_6]$, $[C_6C_1im][BF_4]$, and $[C_8C_1im][BF_4]$) were also tested for the recovery of tryptophan, phenylalanine, tyrosine, leucine, valine, lysine, alanine, and glutamic acid [20–22]. The authors [20] proposed hydrophobic interactions as the driving forces responsible for the enrichment of amino acids in the IL-phase, supported by a demonstrated correlation between the logarithm function of the partition coefficient and the hydrophobicity of the studied amino acids. In these works a pH-dependent partitioning behavior was shown, where higher extraction efficiencies to the IL-rich phase are obtained at lower pH values [20, 21]. These results therefore confirm the occurrence of electrostatic interactions between the cationic form of the amino acids and the IL anions in addition to the proposed dispersive-type interactions. Besides the well-studied imidazolium-based ILs, Tomé et al. [22] investigated the partition coefficients of L-tryptophan between water and different ILs, including pyrrolidinium-, piperidinium-, and pyridinium-based. The authors demonstrated that pyrrolidinium-based ILs are the better extraction phases amongst the ILs investigated [22]. These authors also demonstrated that the pH of the aqueous phase strongly influences the success of the amino acids separation [22].

Huaxi et al. [23] synthesized a new hydrophobic amide-based functionalized IL, $[(CH_2CONHC_4H_9)C_2im][NTf_2]$, and evaluated its performance in extracting amino acids. Again, the partition coefficients were shown to be dependent on the medium pH, and the new functionalized IL allowed a higher partition coefficient and selectivity for L-tryptophan than more traditional ILs. Although contradicting in part the previous findings on the relevance of dispersive-type and electrostatic interactions [20–22], the IL-enhanced performance was explained by the favorable hydrogen bonding interactions established between the acetyl group of the IL and the amino acid NH_2 group. All of the described works thus suggest that there are a large

Table 1 Description, abbreviation, and chemical structure of non-water soluble ILs used in liquid–liquid extractions

Description	Abbreviation	Chemical structure
1-Alkyl-3-methylimidazolium hexafluorophosphate	$[C_nC_1im][PF_6]$ ($n = 4, 6, 8$)	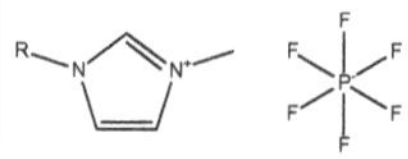
1-Alkyl-3-methylimidazolium tetrafluoroborate	$[C_nC_1im][BF_4]$ ($n = 6, 8$)	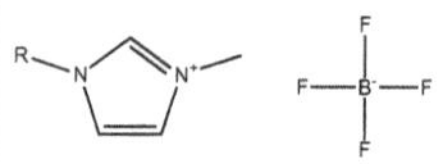
1-Butyl-3-trimethylsilylimidazolium hexafluorophosphate	$[C_4(C_1C_1C_1Si)im]$ $[PF_6]$	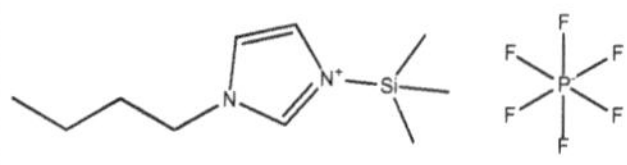
1-Ethyl-3-methylimidazolium bis (trifluoromethylsulfonyl)imide	$[C_2C_1im][Tf_2N]$	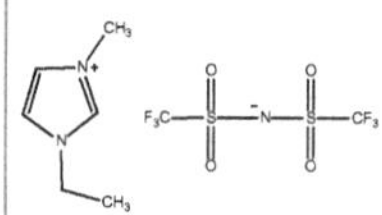
1-Butyl-2,3-dimethylimidazolium bis(trifluoromethylsulfonyl)imide	$[C_4C_1C_1im][Tf_2N]$	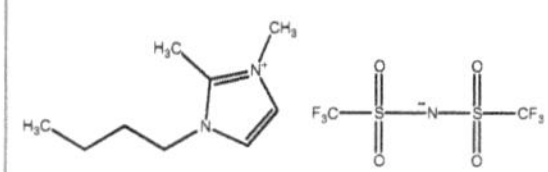
1,3-Diethylimidazolium bis (trifluoromethylsulfonyl)imide	$[C_2C_2im][Tf_2N]$	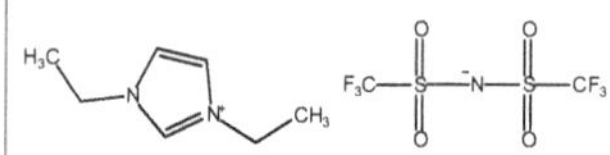
1-Propyl-3-methylpiperidinium bis (trifluoromethylsulfonyl)imide	$[C_3C_1pip][Tf_2N]$	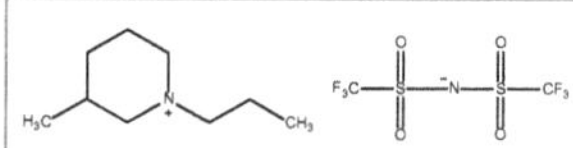
1-Butyl-3-methylpyrrolidinium bis (trifluoromethylsulfonyl)imide	$[C_4C_1pyrr][Tf_2N]$	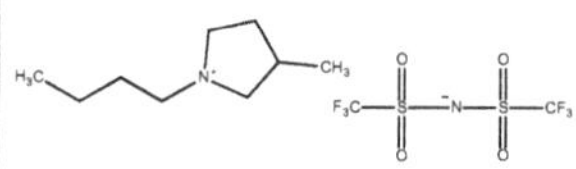
3-(2-(Butylamino)-2-oxoethyl)-1-ethylimidazolium bis (trifluoromethylsulfonyl)imide	$[(CH_2CONHC_4H_9)$ $C_2im][NTf_2]$	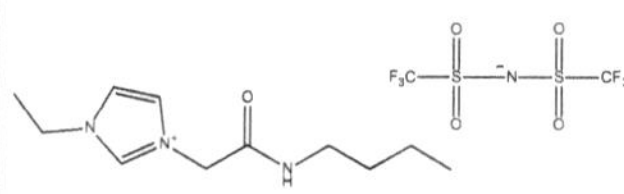
1-Ethyl-3-methylimidazolium prolinate	$[C_2C_1im][Pro]$ ($n = 2, 6, 8$)	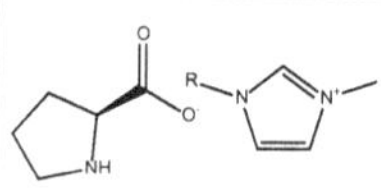
Tetrabutylphosphonium *N*-trifluoromethanesulfonylleucine	$[P_{4444}][Tf\text{-}Leu]$	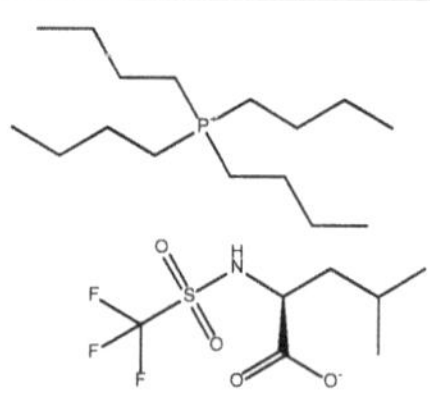

number of interactions taking place, and that all are important for the separation of amino acids. Tang et al. [24] further employed functional hydrophobic amino acid-based ILs as both solvents and selectors for the extraction of racemic mixtures of amino acids, being able to recover the L-enantiomer at the IL-rich phase. Most of the works described before addressed the pH effect on the amino acids partition coefficients, demonstrating that these biomolecules are better extracted to the IL-rich phase at acidic conditions. This pH dependence was additionally found to be useful to recover amino acids from the IL-rich phase [21, 23], which can be considered a remarkable advantage in the recovery of target products from the non-volatile IL solvents studied.

In addition to amino acids, hydrophobic IL-water systems have been investigated for the separation of proteins. It should, however, be noted that only a limited number of works on proteins extraction has been reported, which is probably because of the labile nature of proteins and the requirement for a water-rich medium. Shimojo et al. [25] were the first to investigate the extraction of heme proteins using hydrophobic ILs from an aqueous phase through the addition of dicyclohexano-18-crown-6 (DCH18C6), demonstrating a high peroxidase activity of the cytocrome-c-DCH18C6 complex in the IL-phase. Tzeng et al. [26] used an imidazolium-based IL ($[C_4C_1im]Cl$) merged with a dye [silver salt Cibacron Blue 3GA (CB)] for the extraction of lysozyme, cytochrome-c, ovalbumin, and bovine serum albumin (BSA). As observed with amino acids, the authors demonstrated that the proteins partitioning behavior is pH-dependent, where an increase in pH leads to lower extraction efficiencies [26]. Accordingly, the authors discussed the relevance of electrostatic interactions between the solvent and the positive surface charge of lysozyme [26]. Finally, the recyclability of the ILs used was addressed, showing that the IL phase could be reused at least eight times without losses on the system extraction performance [26].

Kohno et al. [27] studied an IL with an anion derived from leucine, $[P_{4444}]$ [Tf-Leu]. Although hydrophobic, this IL still dissolves a high water content and, with 21 wt% of water, cytochrome-c was completely extracted into the IL-rich phase. Other proteins (lysozyme, myoglobin, chymotrypsin, laccase, hemoglobin, horseradish peroxidase, and BSA) were also investigated, leading to the conclusion that this IL is not capable of extracting laccase and horseradish peroxidase. Ohno and co-workers [28] investigated the cytochrome-c extraction using a phosphonium-type zwitterion, attempting to control the water content at the zwitterion phase. By combining different zwitterions, the authors [28] were able to increase the water content from 0.4% to 62.7%, helping to increase the dissolution and partition of the target protein. Finally, the authors demonstrated back-extraction of the protein by the addition of an inorganic salt, without significant changes in the structure of the recovered protein [28]. Two works regarding the extraction of recombinant proteins using a triazacyclononane-IL phase were recently published [29, 30]. The authors demonstrated that the selective partitioning of hexahistidine-tagged (His-tagged) proteins between the IL and aqueous phases is governed by the proteins' affinity to the IL, the presence and nature of coordinated metal ions, and the medium ionic strength.

In addition to the studies carried out with pure and model proteins discussed above, Cheng et al. [31] reported the direct extraction of hemoglobin from human whole blood using $[C_4(C_1C_1C_1Si)im][PF_6]$. The same non-water miscible IL was used in the extraction of cytochrome-c from an aqueous solution [32]. In both works [31, 32], a pH-dependent back-extraction process to recover proteins and the IL was developed. In the same line, Hu et al. [33] addressed the separation of bacteriorhodopsin from *Halobacterium salinarium* using the IL $[C_6C_1im][PF_6]$, where it was shown that this IL displays a strong ability to remove the main contaminants (lipids, proteins, and sugars). Imidazolium-based hydrophobic ILs ($[C_4C_1im][NTf_2]$ and $[C_4C_1im][PF_6]$) were also used to separate lactoferrin from bovine whey [34]. It was found that the IL extraction efficiency is enhanced at neutral pH, low ionic strength, and low concentration. Even so, the maximum extraction efficiencies reported to the IL-rich were around 20%. Hydrophobic ILs therefore seem more attractive to tailor the extractions selectivity than to obtain enhanced extraction efficiencies.

Overall, the aforementioned works show that, under appropriate conditions and using adequate IL chemical structures, hydrophobic ILs are enhanced extraction media for recovering amino acids and proteins from aqueous solutions. The range of hydrophobic ILs is, however, more limited than hydrophilic (water-soluble) ILs typically used in the formation of IL-based ABS, discussed in the next section, which may impose some constraints in the design of effective separation processes. Furthermore, most of the studies described above focused on ILs formed by imidazolium-based cations, mainly combined with fluorinated anions, thus raising concerns in terms of the ILs biocompatibility. When dealing with proteins, water content seems to be a relevant issue, which may also justify the scarce number of works dealing with hydrophobic ILs to separate proteins.

2.2 *Water-Soluble ILs in Liquid–Liquid Extractions*

The limited number of hydrophobic ILs when compared to hydrophilic (water-soluble) ILs and its impact in terms of chemical structure design, coupled with the low water content in the first set, are the main reasons for the significantly larger number of studies found with IL-based ABS for the separation of amino acids and proteins. In fact, a much greater number of water-miscible ILs can be considered for the formation of ABS, and their environmental and toxicity impact is also reduced [16]. The fundamentals and applications of IL-based ABS were recently reviewed [4]. In this section we only focus on their use for the separation of amino acids and proteins, highlighting some recent and innovative works. The descriptions, abbreviations, and chemical structures of the ILs used and discussed in this section are given in Table 2.

The extraction of a wide range of amino acids (Fig. 2) has been investigated using IL-based ABS, and the pioneering works in this field were demonstrated in 2009 [35, 36]. ABS formed by $[C_nC_1im][C_1CO_2]$ ($n = 4, 6, 8$) or $[C_nC_1im]Br$ ($n = 4, 6, 8$)

Table 2 Description, abbreviation, and chemical structure of the ILs used in the formation of ABS for liquid–liquid extractions

Description	Abbreviation	Chemical structure
1-Alkyl-3-methylimidazolium bromide	$[C_nC_1im]Br$ ($n = 2, 4, 6, 8$)	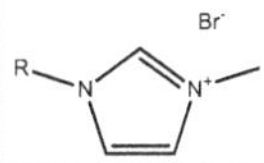
1-Alkylimidazolium chloride	$[C_nim]Cl$ ($n = 1, 2$)	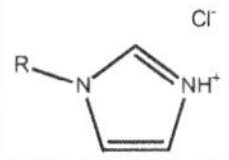
1-Alkyl-3-methylimidazolium chloride	$[C_nC_1im]Cl$ ($n = 2, 4, 6, 8$)	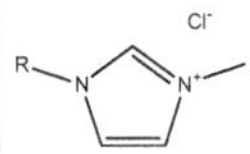
1-Butyl-2,3-dimethylimidazolium chloride	$[C_4C_1C_1im]Cl$	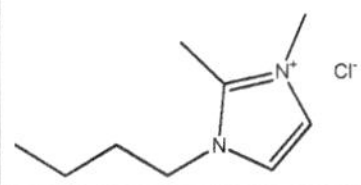
1-Allyl-3-methylimidazolium chloride	[amim]Cl	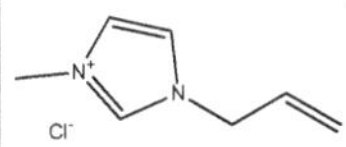
1-Hydroxyethyl-3-methylimidazolium chloride	$[OHC_2mim]Cl$	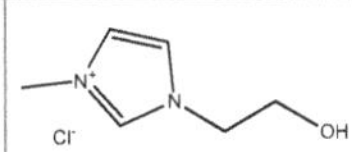
1-Benzyl-3-methylimidazolium chloride	$[C_7H_7C_1im]Cl$	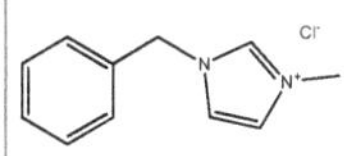
1-Alkyl-3-methylimidazolium acetate	$[C_nC_1im][CH_3CO_2]$ ($n = 2, 4, 6$)	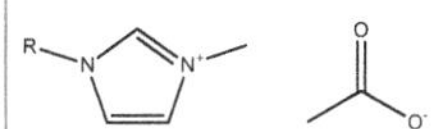
1-Ethyl-3-methylimidazolium alkylsulfate	$[C_2C_1im][C_nSO_4]$ ($n = 1, 2$)	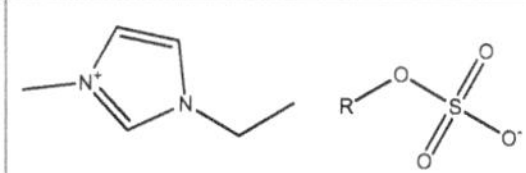
1-Alkyl-3-methylimidazolium trifluoromethanesulfonate	$[C_nC_1im][CF_3SO_3]$ ($n = 2, 4, 6$)	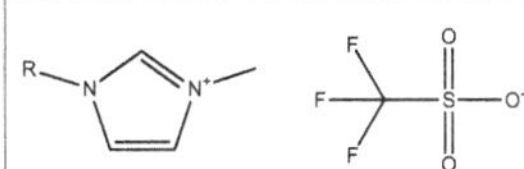
1-Alkyl-3-methylimidazolium methanesulfonate	$[C_nC_1im][CH_3SO_3]$ ($n = 2, 4$)	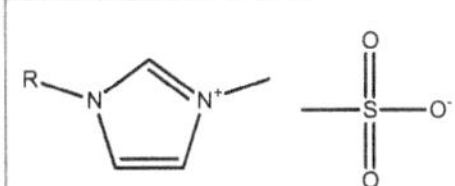
1-Butyl-3-methylimidazolium hydrogensulfate	$[C_4C_1im][HSO_4]$	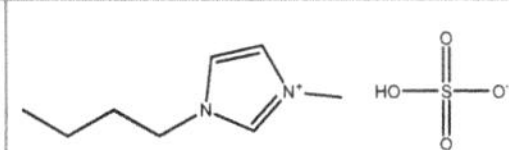
Tetraalkylammonium chloride	$[N_{nnnn}]Cl$ ($n = 2, 4$)	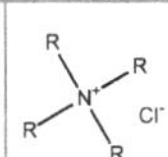
Tetraalkylammonium bromide	$[N_{nnnn}]Br$ ($n = 2, 4$)	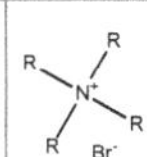

(continued)

Table 2 (continued)

Description	Abbreviation	Chemical structure
Tetrabutylphosphonium chloride	$[P_{4444}]Cl$	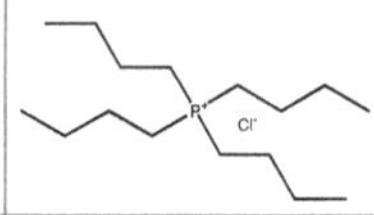
Tetrabutylphosphonium bromide	$[P_{4444}]Br$	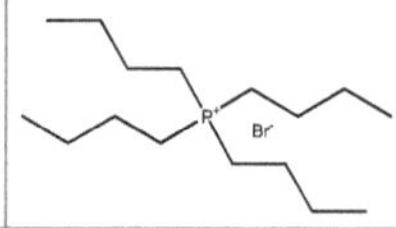
Triisobutyl(methyl)phosphonium tosylate	$[P_{i(444)4}]$ [Tos]	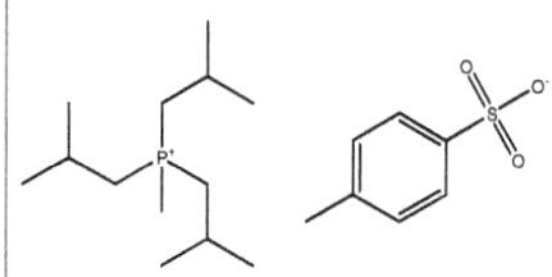
Tributyl(methyl)phosphonium methylsulfate	$[P_{4441}]$ $[MeSO_4]$	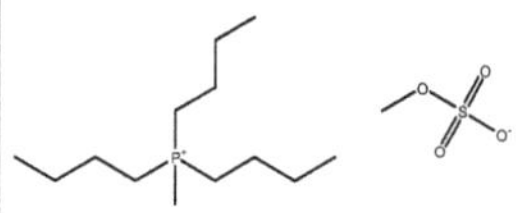
1-Butyl-3-methylimidazolium tetrafluoroborate	$[C_4C_1im]$ $[BF_4]$	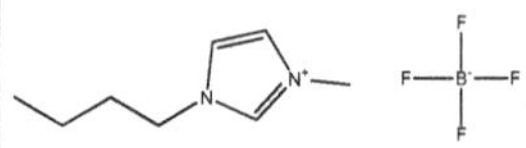
1-Butyl-3-methylimidazolium thiocyanate	$[C_4C_1im]$ [SCN]	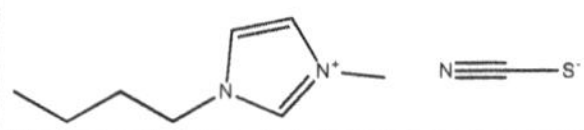
1-Butyl-3-methylimidazolium dicyanamide	$[C_4C_1im][N(CN)_2]$	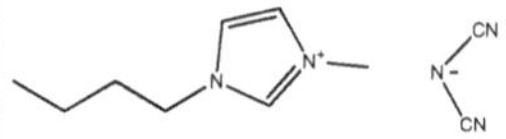
1-Butyl-3-methylimidazolium tosylate	$[C_4C_1im]$ [TOS]	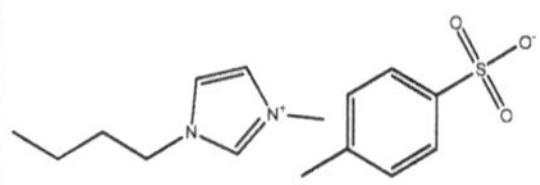
1-Ethyl-1-methylpyrrolidinium trifluoromethanesulfonate	$[C_2C_1pyr]$ $[CF_3SO_3]$	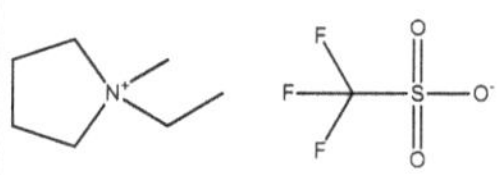
1-Ethyl-3-methylimidazolium butylsulfate	$[C_2C_1im]$ $[C_4SO_4]$	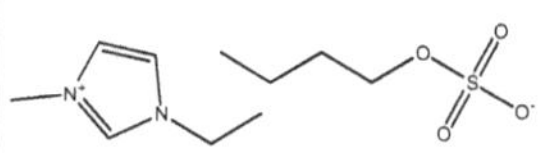
1-Ethyl-3-methylimidazolium dimethylphosphate	$[C_2C_1im]$ [DMP]	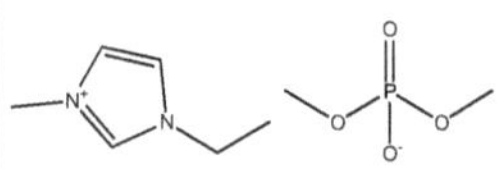
Tri(butyl)methylphosphonium methylsulfate	$[P_{4441}]$ $[MeSO_4]$	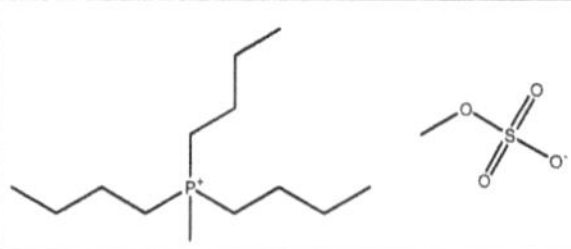
N,N-Dimethylethanolamine alkanoate	$[N_{011(2OH)}]$ $[C_nCO_2]$ ($n = 3, 4, 5, 6$)	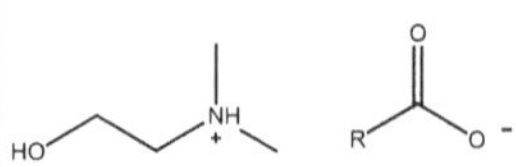

(continued)

Table 2 (continued)

Description	Abbreviation	Chemical structure
N,*N*-Diethylethylenediamine alkanoate	$[N_{022(2OH)}]$ $[C_nCO_2]$ (n = 3, 4, 5, 6)	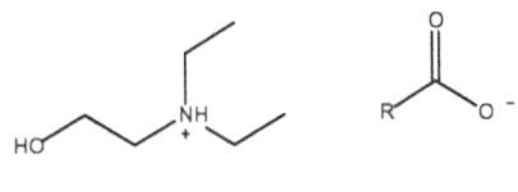
Tetraalkylammonium glycine	$[N_{nnnn}]$[Gly] (n = 1, 2, 4, 5)	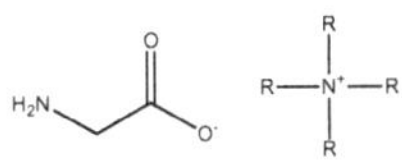
Tetrabutylphosphonium glycine	$[P_{4444}]$[Gly]	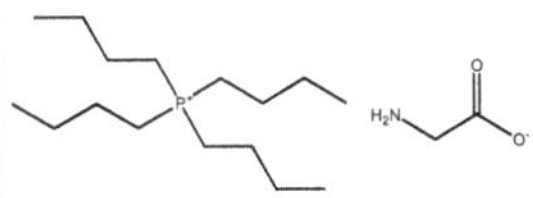
Cholinium 2-[4-(2-hydroxyethyl) piperazin-1-yl]ethane sulfonate	$[N_{111(2OH)}]$ [HEPES]	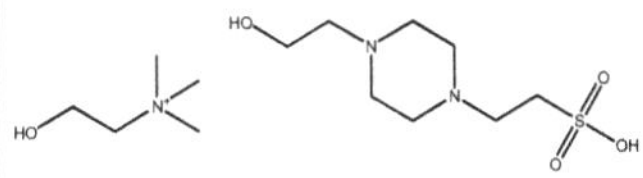
Cholinium *N*-[tris(hydroxymethyl) methyl]glycine	$[N_{111(2OH)}]$ [Tricine]	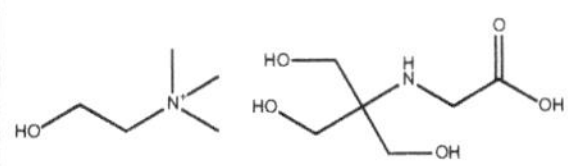
Cholinium 2-(cyclohexylamino) ethanesulfonate	$[N_{111(2OH)}]$ [CHES]	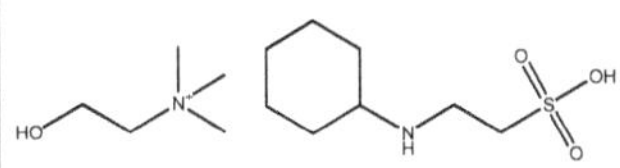
Cholinium 2-(*N*-morpholino) ethanesulfonate	$[N_{111(2OH)}]$ [MES]	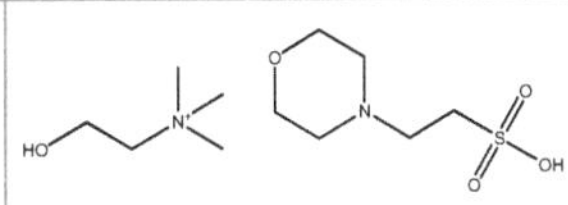
Cholinium 2-[(2-hydroxy-1,1-bis (hydroxymethyl) ethyl)amino] ethanesulfonate	$[N_{111(2OH)}]$ [TES]	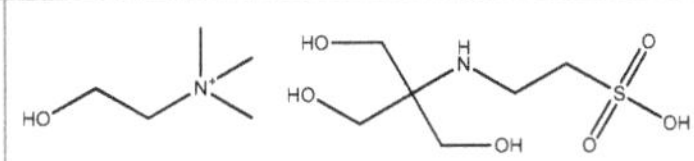
Cholinium chloride	$[N_{111(2OH)}]$ Cl	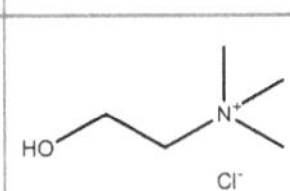
Cholinium alkanoate	$N_{111(2OH)}]$ $[C_nCO_2]$ (n = 2, 3, 4)	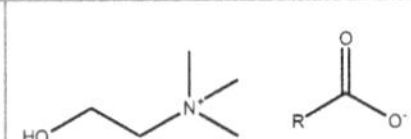
Cholinium bitartrate	$[N_{111(2OH)}]$ [Bit]	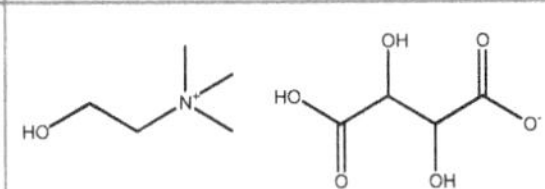
Cholinium dihydrogen citrate	$[N_{111(2OH)}]$ [DHCit]	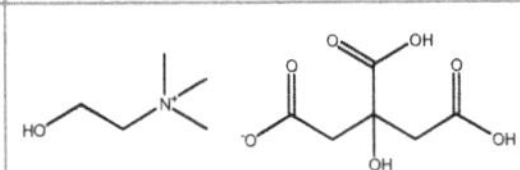
Cholinium dihydrogen phosphate	$[N_{111(2OH)}]$ [DHP]	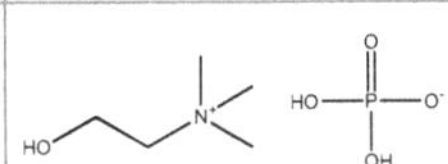

(continued)

Table 2 (continued)

Description	Abbreviation	Chemical structure
Cholinium glycolate	$[N_{111(2OH)}]$ [Gly]	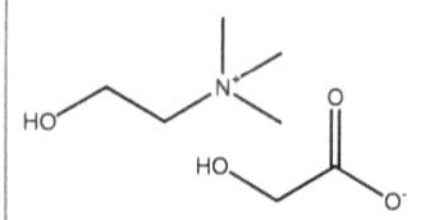
Cholinium lactate	$[N_{111(2OH)}]$ [Lac]	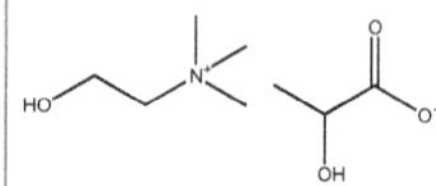
Cholinium benzoate	$[N_{111(2OH)}]$ [BE]	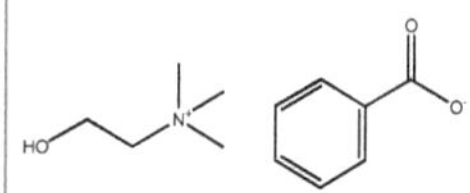
Di-cholinium oxalate	$[N_{111(2OH)}]_2$[Ox]	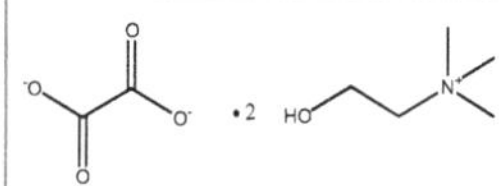
Tri-cholinium citrate	$[N_{111(2OH)}]_3$[cit]	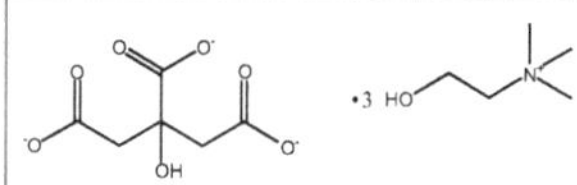

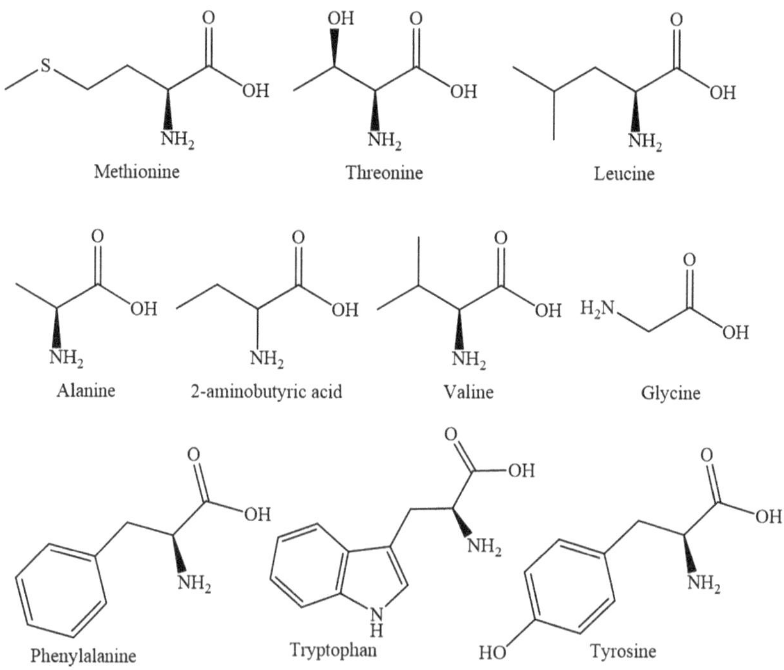

Fig. 2 Chemical structures of amino acids separated with IL-based ABS

and inorganic salts (K_3PO_4, K_2HPO_4, and K_2CO_3) were further explored for the separation of tryptophan, glycine, alanine, 2-aminobutyric acid, valine, leucine, threonine, methionine, and tyrosine [37, 38]. In these works [37, 38] it was found that ILs with more hydrophobic cations lead to the best extraction results, which contradicts the results of Louros et al. [39] with ABS formed by phosphonium-based ILs. Additional investigations on the separation of amino acids (L-tryptophan, L-phenylalanine, L-tyrosine, L-leucine, and L-valine) by IL-based ABS were carried out by Zafarani-Moattar and Hamzehzadeh [40]. The authors concluded that hydrophobic interactions are the main forces responsible for the amino acids' preferential migration to the IL-rich phase, although other parameters, such as amino acid size, accessible surface area, and polarizability, were also described as relevant. ABS formed by several types of ILs, namely imidazolium-, pyrrolidinium-, phosphonium-, and ammonium-based ILs, for the separation of L-tryptophan, were then investigated [41]. A high extraction performance was displayed by these systems, with extraction efficiencies of 72–99% [41]. In contrast to the trend observed with hydrophobic ILs, with IL-based ABS lower partition coefficients were observed at lower pH values. This trend is, however, predictable because in the studied ABS a citrate-based salt is used, which at lower pH values is protonated and thus acts as a weaker salting-out species. Furthermore, two imidazolium-based chiral ILs were synthesized to separate racemic amino acids [42]. Results showed that D-enantiomer amino acids partition to the IL-rich phase, whereas L-enantiomer amino acids are transferred into the Na_2SO_4-rich phase. Overall, all these works combining ABS formed by ILs and salts lead to significantly higher extraction efficiencies than those observed with more conventional ABS formed by polymers and salts [43].

Aiming at the design of biocompatible ABS, Zafarani-Moattar and Hamzehzadeh [44] proposed ABS formed by polypropylene glycol (PPG) 400 (instead of salts) and hydrophilic ILs, $[C_nC_1im]Br$ ($n = 2, 4$), for the extraction of L-tryptophan and L-tyrosine. In this type of polymer-IL-based ABS, and in contrast to the behavior observed in IL-salt ABS, L-tyrosine displays a preferential partition toward the polymer-rich phase, whereas L-tryptophan preferentially migrates to the IL-rich phase. This difference was explained by the lack of one pyrrole ring in L-tyrosine when compared to L-tryptophan [44]. Nevertheless, it should be noted that in this type of systems the strong salting-out species (i.e., salts) were replaced by polymers, and thus there is a more tailored balance between the nature and properties of the phase-forming components and the target amino acids partition between the coexisting phases. Focused on the same goal, Xie et al. [45] investigated novel ABS formed by biocompatible ILs, composed of the cholinium cation and anions derived from long-chain carboxylic acids. The authors [45] showed that ILs with long-chain carboxylate anions are suitable for improving the systems' extraction performance. Freire et al. [46] also replaced the typically used inorganic/organic salts by mono- and disaccharides combined with $[C_4C_1im][CF_3SO_3]$ to form ABS, and evaluated the performance of these systems in extracting amino acids. Despite the advantages of the biodegradable and nontoxic nature of carbohydrates, the extraction efficiencies obtained in this work were significantly lower (around 50%)

than those previously discussed with ABS formed by ILs and salts. Organic biological buffers, namely Good's Buffers, were later used in combination with ILs to form ABS to recover L-tryptophan and L-phenylalanine from aqueous media [47]. In these systems, L-phenylalanine completely partitions to the GB-rich phase, whereas L-tryptophan shows a preferential migration to the opposite phase. These systems [47] and those reported by Zafarani-Moattar and Hamzehzadeh [44] can thus be used in the fractionation of amino acids from complex mixtures, such as fermentation broths or hydrolyzed peptide mixtures, and should be explored in such a direction. With this idea in mind, recently, Capela et al. [48] demonstrated the formation of IL-amino acid-based systems, which allow the selective separation of mixtures of aliphatic and aromatic amino acids in one step (Fig. 3). These systems were designed based on the particular ability of aliphatic amino acids to form ABS with ILs. Finally, the authors [48] used a solid-phase extraction approach, by means of a cation exchange column, to retain the ILs used in ABS formation and thus recover the target amino acids.

In addition to IL-salt and IL-polymer ABS, the use of ILs as additives in polymer-salt systems have been explored to tailor the separation of amino acids. Pereira et al. [13] investigated several imidazolium-based ILs added at 5 wt% to an ABS formed by polyethylene glycol (PEG 600) and $NaSO_4$. The authors [13] demonstrated that both an increase and decrease in the extraction efficiencies can be obtained, and that these depend on the chemical structure of the IL employed. $[C_4C_1im]Br$ was also investigated as an adjuvant in PEG 400 + potassium citrate ABS to tailor the partition of amino acids toward the polymer-rich phase [49].

In summary, a large number of IL-based ABS and amino acids have been investigated in the past decade, and in most cases high extraction efficiencies to the IL-rich phase are obtained in a single step. However, with ABS not employing salts, but instead polymers or biological buffers, there is a more suitable amino acid partitioning, which could be useful to separate mixtures of amino acid from real matrices. With systems not composed of strong salting-out species there is a better balance between the nature and properties of the phase-forming components and the target amino acids partition between the coexisting phases. In terms of effectiveness and cost, the most effective systems for extracting amino acids are those using ILs as

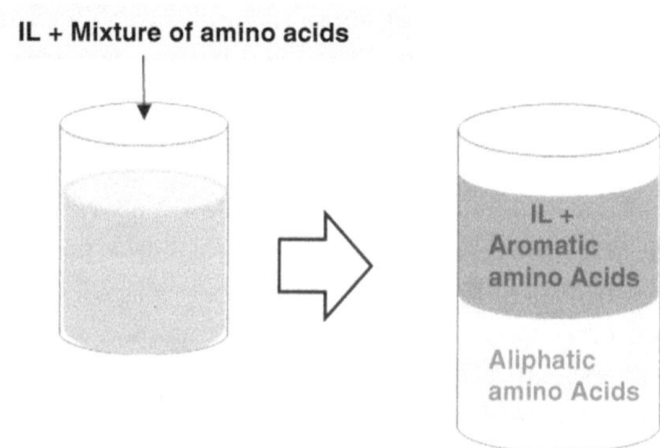

Fig. 3 Schematic representation of the selective separation of aliphatic and aromatic amino acids mixtures (adapted from [48])

adjuvants, mainly because they also allow high extraction efficiencies associated with reduced costs by the low amounts of ILs used. Although poorly investigated, Capela et al. [48] demonstrated the potential of ILs to be added directly to aqueous mixtures of amino acids and allowing their direct fractionation. The same authors demonstrated the amino acid recovery from the IL-rich phase by applying a cation exchange column, allowing the IL reuse, which is a scarcely investigated strategy [48].

In addition to amino acids, IL-based ABS have been investigated as separation strategies for several proteins, namely BSA, hemoglobin, ovalbumin, lysozyme, myoglobin, immunoglobulin Y, immunoglobulin G, cytochrome c, trypsin, pepsin, and rubisco. Because of the labile nature of proteins, they are easily denatured by organic solvents or hydrophobic ILs, and therefore ABS are more appropriate media because of their high water content. Proteins are highly sensitive biomolecules and can lose their native structure through slight changes in the surrounding environment, such as changes in pH, temperature, and ABS phase-forming component compositions. Several works have been conducted in this direction, that is, investigating the effect of several parameters and phase-forming components on the partition behavior of proteins in IL-based ABS. For instance, the influence of pH, IL concentration, temperature, protein size, conformation, and surface structure was studied for large number of proteins. Studies at variable pH values showed that the extraction efficiency of cytochrome-c decreases as the pH increases [50], but the water content in the coexisting phases also plays a role [51]. Pei et al. [52] explored IL-based ABS to recover proteins from wastewater, using BSA, lysozyme, and hemoglobin as model proteins. In this work, hydrophobic interactions were identified as the driving forces in the proteins partitioning between the coexisting phases, as the complete extraction into the IL-rich phase was only achieved when the pH of the systems was close to the proteins pI [52]. The authors [52] further demonstrated that higher temperatures favor protein extraction, contradicting previous results on the extraction of BSA using $[C_{10}C_1im]Cl$-based ABS [53], meaning that both endothermic and exothermic processes may occur, and that these depend on the protein and phase-forming components used. Lin et al. [54] described the effect of IL-protein complexes on the extraction performance of IL-based ABS, in agreement with the findings of Pei and co-authors [55]. An additional study on the partitioning of model proteins (lysozyme, myoglobin, BSA) was performed in IL + salt ABS, although using a dye (Reactive Red-120) as ligand to afford affinity-induced partitions [56]. The authors demonstrated that the nature of the proteins, pH, temperature and ABS composition are parameters that influence the partition coefficients. In this work [56], the hydrophilic IL used was recovered by adding a hydrophobic IL $[C_4C_1im][PF_6]$ that acted as extractant.

In addition to the well-studied imidazolium-based ILs, the separation of BSA, ovalbumin, and hemoglobin was addressed using a series of protic ammonium-based ILs [57]. Under optimum conditions, *N,N*-dimethylethanolamine propionate $[N_{11(2OH)}][C_2CO_2]$ allows extraction efficiencies up to 99.5% in a single step [57]. However, it has been reported that the stability of lysozyme in ammonium-based ILs decreases with IL concentration [58]. The separation of cytochrome-c was investigated by Santos et al. [59] using a quaternary system formed by PEG 8000, sodium poly(acrylate) (NaPA8000), water, and ILs as electrolytes. With these systems, cytochrome-c is enriched in the NaPA8000-rich phase (extraction efficiency >96%).

Foreseeing the development of biocompatible IL-based ABS, Wu et al. [60] employed amino acid-based ILs combined with K_2HPO_4 to separate cytochrome-c, which preferentially partitions to the IL-rich phase. In the same line, $[N_{111(2OH)}]Cl$ and related ILs were used in the design of ABS to extract and purify proteins (BSA, trypsin, papain, and lysozyme) [61, 62]. With an ABS formed by $[N_{11(2OH)(3OH)}]Cl$ and KH_2PO_4, BSA is mainly enriched in the IL-rich phase (extraction yield of 84.3%), thus representing better alternatives to ABS formed by imidazolium-based fluids (extraction yields of 60–68%) [61]. ILs formed by the cholinium cation and anions derived from carboxylic acids, namely $[N_{111(2OH)}][C_1CO_2]$, $[N_{111(2OH)}][C_2CO_2]$, $[N_{111(2OH)}][C_3CO_2]$, $[N_{111(2OH)}][Glyc]$, $[N_{111(2OH)}][Lac]$, $[N_{111(2OH)}]_2[Oxa]$, and $[N_{111(2OH)}]_3[Cit]$, were combined with PPG400 (a non-toxic, biodegradable and thermo-sensitive polymer) to form ABS, demonstrating how these are enhanced systems to separate proteins [62]. Cholinium-based ILs and PPG400 were used in three other works to extract BSA [63–65]. The preferential partition of BSA to the IL-rich phase was always observed, with two of these works reporting the complete extraction of BSA into the IL phase in a single step [63, 64]. In both works, the partition of BSA seems to be dictated by specific interactions occurring between the protein and the IL, such as hydrogen bonding and dispersive interactions. The remarkable extraction efficiencies obtained (92–100% for the IL-rich phase) were far superior to those observed with typical polymer-based ABS, and without evidence of protein denaturation in concentrations at the aqueous media up to 10 g/L [64]. The authors further corroborated their results by extracting BSA from bovine serum, where the extraction efficiencies for the target protein were kept at 100% in a single step. The BSA partition behavior in ABS formed by ILs composed of cholinium as cation and amino acid-derived anions was then investigated by Song et al. [65]. In this work, the authors manipulated the interactions between the phase-forming components and the proteins by changing their surface charge through pH variations. When the pH of the system is higher than the pI of the proteins and the amino acid anions, proteins mainly partition to the IL phase, whereas the reverse occurs at lower pH values [65].

In most of the works described, a buffered salt aqueous solution was used to maintain the medium pH. Recently, ILs with self-buffering characteristics, namely Good's buffer ILs (GB-ILs), were suggested, and their ability to form ABS and to extract BSA evaluated [66]. Remarkably, extraction efficiencies of 100% of BSA to the GB-IL rich phase were obtained in a single step. Authors also found that GB-ILs display a higher stabilizing effect over the studied protein when compared to conventional ILs [66]. Recently, protic ILs were proposed to form thermoreversible ABS when combined with polymers [67]. In contrast to imidazolium-based ILs, the phase diagrams of ABS formed by protic ILs are highly dependent on temperature, thus allowing monophasic–biphasic transitions through small changes in temperature. The complete extraction of cytochrome-c and azocasein was accomplished in a single step using these systems and maintained along three cooling-heating cycles, with no stability losses [67].

Most studies discussed above were carried out with pure/model proteins, and only the extraction efficiency of IL-based ABS and their impact on protein stability and enzyme activity were addressed. Nevertheless, proteins of biological, clinical, pharmaceutical, and industrial relevance are produced in complex media, therefore requiring downstream processing for their purification and recovery. Even so, few works are to be found on the application of IL-based ABS to separate proteins from real matrices. Among these, Taha et al. [68] prepared novel ABS by using self-buffering cholinium-based ILs and PPG 400 for the separation of immunoglobulin Y (IgY) from chicken egg yolk. The combination of the investigated ILs with PPG 400 to form ABS allows the preferential partitioning of IgY to the IL-rich phase, with extraction efficiencies of 79–94% in a single step. However, the greatest challenge was to purify IgY, as the total separation of IgY from the major contaminant proteins was not achieved [68]. Still focused on antibodies and biopharmaceuticals, Ferreira et al. [69] investigated the separation of immunoglobulin G (IgG) from a rabbit serum source, using ABS composed of PEG of different molecular weights and potassium citrate, with 5 wt% of ILs as adjuvants. The tuning ability of ILs was confirmed because the complete extraction (100%) of the target antibody was obtained in a single step, as well as higher recovery yields and enhancements in the IgG purity, by applying ILs (as compared to the IL-free ABS). More recently, ABS formed by cholinium-based ILs + PPG 400 were investigated in the separation of IgG from rabbit serum [70, 71]. These systems allowed an efficient extraction performance and recovery yields >80%, with a purity level up to 66% [71].

Still dealing with more complex matrices, Desai and co-workers [72] addressed the recovery of rubisco from plant extracts using ABS formed by ammonium-based ILs (Ammoeng 110) and phosphate salts. Rubisco migrates to the IL-rich phase with partition coefficients three to four times higher than those obtained with PEG-based systems. However, at high concentrations of ILs (>50%), the studied protein starts to aggregate, demonstrating that there is a limited range of appropriate ILs concentrations that can be used in protein extraction [72]. Pei and co-workers applied IL-based ABS to separate proteins from polysaccharides [55]. BSA was successively separated into the IL-rich phase, with saccharides preferentially enriched in the opposite phase. Similar studies were further published, aiming at the fractionation of sugars and proteins from plant extracts [57, 73, 74]. In order to decrease the environmental and economic impact of IL-based ABS as separation strategies for proteins, some studies attempted to recover the ABS phase-forming components after the extraction of proteins [73–75]. For instance, Pereira et al. studied the recyclability and reusability of phosphonium- and ammonium-based ILs after the extraction of BSA into the IL-rich phase by recovering the protein by dialysis [75]. Extraction efficiencies were kept at 100% in three-step sequential extractions comprising both the BSA recovery and the IL reusability. With the same aim, Li et al. used a thermosensitive polymer in the formulation of ABS, allowing its recovery by temperature increase (Fig. 4) [62].

IL-based ABS were also investigated in the separation of lipases [76–82]. Deive et al. [76] explored the partition behavior of *Thermomyces lanuginosus* lipase (TLL)

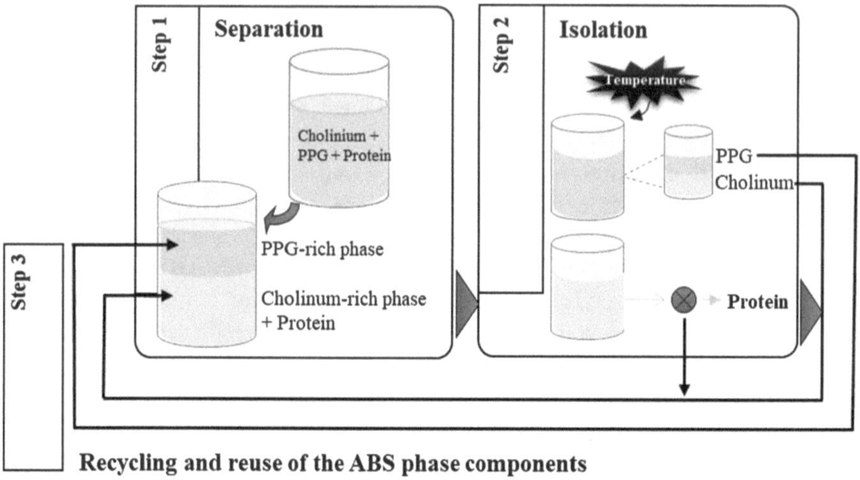

Fig. 4 Schematic representation of the separation and recovery steps of the ABS phase-forming components and proteins, by temperature increase (adapted from [62])

where, under optimized conditions, the authors recovered 99% of TLL from aqueous solutions, preserving the enzyme biocatalytic activity [76]. The same research group employed ABS formed by ILs and inorganic salts to separate *Candida antarctica* lipase A (CaLA) [77]. The results revealed that ABS formed by $[C_2C_1im][C_4SO_4]$ allow extraction efficiencies higher than 99%. Additional investigations on lipase (*C. antarctica* lipase B, CaLB) partitioning were carried out by Ventura et al. using a wide variety of IL-based ABS [78]. With IL + salt ABS, the enzyme recovery efficiencies are higher than 97%, with a maximum purification factor of 2.6. The same group of researchers then applied IL-ABS as downstream processes to purify an extracellular lipolytic enzyme produced by *Bacillus* sp. ITP-001 from its fermentation broth [79]. The best results were obtained with ABS formed by $[C_8C_1im]Cl$ (recovery = 92.2% and purification factor (PF) = 51). Finally, the authors applied ABS formed by polymers and salts, using ILs as adjuvants [80]. With this type of system, a significant increase in the PF was achieved (PF = 245) when compared with IL-based ABS (PF ≈ 51–137) [78, 81] and with polymer-based ABS (PF = 202) [79]. This last type of ABS also allows a decrease of the concentration of the IL employed, thus resulting in a more sustainable separation process. Recently, the same group of researchers [82] investigated the impact of different cations and anions of self-buffering ILs to form ABS when conjugated with salts or polymers aiming at recovering lipase produced via submerged fermentation by *Burkholderia cepacia* ST8. The authors showed that lipase preferentially partitions toward the IL-rich phase in both types of systems investigated, with a recovery yield of 94% [82].

In addition to lipases, other enzymes have been investigated with IL-based ABS. ABS formed by $[C_4C_1im]Cl$ or $[C_4C_1im]Br$ + K_2HPO_4 have been investigated in the separation of papain [83]. The increase of $[C_4C_1im]Br$ and K_2HPO_4 concentrations allow a maximum extraction of 98% to the IL-rich phase [83]. A similar ABS has been applied to the extraction of horseradish peroxidase, with a recovery of 80% into the IL-rich phase and preservation of more than 90% of the enzyme activity [84]. Several ABS (polymer-polymer, polymer-salt, alcohol-salt, and IL-salt) were studied for the recovery of superoxide dismutases (SOD) by Simental-Martinez et al. [85]. The authors found that, although IL-based ABS allow a better enzyme recovery, a polymer-salt ABS is more useful in terms of SOD recovery, enzyme specific activity maintenance, and purification. Santos et al. recently demonstrated an efficient integrated downstream process for the purification of L-asparaginase (ASNase) using IL-based ABS (Fig. 5) [86]. This work evaluated the use of polymer-salt ABS employing ILs as adjuvants, combined with the permeabilization of cell membrane using *n*-dodecane and glycine for the in situ purification of periplasmatic ASNase from *Escherichia coli* cells. The results show that ASNase partitions mostly to the PEG-rich phase, which was explained based on hydrophobic interactions established between the polymer and the enzyme. With the addition of 5 wt% of $[C_4C_1im][CH_3SO_3]$, recoveries of the enzyme up to 88% have been reported.

Jiang et al. evaluated IL-based ABS for the extraction and purification of wheat esterase [87]. It was demonstrated that the enzyme preferentially partitions to the IL-rich phase, and under optimized conditions an extraction yield of 88.93% and a purification factor of 4.23 were obtained [87]. Aiming at the purification of proteins in continuous mode, Novak et al. [88] addressed the intensification of BSA purification by IL-based ABS using microfluidic devices. By adjusting the fluid flow pattern, the BSA extraction within the microchannels was successfully carried out, yielding partition coefficients of 14.6, 20-fold greater than those achieved with conventional ABS composed of polymers and salts [88]. Among the several

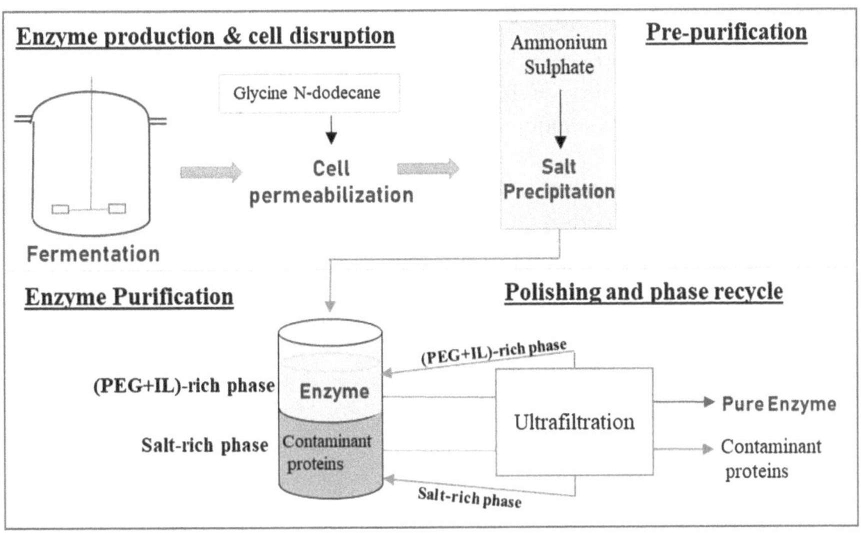

Fig. 5 Schematic representation of the in situ purification of enzymes with the recycling of the ABS phase-forming components (adapted from [86])

advantages of IL-based ABS over more conventional systems, the low viscosity of the coexisting phases was reported as the most significant to the microfluidics approach [88].

With regard to IL-based ABS for the separation of proteins, it is clear that the largest fraction of published works is devoted to studies on the partition of model proteins and enzymes, which give no indications of the behavior of contaminants and how these systems are useful to purify proteins from complex matrices. However, some works started to appear on the application of IL-based ABS to separate proteins from biological-based media, showing that IL-based ABS represent a promising alternative to deal with complex matrices. Nevertheless, the major challenge is still to optimize the ILs chemical structures to separate target proteins, which differ for each protein and matrix.

3 ILs in Solid–Liquid Bioseparation Processes

Solid-phase extraction (SPE) is the most widely used sample-preparation technique for liquid samples, belonging to the group of sorptive-based extraction techniques [89]. The development of new materials as effective and selective adsorbents in sample preparation has been explored in the past few years [89, 90]. Taking advantage of the chemical versatility and designer ability that ILs can impart, in recent years IL-modified materials (also known as supported IL phases – SILPs) have been the focus of a great deal of attention [17]. However, their use for the separation of amino acids and proteins is still limited.

IL-based SPE methods aimed at extracting and separating amino acids, although scarce, comprise ILs immobilized on silica and the use of ILs on the preparation of molecularly imprinted polymers (MIPs). Marwani et al. [91] immobilized a chiral IL $[C_2(L\text{-Phe})][NTf_2]$ on silica, envisioning the enantioselective separation of D-phenylalanine from aqueous solutions. The determined adsorption isotherms revealed that the adsorption capacity of the solid support for D-phenylalanine was 97.35% at pH 3.0 for an initial concentration of the amino acid at 0.1 g/L. The authors examined the enhanced performance of the prepared material by applying it to real water samples (ground water, lake water, seawater, waste water) with satisfactory results [91].

Yang et al. applied the IL 1-butyl-3-methylimidazolium aminohydrocinnamic acid in acetonitrile to prepare surfaces of MIPs for the selective recovery of L-phenylalanine [92]. Studies on the adsorption kinetics, SPE application, and the chiral resolution of racemic phenylalanine mixtures were performed. The IL-based copolymerizing process carried out in acetonitrile, when compared with the traditional imprinting process with acetonitrile/H_2O, allows a higher adsorption of L-phenylalanine, further resulting in the selective separation of L-phenylalanine from other amino acids (L-tryptophan and L-histidine). Although these materials appear promising for recovering amino acids, with recovery values above 91% as reported [92], no further studies for amino acids are so far available.

In the field of proteins, Shu and co-workers immobilized $[C_1im]^+$ moieties onto polyvinyl chloride (PVC), forming $[C_1im]Cl$–PVC materials [93]. The authors demonstrated that the immobilization of the IL strongly depends on the $[C_1im]$/PVC molar ratio during the reaction step. A maximum immobilization ratio of 15.1% was obtained with a 4:1 M ratio of $[C_1im]$:PVC, further showing that these materials adsorb lysozyme, cytochrome-c, and hemoglobin with efficiencies of 97%, 98%, and 94%, respectively. On the other hand, this material was found not to be promising for the removal of BSA, transferrin and IgG [93]. These results were justified based on the pH and pI of the proteins and ionic strength of the solutions.

The selective isolation of hemoglobin was also studied using imidazolium-modified polystyrene materials [94]. In this work, imidazolium cations were grafted onto the surface of a chloromethyl polystyrene, forming PS-CH_2-$[C_1im]^+Cl^-$, which is a crosslinked rigid polymer that can act directly as a support [94]. Adsorption efficiencies of this material to hemoglobin reached values up to 91% [94]. Two other polymer materials were synthesized and used for protein separation, in which 1-allyl-3-butylimidazolium chloride and 1-vinyl-3-octylimidazolium bromide were used as functional monomers, and acrylamide as a co-functional monomer [95]. The first ionic liquid polymeric material has a high binding capacity for hemoglobin, whilst the former possesses a high binding capacity for BSA. These results suggest that these materials can be tailored to adsorb specific proteins, and thus can be envisioned as promising strategies to separate target proteins from complex matrices. IL-modified magnetic nanoparticles (ILs-MNPs) were recently proposed to recover BSA (Fig. 6), leading to extraction efficiencies of 86.9% [96]. The recovery of the protein from the material was also addressed; with NaCl at concentrations higher than 1.1 mol/L the desorption ratio of BSA reaches 95.3%. Moreover, almost 95% of the prepared MNPs were recovered, with no significant losses on their extraction efficiency over four cycles [96]. Recently, Wen et al. [97] investigated the performance of an amino functional dicationic ionic liquid (AFDCIL) coated on the surface of magnetic graphene oxide (Fe@GO) as a magnetic adsorbent (Fe@GO@AFDCIL) for proteins. The authors reported that, compared with conventional Fe@GO@IL composites, the Fe@GO@AFDCIL composite exhibits the highest extraction capacity for bovine hemoglobin. It was also suggested that this magnetic adsorbent could be successfully employed in extraction of bovine hemoglobin from real samples [97].

Although all described works highlight the potential of SILPs to recover proteins, there are no indications regarding their selective nature or performance when applied to real and complex matrices, in which a large number of proteins or other metabolites may be present. The use of SILPs for the separation of proteins is at an early stage. Based on the versatility of ILs transferred to SILPs, it is predicted that further work in this arena should confirm the potential of these materials in protein downstream processing.

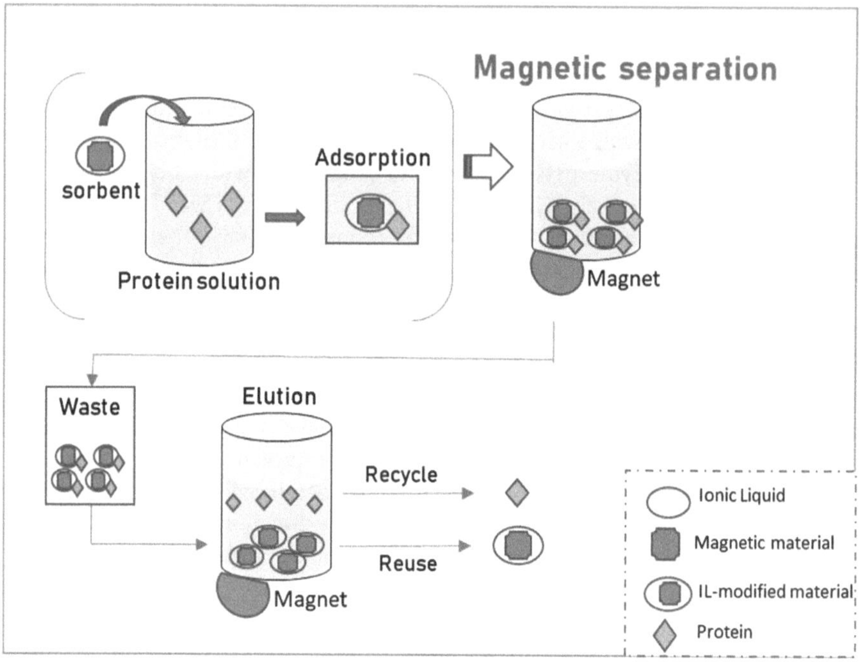

Fig. 6 Schematic representation of magnetic solid phase extractions of proteins [96]

4 Conclusions

In the past decade increased attention has been given to the use of ILs in bioseparation processes. This chapter provides an overview of the potential and suitability of IL-based liquid–liquid extractions, employing hydrophobic ILs and ABS, and solid–liquid extractions, based on the use of IL-modified materials, for the separation of amino acids and proteins, including enzymes and biopharmaceuticals. Most published works show that IL-based strategies lead to improved extraction and separation performance when compared to more traditional approaches. This trend is a direct result of the ILs tailoring ability, which is transferrable to bioprocesses in which they are used, thus allowing the design of specific ILs and processes for target separations.

Most of the works discussed are based on imidazolium-based ILs, majorly combined with fluorinated anions. However, more recently, other classes of ILs have been considered, such as with cholinium- and amino acid-based ILs, bringing additional biocompatible features to bioseparation processes. Furthermore, most works dealing with the separation of amino acids and proteins employing ILs are focused on ABS, which may be explained by the advantages of these liquid–liquid systems over the application of hydrophobic ILs. ABS are water-rich and usually lead to improved extraction efficiencies and recovery yields – a result also directly connected to the second phase-forming component of ABS that most of the time is a strong salting-out species. In addition to liquid–liquid extractions, SILPs seem to be a promising option. However, works on SILPs are still scarce, with no reports on their use to purify amino acids and proteins from complex/real matrices.

In summary, this chapter aims to provide an overview of the application of ILs in bioseparation processes. From the works discussed it is clear that the use of appropriate ILs leads to high extraction efficiencies, recovery yields, and purification factors. However, a number of requirements need to be addressed before their application in scaled-up separation processes becomes a reality, namely to: (1) find efficient and low-cost ILs able to compete with common organic solvents; (2) develop integrated processes, particularly focused on decreasing the number of steps involved and the process cost; (3) develop strategies for biomolecules recovery and IL recycling; (4) perform scale-up studies of the optimized processes; and (5) carry out economic and life cycle analysis of the developed processes. Although there is still a path to follow, IL-based separation processes display several advantages to become an industrial reality in the following decades.

Acknowledgments This work was developed in the scope of projects CICECO – Aveiro Institute of Materials (Ref. FCT UID/CTM/50011/2013) and MultiBiorefinery (POCI-01-0145-FEDER-016403), financed by national funds through Fundação para a Ciência e a Tecnologia (FCT, Portugal)/MEC and co-financed by FEDER under the PT2020 Partnership agreement. The authors acknowledge financial support from the European Union Framework Programme for Research and Innovation HORIZON 2020, under the TEAMING Grant agreement No 739572 – The Discoveries CTR. D.C.V. Belchior acknowledges financial support from Conselho Nacional de Desenvolvimento Científico e Tecnológico – CNPq for the PhD grant [202337/2015-4]). I.F. Duarte acknowledges the FCT/MCTES for a research contract under the Program 'Investigador FCT'. M.G. Freire acknowledges the European Research Council under the European Union's Seventh Framework Programme (FP7/2007-2013) /ERC grant agreement n° 337753.

References

1. Martínez-Aragón M, Burghoff S, Goetheer ELV, de Haan AB (2009) Guidelines for solvent selection for carrier mediated extraction of proteins. Sep Purif Technol 65:65–72. https://doi.org/10.1016/j.seppur.2008.01.028
2. Ahamed T, Ottens M, Nfor BK et al (2006) A generalized approach to thermodynamic properties of biomolecules for use in bioseparation process design. Fluid Phase Equilib 241: 268–282. https://doi.org/10.1016/j.fluid.2005.12.011
3. Bhawsar CM, Pandit B, Sawant B, Joshi B (1994) Enzyme mass transfer coefficient in a sieve plate extraction column. Chem Eng J 55:B1–B17. https://doi.org/10.1016/0923-0467(94)87012-8
4. Ventura SPM, Silva FA, Quental MV et al (2017) Ionic-liquid-mediated extraction and separation processes for bioactive compounds: past, present, and future trends. Chem Rev 117: 6984–7052. https://doi.org/10.1021/acs.chemrev.6b00550
5. Plechkova NV, Seddon KR (2008) Applications of ionic liquids in the chemical industry. Chem Soc Rev 37:123–150. https://doi.org/10.1039/B006677J

6. Albertsson P (1986) Partition of cell particles and macromolecules: separation and purification of biomolecules, cell organelles, membranes, and cells in aqueous polymer two-phase systems and their use in biochemical analysis and biotechnology.3rd edn. Wiley, Chichester
7. Iqbal M, Tao Y, Xie S et al (2016) Aqueous two-phase system (ATPS): an overview and advances in its applications. Biol Proced Online 18:1–18. https://doi.org/10.1186/s12575-016-0048-8
8. Grilo AL, Aires-Barros MR, Azevedo AM (2014) Partitioning in aqueous two-phase systems: fundamentals, applications and trends. Sep Purif Rev 45:68–80. https://doi.org/10.1080/15422119.2014.983128
9. Azevedo AM, Fonseca LP, Prazeres DMF (1999) Stability and stabilisation of penicillin acylase. J Chem Technol Biotechnol 74:1110–1116. https://doi.org/10.1002/(SICI)1097-4660(199911)74:11<1110::AID-JCTB149>3.0.CO;2-B
10. Pereira JFB, Rebelo LPN, Rogers RD et al (2013) Combining ionic liquids and polyethylene glycols to boost the hydrophobic-hydrophilic range of aqueous biphasic systems. Phys Chem Chem Phys 15:19580–19583. https://doi.org/10.1039/c3cp53701c
11. Li J, Kao WJ (2003) Synthesis of polyethylene glycol (PEG) derivatives and PEGylated - peptide biopolymer conjugates. Biomacromolecules 4:1055–1067. https://doi.org/10.1021/bm034069l
12. Rosa PAJ, Azevedo AM, Ferreira IF et al (2007) Affinity partitioning of human antibodies in aqueous two-phase systems. J Chromatogr A 1162:103–113. https://doi.org/10.1016/j.chroma.2007.03.067
13. Pereira JFB, Lima ÁS, Freire MG, Coutinho JAP (2010) Ionic liquids as adjuvants for the tailored extraction of biomolecules in aqueous biphasic systems. Green Chem 12:1661. https://doi.org/10.1039/c003578e
14. Dhadge VL, Rosa SASL, Azevedo AM et al (2014) Magnetic aqueous two-phase fishing: a hybrid process technology for antibody purification. J Chromatogr A 1339:59–64. https://doi.org/10.1016/j.chroma.2014.02.069
15. Gutowski KE, Broker GA, Willauer HD et al (2003) Controlling the aqueous miscibility of ionic liquids: aqueous biphasic systems of water-miscible ionic liquids and water-structuring salts for recycle, metathesis, and separations. J Am Chem Soc 125:6632–6633. https://doi.org/10.1021/ja0351802
16. Freire MG, Cláudio AFM, Araújo JMM et al (2012) Aqueous biphasic systems: a boost brought about by using ionic liquids. Chem Soc Rev 41:4966–4995. https://doi.org/10.1039/c2cs35151j
17. Abd A, Ahmed A, Xiashi Z (2017) Developments/application of ionic liquids/poly ionic liquids in magnetic solid-phase extraction and solid phase microextraction. Colloid Surf Sci 2:162–170. https://doi.org/10.11648/j.css.20170204.15
18. Carda-Broch S, Berthod A, Armstrong DW (2003) Solvent properties of the 1-butyl-3-methylimidazolium hexafluorophosphate ionic liquid. Anal Bioanal Chem 375:191–199. https://doi.org/10.1007/s00216-002-1684-1
19. Smirnova SV, Torocheshnikova II, Formanovsky AA, Pletnev IV (2004) Solvent extraction of amino acids into a room temperature ionic liquid with dicyclohexano-18-crown-6. Anal Bioanal Chem 378:1369–1375. https://doi.org/10.1007/s00216-003-2398-8
20. Wang J, Pei Y, Zhao Y, Hu Z (2005) Recovery of amino acids by imidazolium based ionic liquids from aqueous media. Green Chem 7:196–202. https://doi.org/10.1039/b415842c
21. Absalan G, Akhond M, Sheikhian L (2010) Partitioning of acidic, basic and neutral amino acids into imidazolium-based ionic liquids. Amino Acids 39:167–174. https://doi.org/10.1007/s00726-009-0391-z
22. Tomé LIN, Catambas VR, Teles ARR et al (2010) Tryptophan extraction using hydrophobic ionic liquids. Sep Purif Technol 72:167–173. https://doi.org/10.1016/j.seppur.2010.02.002
23. Huaxi L, Zhuo L, Jingmei Y et al (2012) Liquid–liquid extraction process of amino acids by a new amide-based functionalized ionic liquid. Green Chem 14:172–1727. https://doi.org/10.1039/c2gc16560k

24. Tang F, Zhang Q, Ren D et al (2010) Functional amino acid ionic liquids as solvent and selector in chiral extraction. J Chromatogr A 1217:4669–4674. https://doi.org/10.1016/j.chroma.2010.05.013
25. Shimojo K, Kamiya N, Tani F et al (2006) Functional conversion of cytochrome c in ionic liquids via crown ether Complexation. Anal Chem 78:7735–7742. https://doi.org/10.1021/ac0612877
26. Tzeng YP, Shen CW, Yu T (2008) Liquid-liquid extraction of lysozyme using a dye-modified ionic liquid. J Chromatogr A 1193:1–6. https://doi.org/10.1016/j.chroma.2008.02.118
27. Kohno Y, Saita S, Murata K et al (2011) Extraction of proteins with temperature sensitive and reversible phase change of ionic liquid/water mixture. Polym Chem 2:862–867. https://doi.org/10.1039/c0py00364f
28. Ito Y, Kohno Y, Nakamura N, Ohno H (2013) Design of phosphonium-type zwitterion as an additive to improve saturated water content of phase-separated ionic liquid from aqueous phase toward reversible extraction of proteins. Int J Mol Sci 14:18350–18361. https://doi.org/10.3390/ijms140918350
29. Xu W, Cao H, Ren G et al (2014) An AIL/IL-based liquid/liquid extraction system for the purification of His-tagged proteins. Appl Microbiol Biotechnol 98:5665–5675. https://doi.org/10.1007/s00253-014-5737-0
30. Ren G, Gong X, Wang B et al (2015) Affinity ionic liquids for the rapid liquid-liquid extraction purification of hexahistidine tagged proteins. Sep Purif Technol 146:114–120. https://doi.org/10.1016/j.seppur.2015.03.025
31. Cheng DH, Chen XW, Shu Y, Wang JH (2008) Selective extraction/isolation of hemoglobin with ionic liquid 1-butyl-3-trimethylsilylimidazolium hexafluorophosphate (BtmsimPF6). Talanta 75:1270–1278. https://doi.org/10.1016/j.talanta.2008.01.044
32. Cheng D-H, Chen X-W, Shu Y, Wang J-H (2008) Extraction of cytochrome C by ionic liquid 1-butyl-3-trimethylsilylimidazolium hexafluorophosphate. Chinese J Anal Chem 36:1187–1190. https://doi.org/10.1016/S1872-2040(08)60066-3
33. Huh YS, Jeong CM, Chang HN et al (2010) Rapid separation of bacteriorhodopsin using a laminar-flow extraction system in a microfluidic device. Biomicrofluidics 4:14103(10)–14103(1). https://doi.org/10.1063/1.3298608
34. Alvarez-Guerra E, Irabien A (2012) Extraction of lactoferrin with hydrophobic ionic liquids. Sep Purif Technol 98:432–440. https://doi.org/10.1016/j.seppur.2012.08.010
35. Ventura PM, Neves CMSS, Freire MG et al (2009) Evaluation of anion influence on the formation and extraction capability of ionic-liquid-based aqueous biphasic systems. J Phys Chem B 113:9304–9310. https://doi.org/10.1021/jp900293v
36. Neves CMSS, Ventura SPM, Freire MG et al (2009) Evaluation of cation influence on the formation and extraction capability of ionic-liquid-based aqueous biphasic systems. J Phys Chem B 113:5194–5199. https://doi.org/10.1021/jp900293v
37. Li Z, Pei Y, Liu L, Wang J (2010) (Liquid+liquid) equilibria for (acetate-based ionic liquids + inorganic salts) aqueous two-phase systems. J Chem Thermodyn 42:932–937. https://doi.org/10.1016/j.jct.2010.03.010
38. Pei Y, Li Z, Liu L, Wang J (2012) Partitioning behavior of amino acids in aqueous two-phase systems formed by imidazolium ionic liquid and dipotassium hydrogen phosphate. J Chromatogr A 1231:2–7. https://doi.org/10.1016/j.chroma.2012.01.087
39. Louros CLS, Claudio AFM, Neves CMSS et al (2010) Extraction of biomolecules using phosphonium-based ionic liquids + K_3PO_4 aqueous biphasic systems. Int J Mol Sci 11:1777–1791. https://doi.org/10.3390/ijms11041777
40. Zafarani-Moattar MT, Hamzehzadeh S (2011) Partitioning of amino acids in the aqueous biphasic system containing the water-miscible ionic liquid 1-butyl-3-methylimidazolium bromide and the water-structuring salt potassium citrate. Biotechnol Prog 27:986–997. https://doi.org/10.1002/btpr.613

41. Passos H, Ferreira AR, Cláudio AFM et al (2012) Characterization of aqueous biphasic systems composed of ionic liquids and a citrate-based biodegradable salt. Biochem Eng J 67:68–76. https://doi.org/10.1016/j.bej.2012.05.004
42. Wu D, Zhou Y, Cai P et al (2015) Specific cooperative effect for the enantiomeric separation of amino acids using aqueous two-phase systems with task-specific ionic liquids. J Chromatogr A 1395:65–72. https://doi.org/10.1016/j.chroma.2015.03.047
43. Salabat A, Abnosi MH, Motahari A (2008) Investigation of amino acid partitioning in aqueous two-phase systems containing polyethylene glycol and inorganic salts. J Chem Eng Data 53: 2018–2021. https://doi.org/10.1021/je700727u
44. Zafarani-Moattar MT, Hamzehzadeh S, Nasiri S (2011) A new aqueous biphasic system containing polypropylene glycol and a water-miscible ionic liquid. Biotechnol Prog 28: 146–156. https://doi.org/10.1002/btpr.718
45. Xie Y, Xing H, Yang Q et al (2015) Aqueous biphasic system containing long chain anion-functionalized ionic liquids for high-performance extraction. ACS Sustain Chem Eng 3: 3365–3372. https://doi.org/10.1021/acssuschemeng.5b01068
46. Freire MG, Louros CLS, Rebelo LPN, Coutinho JAP (2011) Aqueous biphasic systems composed of a water-stable ionic liquid + carbohydrates and their applications. Green Chem 13:1536–1545. https://doi.org/10.1039/c1gc15110j
47. Luis A, Dinis TBV, Passos H et al (2015) Good's buffers as novel phase-forming components of ionic-liquid-based aqueous biphasic systems. Biochem Eng J 101:142–149. https://doi.org/10.1016/j.bej.2015.05.008
48. Capela EV, Quental MV, Domingues P et al (2017) Effective separation of aromatic and aliphatic amino acid mixtures using ionic-liquid-based aqueous biphasic systems. Green Chem 19:1850–1854. https://doi.org/10.1039/C6GC03060B
49. Hamzehzadeh S, Vasiresh M (2014) Ionic liquid 1-butyl-3-methylimidazolium bromide as a promoter for the formation and extraction capability of poly(ethylene glycol)-potassium citrate aqueous biphasic system at T=298.15K. Fluid Phase Equilib 382:80–88. https://doi.org/10.1016/j.fluid.2014.08.029
50. Lu Y, Lu W, Wang W et al (2011) Thermodynamic studies of partitioning behavior of cytochrome c in ionic liquid-based aqueous two-phase system. Talanta 85:1621–1626. https://doi.org/10.1016/j.talanta.2011.06.058
51. Dreyer S, Salim P, Kragl U (2009) Driving forces of protein partitioning in an ionic liquid-based aqueous two-phase system. Biochem Eng J 46:176–185. https://doi.org/10.1016/j.bej.2009.05.005
52. Pei Y, Li L, Li Z et al (2012) Partitioning behavior of wastewater proteins in some ionic liquids-based aqueous two-phase systems. Sep Sci Technol 47:277–283. https://doi.org/10.1080/01496395.2011.609241
53. Yan H, Wu J, Dai G et al (2012) Interaction mechanisms of ionic liquids [Cnmim]Br (n = 4, 6, 8, 10) with bovine serum albumin. J Lumin 132:622–628. https://doi.org/10.1016/j.jlumin.2011.10.026
54. Lin X, Wang Y, Zeng Q et al (2013) Extraction and separation of proteins by ionic liquid aqueous two-phase system. Analyst 138:6445–6453. https://doi.org/10.1039/c3an01301d
55. Pei Y, Li Z, Liu L et al (2010) Selective separation of protein and saccharides by ionic liquids aqueous two-phase systems. Sci China Chem 53:1554–1560. https://doi.org/10.1007/s11426-010-4025-9
56. Sheikhian L, Akhond M, Absalan G, Goltz DM (2013) Dye-affinity partitioning of acidic, basic, and neutral proteins in ionic liquid-based aqueous biphasic systems. Sep Sci Technol 48: 2372–2380. https://doi.org/10.1080/01496395.2013.804086
57. Chen J, Wang Y, Zeng Q et al (2014) Partition of proteins with extraction in aqueous two-phase system by hydroxyl ammonium-based ionic liquid. Anal Methods 6:4067–4076. https://doi.org/10.1039/c4ay00233d

58. Bisht M, Kumar A, Venkatesu P (2015) Analysis of the driving force that rule the stability of lysozyme in alkylammonium-based ionic liquids. Int J Biol Macromol 81:1074–1081. https://doi.org/10.1016/j.ijbiomac.2015.09.036
59. Santos JHPM, E Silva FA, Coutinho JAP et al (2015) Ionic liquids as a novel class of electrolytes in polymeric aqueous biphasic systems. Process Biochem 50:661–668. https://doi.org/10.1016/j.procbio.2015.02.001
60. Wu C, Wang J, Li Z et al (2013) Relative hydrophobicity between the phases and partition of cytochrome-c in glycine ionic liquids aqueous two-phase systems. J Chromatogr A 1305:1–6. https://doi.org/10.1016/j.chroma.2013.06.066
61. Huang S, Wang Y, Zhou Y et al (2013) Choline-like ionic liquid-based aqueous two-phase extraction of selected proteins. Anal Methods 5:3395–3402. https://doi.org/10.1016/B978-1-895198-85-0.50012-1
62. Li Z, Liu X, Pei Y et al (2012) Design of environmentally friendly ionic liquid aqueous two-phase systems for the efficient and high activity extraction of proteins. Green Chem 14: 2941. https://doi.org/10.1039/c2gc35890e
63. Taha M, Quental MV, Correia I et al (2015) Extraction and stability of bovine serum albumin (BSA) using cholinium-based Good's buffers ionic liquids. Process Biochem 50:1158–1166. https://doi.org/10.1016/j.procbio.2015.03.020
64. Quental MV, Caban M, Pereira MM et al (2015) Enhanced extraction of proteins using cholinium-based ionic liquids as phase-forming components of aqueous biphasic systems. Biotechnol J 10:1457–1466. https://doi.org/10.1002/biot.201500003
65. Song CP, Ramanan RN, Vijayaraghavan R et al (2015) Aqueous two-phase systems based on cholinium aminoate ionic liquids with tunable hydrophobicity and charge density. ACS Sustain Chem Eng 3:3291–3298. https://doi.org/10.1021/acssuschemeng.5b00881
66. Taha M, e Silva FA, Quental MV et al (2014) Good's buffers as a basis for developing self-buffering and biocompatible ionic liquids for biological research. Green Chem 16:3149–3159. https://doi.org/10.1039/C4GC00328D
67. Passos H, Luís A, Coutinho JAP, Freire MG (2016) Thermoreversible (ionic-liquid- based) aqueous biphasic systems. Sci Rep 6:1–7. https://doi.org/10.1038/srep20276
68. Taha M, Almeida MR, Francisca A, Domingues P (2015) Novel biocompatible and self-buffering ionic liquids for biopharmaceutical applications. Chem Eur J 21:4781–4788. https://doi.org/10.1002/chem.201405693
69. Ferreira AM, Faustino VFM, Mondal D et al (2016) Improving the extraction and purification of immunoglobulin G by the use of ionic liquids as adjuvants in aqueous biphasic systems. J Biotechnol 236:166–175. https://doi.org/10.1016/j.jbiotec.2016.08.015
70. Ramalho CC, Neves CMSS, Quental MV et al (2018) Separation of immunoglobulin G using aqueous biphasic systems composed of cholinium-based ionic liquids and poly(propylene glycol). J Chem Technol Biotechnol. https://doi.org/10.1002/jctb.5594
71. Mondal D, Sharma M, Quental MV et al (2016) Suitability of bio-based ionic liquids for the extraction and purification of IgG antibodies. Green Chem 18:6071–6081. https://doi.org/10.1039/C6GC01482H
72. Desai RK, Streefland M, Wijffels RH, Eppink MHM (2014) Extraction and stability of selected proteins in ionic liquid based aqueous two-phase systems. Green Chem 16:2670–2679. https://doi.org/10.1039/C3GC42631A
73. Tan ZJ, Li FF, Xu XL, Xing JM (2012) Simultaneous extraction and purification of aloe polysaccharides and proteins using ionic liquid based aqueous two-phase system coupled with dialysis membrane. Desalination 286:389–393. https://doi.org/10.1016/j.desal.2011.11.053
74. Yan JK, Ma HL, Pei JJ et al (2014) Facile and effective separation of polysaccharides and proteins from Cordyceps sinensis mycelia by ionic liquid aqueous two-phase system. Sep Purif Technol 135:278–284. https://doi.org/10.1016/j.seppur.2014.03.020
75. Pereira MM, Pedro SN, Quental MV et al (2015) Enhanced extraction of bovine serum albumin with aqueous biphasic systems of phosphonium- and ammonium-based ionic liquids. J Biotechnol 206:17–25. https://doi.org/10.1016/j.jbiotec.2015.03.028

76. Deive FJ, Rodríguez A, Pereiro AB et al (2011) Ionic liquid-based aqueous biphasic system for lipase extraction. Green Chem 13:390–396. https://doi.org/10.1039/C0GC00075B
77. Deive FJ, Rodríguez A, Rebelo LPN, Marrucho IM (2012) Extraction of Candida antarctica lipase A from aqueous solutions using imidazolium-based ionic liquids. Sep Purif Technol 97:205–210. https://doi.org/10.1016/j.seppur.2011.12.013
78. Ventura SPM, Sousa SG, Freire MG et al (2011) Design of ionic liquids for lipase purification. J Chromatogr B Anal Technol Biomed Life Sci 879:2679–2687. https://doi.org/10.1016/j.jchromb.2011.07.022
79. Ventura SPM, de Barros RLF, de Pinho Barbosa JM et al (2012) Production and purification of an extracellular lipolytic enzyme using ionic liquid-based aqueous two-phase systems. Green Chem 14:734–740. https://doi.org/10.1039/c2gc16428k
80. Souza RL, Ventura SPM, Soares CMF et al (2015) Lipase purification using ionic liquids as adjuvants in aqueous two-phase systems. Green Chem 17:3026–3034. https://doi.org/10.1039/C5GC00262A
81. Souza RL, Lima RA, Coutinho JAP et al (2015) Aqueous two-phase systems based on cholinium salts and tetrahydrofuran and their use for lipase purification. Sep Purif Technol 155:118–126. https://doi.org/10.1016/j.seppur.2015.05.021
82. Lee SY, Khoiroh I, Coutinho JAP et al (2017) Lipase production and purification by self-buffering ionic liquid-based aqueous biphasic systems. Process Biochem 63:221–228. https://doi.org/10.1016/j.procbio.2017.08.020
83. Bai Z, Chao Y, Zhang M et al (2013) Partitioning behavior of papain in ionic liquids-based aqueous two-phase systems. J Chem 2013:1–6. https://doi.org/10.1155/2013/938154
84. Cao Q, Quan L, He C et al (2008) Talanta partition of horseradish peroxidase with maintained activity in aqueous biphasic system based on ionic liquid. Talanta 77:160–165. https://doi.org/10.1016/j.talanta.2008.05.055
85. Simental-Martínez J, Rito-Palomares M, Benavides J (2014) Potential application of aqueous two-phase systems and three-phase partitioning for the recovery of superoxide dismutase from a clarified homogenate of Kluyveromyces marxianus. Biotechnol Prog 30:1326–1334. https://doi.org/10.1002/btpr.1979
86. Santos JHPM, Santos JC et al (2018) In situ purification of periplasmatic L-asparaginase by aqueous two phase systems with ionic liquids (ILs) as adjuvants. J Chem Technol Biotechnol. https://doi.org/10.1002/jctb.5455
87. Jiang B, Feng Z, Liu C et al (2015) Extraction and purification of wheat-esterase using aqueous two-phase systems of ionic liquid and salt. J Food Sci Technol 52:2878–2885. https://doi.org/10.1007/s13197-014-1319-5
88. Novak U, Pohar A, Plazl I, Žnidaršič-Plazl P (2012) Ionic liquid-based aqueous two-phase extraction within a microchannel system. Sep Purif Technol 97:172–178. https://doi.org/10.1016/j.seppur.2012.01.033
89. Vidal L, Riekkola ML, Canals A (2012) Ionic liquid-modified materials for solid-phase extraction and separation: a review. Anal Chim Acta 715:19–41. https://doi.org/10.1016/j.aca.2011.11.050
90. Fumes BH, Silva MR, Andrade FN et al (2015) Recent advances and future trends in new materials for sample preparation. TrAC Trends Anal Chem 71:9–25. https://doi.org/10.1016/j.trac.2015.04.011
91. Marwani HM, Bakhsh EM, Al-Turaif HA et al (2014) Enantioselective separation and detection of D-phenylalanine based on newly developed chiral ionic liquid immobilized silica gel surface. Int J Electrochem Sci 9:7948–7964
92. Yang L, Hu X, Guan P et al (2015) Molecularly imprinted polymers for the selective recognition of L-phenylalanine based on 1-buty-3-methylimidazolium ionic liquid. J Appl Polym Sci 132:42485(1)–42485(9). https://doi.org/10.1002/app.42485
93. Shu Y, Chen XW, Wang JH (2010) Ionic liquid-polyvinyl chloride ionomer for highly selective isolation of basic proteins. Talanta 81:637–642. https://doi.org/10.1016/j.talanta.2009.12.059

94. Zhao G, Chen S, Chen XW, He RH (2013) Selective isolation of hemoglobin by use of imidazolium-modified polystyrene as extractant. Anal Bioanal Chem 405:5353–5358. https://doi.org/10.1007/s00216-013-6889-y
95. Liu Y, Ma R, Deng Q et al (2014) Preparation of ionic liquid polymer materials and their recognition properties for proteins. RSC Adv 4:52147–52154. https://doi.org/10.1039/C4RA05713A
96. Chen J, Wang Y, Ding X et al (2014) Magnetic solid-phase extraction of proteins based on hydroxy functional ionic liquid-modified magnetic nanoparticles. Anal Methods 6:8358–8367. https://doi.org/10.1039/C4AY01786B
97. Wen Q, Wang Y, Xu K et al (2016) Magnetic solid-phase extraction of protein by ionic liquid-coated Fe@graphene oxide. Talanta 160:481–488. https://doi.org/10.1016/j.talanta.2016.07.031

Adv Biochem Eng Biotechnol (2019) 168: 31–60
DOI: 10.1007/10_2018_67

Published online: 24 May 2018

Natural Deep Eutectic Solvents and Their Applications in Biotechnology

Zhen Yang

Contents

Z. Yang (✉)
College of Life Sciences and Oceanography, Shenzhen University, Shenzhen, Guangdong, P. R. China
e-mail: zyang@szu.edu.cn

Abstract Following the appearance of ionic liquids (ILs) and deep eutectic solvents (DESs), natural deep eutectic solvents (NADESs) have emerged as a new type of truly green solvents with many excellent advantages such as cheapness, sustainability, biocompatibility, environmental friendliness, and, in particular, remarkable solubilizing power and outstanding designability. Although only at an early stage, research on NADESs has started to blossom with exponential growth, showing attractive and promising potentials for applications in various areas. In this chapter we focus on an introduction to what is currently known about NADESs: their formation, structure and roles in nature, their physical/chemical properties, their toxicity and biodegradability, and, more importantly, their beneficial applications in biotechnology.

Graphical Abstract

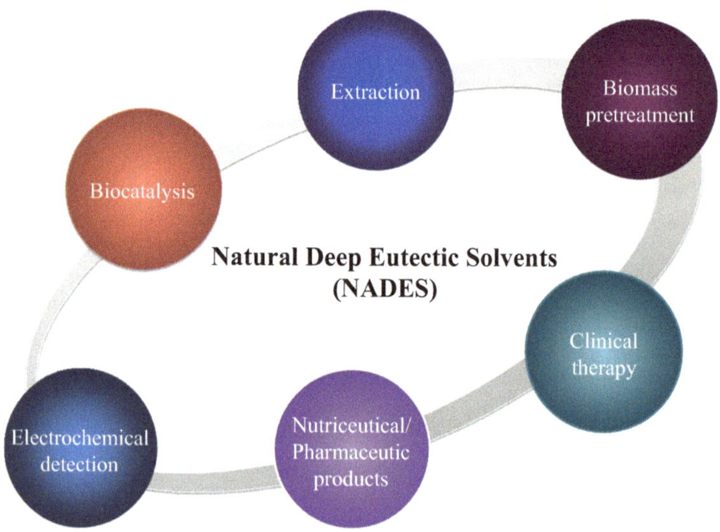

Keywords Biosafety, Biotechnology, Deep eutectic solvents (DESs), Ionic liquids (ILs), Natural deep eutectic solvents (NADESs), Physicochemical properties

Abbreviations

[BMIm][Ac]	1-Butyl-3-methylimidazolium acetate
[BMIm][BF_4]	1-Butyl-3-methylimidazolium tetrafluoroborate
[BMIm][Cl]	1-Butyl-3-methylimidazolium chloride

ChAc	Choline acetate
ChCl	Choline chloride
DES	Deep eutectic solvent
IL	Ionic liquid
NADES	Natural deep eutectic solvent

1 Green Solvents: ILs, DESs, and NADESs

Continuous searching for inexpensive, sustainable, and eco-friendly solvents for chemical and biochemical processes has always been a tremendously important issue. Ionic liquids (ILs) were regarded as potential green alternatives to conventional organic solvents mainly because of their negligible vapor pressures and excellent chemical/thermal inertness [1], but their biotoxicity and biodegradability have been more and more questioned by later research [2].

First described in 2003 by Abbott's group [3], deep eutectic solvents (DESs) make up a new class of nonaqueous solvents derived from ILs. DESs are eutectic mixtures of ammonium salts (such as choline chloride, ChCl) and hydrogen bond donors (HBDs, such as urea and glycerol) with melting points lower than those of any of their individual components. A typical example is given by mixing ChCl ($T_m = 302°C$) and urea ($T_m = 133°C$) at a molar ratio of 1:2, resulting in a liquid-form DES with a rather low freezing point of 12°C [3].

DESs are found to share some attractive IL-like solvent properties such as low melting point, low volatility, high thermal stability, high solubility for various substances, and the "designer solvent" property (i.e., the physicochemical properties of the solvent can be fine tuned by altering the DES components) [4]. Because the components used to prepare DESs are rich in nature and DESs are easily prepared by mixing the components through freeze-drying or thermal mixing, DESs are considered a potentially greener alternative to ILs with some additional advantages such as (1) lower price, (2) easier preparation with higher purity, (3) higher biodegradability and lower toxicity, and (4) outstanding designability with a broader selection (with choices of cations, anions, and HBDs at different salt/HBD molar ratios). Therefore, DESs are currently attracting widespread academic and industrial interest, with a broad range of applications in catalysis, organic synthesis, dissolution and extraction processes, electrochemistry, material chemistry, biocatalysis and biotransformation, etc. [5, 6]. Very recently, the applications of DESs in biocatalysis, biotechnology, and bioengineering have been extensively reviewed in [7, 8].

In 2011, a new type of DESs, natural deep eutectic solvents (NADESs), emerged when Choi et al. [9] noticed that many plant primary metabolites such as choline, sugars, and amino acids, which are also potential ingredients of normal ILs and DESs, can form a DES-like liquid when being mixed in certain combinations. They assigned the term "natural deep eutectic solvents" for these mixtures. Although only at a nascent stage, NADESs have shown great potential for many applications

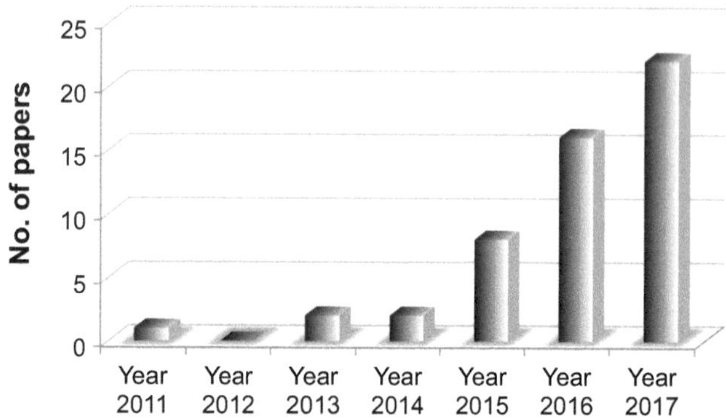

Fig. 1 Number of publications on NADES during the years 2011–2017. Searched from Web of Science with "natural deep eutectic solvents" as the keyword appearing in the title

because of their superior nontoxicity, sustainability, and environmental friendliness relative to ILs and DESs [10, 11], and the numbers of publications on NADESs have grown exponentially (see Fig. 1 for the number of papers published on NADES).

This chapter is focused on the introduction of NADESs: their formation, structure, and roles in nature, their physicochemical properties, their toxicity and biodegradability, and their applications in biotechnology and bioengineering. It has to be noted that there is no clear line between DESs and NADESs; some eutectic mixtures prepared from natural metabolites, such as ChCl/glucose, can be deemed as both a DES and an NADES.

2 An Introduction to NADESs: Formation, Structure, and Roles

2.1 Preparation of NADESs

NADESs can be prepared from ubiquitous natural compounds, including ChCl, sugars, polyols, amino acids, and organic acids (Scheme 1). These compounds in the solid state become liquid when mixed at a specific combination and a proper molar ratio. Similar to DES preparation, three methods can be used for preparing an NADES:

1. Thermal mixing [12]: the components are mixed at about 80°C for 1–2 h on a hot plate with magnetic stirring until a colorless clear liquid is formed
2. Vacuum evaporation [12]: the components are dissolved in water before being evaporated at 50°C with a rotary evaporator
3. Freeze-drying [13]: the components are dissolved in water and then subjected to freeze-drying

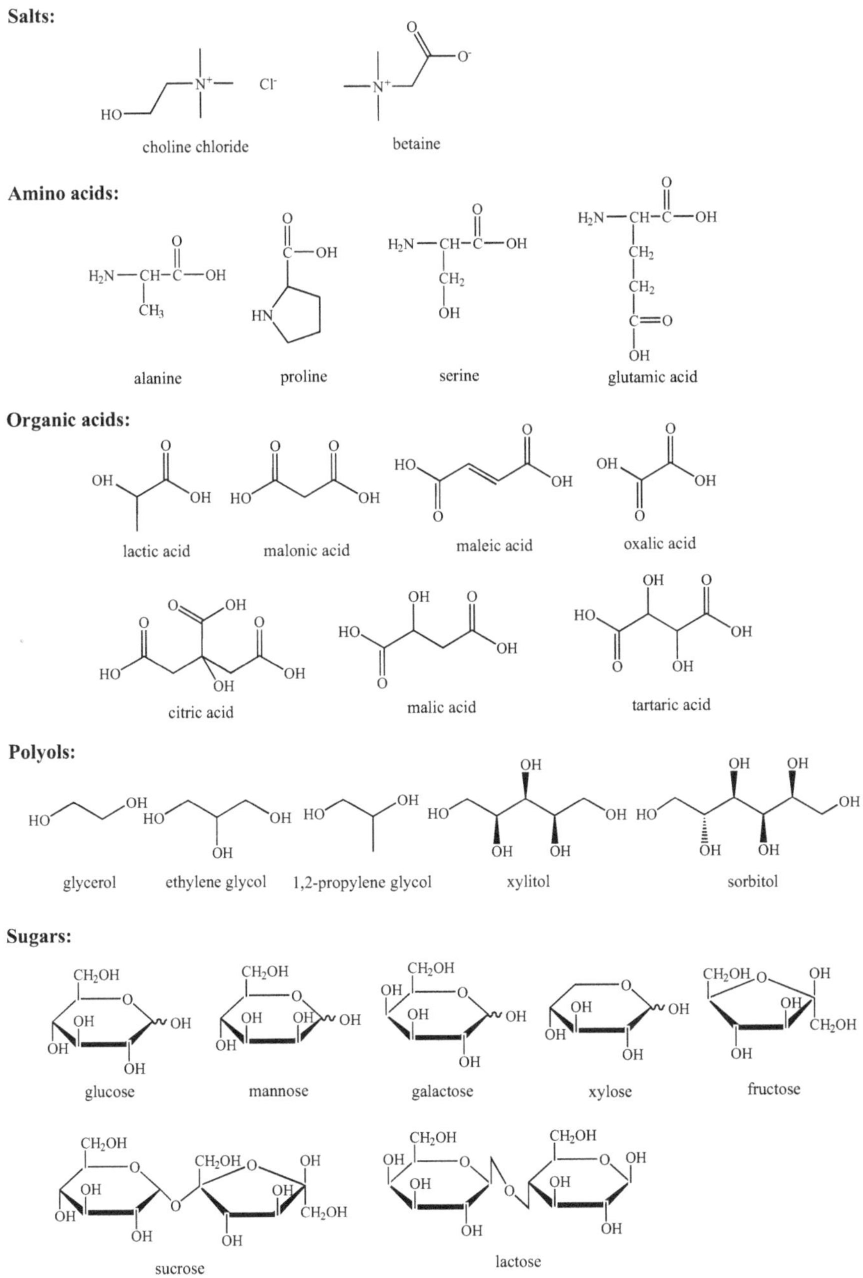

Scheme 1 Typical natural components commonly used for NADES preparation

All three methods result in the formation of the same NADES, as has been demonstrated by the ^{1}H nuclear magnetic resonance (NMR) spectra [12], but comparatively speaking the first one is both cheaper and safer.

2.2 Structure of NADESs

In the case of a DES, the main driving force for the significant depression of the freezing point upon its formation involves the interaction between the HBD and the salt anion through hydrogen bonding, and the existence of this H-bonding has been observed using NMR spectroscopy [3]. Other factors responsible for depression of the freezing point include the lattice energies of the ionic species of the DESs, the nature and asymmetry of the organic salts, and the charge delocalization that occurs through hydrogen bonding.

A similar situation is believed to occur in NADESs. By utilizing ^{1}H-^{1}H-nuclear Overhauser enhancement spectroscopy, heteronuclear Overhauser spectroscopy, and Fourier transform infrared spectroscopy (FTIR), Choi's group [9, 12, 14] has successfully demonstrated the existence of intensive H-bonding interactions between the components within NADESs such as sucrose/malic acid, 1,2-propanediol/ChCl/H_2O, and proline/malic acid (PMH)/H_2O, allowing them to form a supermolecular structure with an extensive H-bond network. This may also account for the fact that in some NADESs water is present as part of the solvent and cannot simply be removed by evaporation. The ^{1}H-NMR and FT-IR results obtained by Xin et al. [15] have also confirmed the existence of intensive H-bonding interactions between trehalose and ChCl in the NADES formed by these two components.

However, experiments also verified that for the NADES formed with ChCl and 1,2-propanediol, the H-bonds between its two components could be progressively ruptured by excessive addition of water, eventually leading to the complete disappearance of the H-bonds and consequent loss of the supramolecular structure when the water content reached 50 vol% [14]. This suggests that the supramolecular complex of the NADES remains intact if the volume of added water is less than 50%. Passing the threshold, the resulting mixture may consist merely of dissociated NADES ingredients. However, this threshold may vary from NADES to NADES, as the eutectic networks of two other NADESs, citric acid/sucrose (1:1) and malic acid/fructose/glucose (1:1:1), appear to be well-maintained upon dilution up to 1:200 [16].

2.3 Roles of NADESs in Organisms

NADESs are actually ubiquitous in nature. For example, maple syrup is a liquid composed mainly of sucrose and minor amounts of other sugars and malic acid, although all these individual components are solid at ambient temperature. In fact, the compounds that can be used to form NADES (Scheme 1) are all abundantly available as metabolites and cellular constituents in all types of cells and organisms, and different combinations of these compounds can yield over 100 different NADESs [12]. In their pioneering paper, Choi and coworkers [9] hypothesized that these NADESs form a third type of liquid phase in cells, in addition to water and lipids. Much circumstantial evidence has been collected to demonstrate that NADESs do play important roles in cellular metabolism. Many biological phenomena become explainable when NADESs are taken into account. For instance, so many poorly water-soluble metabolites and macromolecules can be synthesized, stored, and transported in plants simply because of the presence of NADESs as solvent/co-solvent, and the cells and organisms can survive in extreme drought and/or cold conditions simply because the membranes, enzymes, and metabolites remain stable in a cellular system rich in NADESs, in which water is strongly retained and freezing is prohibited by the very low melting point of the deep eutectic mixtures. In this way, cell rupture caused by the presence of ice crystals is avoided and osmotic effects are overcome [17].

3 Physicochemical Properties of NADESs

Similar to conventional ILs and DESs, NADESs are also chemically tailorable solvents, that is, their physical/chemical properties can be fine-tuned by altering their compositions, such as the choice of the components and their molar ratios. Other factors affecting the properties of NADESs include temperature and water content.

3.1 *Thermal Behavior*

In their early study of NADESs, Francisco et al. [18] reported the glass transition temperatures of 26 NADESs, which varied from -78 to -13°C. No melting points were found. Dai et al. [12] have also examined the thermal behavior of 13 NADESs using thermogravimetric analysis (TGA) and differential scanning calorimetry (DSC). All 13 NADESs have decomposition temperatures higher than 135°C and glass transition temperatures below -50°C, and again no melting points were observed. This implies that NADESs are suitable to be used as solvents in the temperature range 0–100°C.

3.2 *Density*

The densities of NADESs are all higher than that of water [12, 19–21], and they decrease linearly as either the temperature [15, 19, 20] or the water content [14] increases. Among the two sets of NADESs made up of ChCl with fructose [19] or glucose [20], the variation of their densities upon a change in the molar ratios between the two components is not the same: For the fructose-based set, the NADESs have a higher density with a higher ChCl/fructose molar ratio, whereas for the glucose-based set, no such correlation was observed.

3.3 *Viscosity*

Viscosity is related to the molecular cohesive forces in a fluid, determining its fluidity and hence the mass transfer limitations in it. Similar to ILs and DESs, most NADESs exhibit higher viscosity than many conventional organic solvents. For example, among the 13 NADESs tested by Dai et al. [12] only four have their kinetic viscosities falling in the range 33–86 mm^2/s at 40°C and the values for the rest were all higher than 138 mm^2/s, even up to 720 mm^2/s (for comparison, water's viscosity is ~ 1 mm^2/s). The fact that all DESs and NADESs are always so viscous, implying a low fluidity, is believed to result mainly from the extensive H-bonding network within the eutectic mixture and, to a lesser extent, the van der Waals and electrostatic interactions. It is therefore reasonable to expect that an NADES may be more viscous if one of its components has extra hydroxyl groups acting as the H-bond donor, thus creating more H-bonds within the eutectic mixture. For example, the dynamic viscosity of ChCl/glycerol at 30°C (0.177 Pa·s) is much lower than that for ChCl/xylitol (3.867 Pa·s), because xylitol has two more hydroxyl groups compared to glycerol [21]. The low fluidity of NADES can also be explained by the "hole theory" [22].

A comparison of the viscosity data of the two sets of NADESs (ChCl/fructose [19] and ChCl/glucose [20]) reveal that the viscosity of an NADES may be less related to the molar ratio of its components and more to the freezing point of the eutectic mixture. However, according to the study on the ChCl/glycerol DES by Abbott et al. [23], the viscosity increases in the presence of a higher ChCl/glycerol molar ratio.

A high viscosity is always a big obstacle for an NADES to be used as a solvent for any application, but fortunately it can be significantly lowered by raising either the temperature or the water content [12, 15, 19–21]. For instance, the kinetic viscosity of glucose/ChCl/H_2O (2:5:5) at 40°C is 397.4 mm^2/s, which drops to ~1 mm^2/s, similar to that of pure water, when the temperature is enhanced to 60°C or when the water content is increased to 25 vol% [12]. The sharp decrease in viscosity upon dilution with water can be attributed to the breakage of the H-bonds, which has been verified by the ^{1}H-NMR studies [14]. The variation of the viscosity upon a change in the temperature seems to fit the Arrhenius model [15, 20]:

$$\ln \eta = \ln \eta_0 + E_a/RT$$

where η is the viscosity, η_0 is a pre-exponential constant, E_a is the activation energy, R is the ideal gas constant, and T is the temperature in Kelvin.

3.4 Surface Tension

Surface tension is an important fluid property, mostly used in applications of emulsions and surfactants. The values at room temperature of the two sets of sugar-based NADESs prepared by Hayyan et al. [19, 20] fall in the range 70–75 mN/m, and they decrease linearly with an increase in the temperature in Kelvin. A reduction in surface tension upon an increase in temperature was also reported by Abbott et al. [23].

3.5 Refractive Index

Refractive index (RI) is also an important solvent property, which is rarely reported for ILs and DESs. The values for ChCl/fructose [19] and ChCl/glucose [20] lie in the ranges 1.5071–1.5228 and 1.6574–1.6661, respectively within the temperature range 25–85 °C. Again, the RI values decrease linearly with an increase in the temperature in Kelvin.

3.6 Conductivity

Dai et al. [14] have determined the conductivity of 14 NADESs at room temperature and found that all except 1 are much more conductive than water. The data seem to be dependent on the compositions of the NADESs, varying in the order of base/polyalcohol > base/organic acid ~ base/sugar > organic acid/amino acid (non-polar) > organic acid/sugar > sugar/sugar. Abbott et al. [23] have reported that the conductivities of ChCl/glycerol DESs increase as a function of the ChCl composition.

The conductivity of an NADES can also be tailored by changing the temperature or the water content. A linear increase in the conductivity upon an increase in temperature has been reported by Abbott et al. [23] and Zhao et al. [21]. As for the influence of water content, although Dai et al. [14] observed a bell-shaped relationship between conductivity and water content, all five NADESs tested by Aroso et al. [24] showed a decreased conductivity when the water content in the NADES increased from 0 to 5 wt%. On the other hand, it is worth noting that a plot of the conductivity data obtained by Dai et al. [14] against the viscosity data presented by Dai et al. [12] for the same set of 14 NADESs does not show a clear relationship, which is in disagreement with the comments claimed by both Dai et al. [14] and Zhao et al. [21] that the conductivities of NADESs are inversely correlated with their viscosities.

3.7 Polarity

Polarity is a key parameter for a solvent, determining its solubilizing power. The solvent polarity is usually determined based on the shift of the charge-transfer absorption band of a solvatochromic probe, such as Nile Red used by Dai et al. [12], in the presence of the solvent. Dai et al. [12] have measured the polarity of 13 typical NADESs made up of different components, and discovered that organic acid-based NADESs are most polar (44.8 kcal/mol), followed by amino acid-based NADESs (48.0–49.55 kcal/mol), and the NADESs composed of sugars and polyols are the least polar (48.2–50.1 kcal/mol), with a polarity close to that of water (48.21 kcal/mol) and methanol (51.89 kcal/mol). A similar trend was observed by Craveiro et al. [25] for the 11 NADESs they tested, and all these NADESs showed a polarity higher than [BMIm][BF_4], a conventional IL used as a reference. As for ChCl/glycerol DESs, a higher ChCl composition favors an increased polarity [23]. The polarity of an NADES can also be enhanced by water addition [12, 26].

3.8 Solubilizing Power

The excellent solubilization ability is undoubtedly one of the most attractive features of NADESs. They are capable of dissolving a wide range of poorly water-soluble metabolites or natural products with low (e.g., rutin, paclitaxel, ginkgolide B, quercetine) or high (e.g., starch, DNA, proteins) molecular weights. For instance, the flavonoid rutin is 50–100 times more soluble in various NADESs than in water, whereas the non-water-soluble polysaccharide, starch, shows a solubility of 17.2 mg/mL in the NADES ChCl/glucose (1:1) [9]. Faggian et al. [27] determined the solubility of rutin in 30 different NADESs: 28 of them dissolve rutin much better than water, three showing a solubility even higher than ethanol.

The solubility of a compound in NADES may be related to the polarity of the solvent, and can be further increased by adding water or increasing the temperature [12]. In fact, there is a bell-shaped relationship between the solubility of a solute and the water content in the NADES, and the optimal water content varies from solute to solute. An NMR study has suggested that the hydrogen bonding between the NADES and the solute should be responsible for the enhanced solubility of these compounds [12].

Choi's group has demonstrated that NADESs are good solvents for the above-mentioned natural products, not only because of their great solubilizing power but also because of their excellent stabilizing ability. Natural pigments such as carthamin from safflower are found to be more stable against light, high temperature, and storage time in several sugar-based NADES than in water or 40% ethanol solution, which may be ascribed to a reduced water content and an increased viscosity [28]. The same NADES-induced stabilizing effect on anthocyanins extracted from *Catharanthus roseus* was also observed in [29]. NMR and FT-IR spectra [12, 14, 28] have shown that there are intensive H-bonding interactions between solute and NADES, which may be responsible for the strong solubilizing and stabilizing ability of NADES.

3.9 *Reactivity of Glycerol in ChCl/Glycerol DES*

In order to assess the reactivity of glycerol incorporated in the eutectic mixture with ChCl, Abbott et al. [23] carried out the esterification reaction with lauric acid in glycerol and three ChCl/glycerol DESs with different molar ratios (1:2, 1:1.5, and 1:1). The esterification between the acid and glycerol did occur, implying that glycerol actively participated in the reaction, regardless of whether it was present as the pure solvent or incorporated in the DES as the reaction medium. Results show that the content of ChCl in the DES have a marked effect on the product distribution (monoester, diester, and triester), and that the rate of the consumption of lauric acid follows the trend 1:2 DES ~ 1:1.5 DES >1:1 DES > glycerol only. Plausible explanations may include a decrease in viscosity caused by the increasing ChCl content and an increased nucleophilicity of glycerol caused by the H-bond between the chloride anion and the hydroxyl group in the glycerol.

3.10 *Effect of Water*

As mentioned above, water plays an important role in affecting the properties and structure of an NADES. Dilution with water leads to a reduction in both density and viscosity of the NADES and an increase in polarity. Water affects the solubility of various solutes in an NADES and conductivity of the NADES with a bell-shaped relationship. Meanwhile, NMR studies have revealed that addition of water results in the breakage of H-bonds in the eutectic mixture, eventually leading to a rupture of its supramolecular structure. A detailed study regarding the impact of adding water on the physicochemical properties of NADES has been reported by Dai et al. [14]. Basically, the solvent properties of an NADES can be tailored by dilution with water, accompanied by a change in its chemical structure.

4 Toxicity and Biodegradability of NADESs

Because NADESs are prepared solely by using raw materials of natural origin, they are usually considered to be "truly green," and even greener than DESs. However, a lack of toxicological data seriously hinders their application. As the question "are deep eutectic solvents benign or toxic?" was raised by Hayyan's group [30], a few studies have been carried out to investigate the toxicity and biodegradability of DESs and NADESs. Although it is still far from complete, this research can provide us with useful information regarding the biosafety of using DESs and NADESs. Here we focus on a brief review of the research data obtained for NADESs.

4.1 Cytotoxicity

The first cytotoxicity test on NADES was reported by Paiva et al. [10], in which a model cell line of L929 fibroblast-like cells was treated with 11 NADESs in comparison with two typical ILs, [BMIm][Ac] and [BMIm][Cl]. Four of the NADES composed of tartaric acid and citric acid yielded an extremely low viability, the same as for the two ILs. This suggests that the presence of an organic acid may exert a detrimental effect on the cells, which was later confirmed by Zhao et al. [21], Radošević et al. [31], and Hayyan et al. [32]. A major reason for this may be the reduction in pH induced by the organic acid [21].

Zhao et al. [21] carried out the toxicology assessment on NADESs using the bacterial growth inhibition method. Twenty (NA)DESs were prepared as a eutectic mixture of ChCl with amines, alcohols, sugars, and organic acids, and their toxicity was evaluated toward two Gram-positive (*Staphylococcus aureus* and *Listeria monocytogenes*) and two Gram-negative (*Escherichia coli* and *Salmonella enteritidis*) bacteria. All the amine-, alcohol-, and sugar-based DESs did not inhibit the growth of the four bacteria, whereas all seven organic acid-based DESs had a significant inhibitory effect, suggesting that these acid-containing (NA)DESs possess antibacterial properties. The finding that the acid-based DESs were more detrimental to the Gram-negative bacteria than to the Gram-positive bacteria is interesting, because the former possess an extra outer lipopolysaccharide membrane on the cell wall, making it more protective. Huang et al. [26] have also used the above four bacteria to evaluate quantitatively the cytotoxicity of 13 different NADESs prepared with ChCl, sugars, alcohols, and amino acids. All except the NADES arginine/glycerol showed extremely low toxicity toward growth of the four bacteria. On the other hand, the 24 DESs and 21 NADESs tested in our recent studies were shown to possess a significant antibacterial activity [33, 34].

Radošević et al. [31] have evaluated three ChCl-based DESs with glucose, glycerol, and oxalic acid as H-bond donor for their cytotoxicity toward fish (CCO) and human tumor (MCF-7) cell lines and phytotoxicity on wheat. Their results reveal that all three DESs do not inhibit wheat seed germination and show low to moderate cytotoxicity. During treatment of the CCO cells with the ChCl/oxalate DES, formation of calcium oxalate crystals was observed, which may be cell damaging and therefore can partly account for the relatively higher cytotoxicity exhibited by this DES. Later in their further study [35], by testing the cytotoxicity against two human tumor cell lines (HeLa and MCF-7), five ChCl-based NADESs were again verified to be non-toxic.

In their recent study of NADES cytotoxic profiles, Hayyan et al. [32] prepared five NADESs using ChCl as the salt and glucose, fructose, sucrose, glycerol, and malonic acid as the H-bond donor, and tested their cytotoxicity toward three human (HelaS3, CaOV3, and MCF-7) and one mouse (B16F10) cancer cell lines. The IC_{50} values obtained have revealed that all five NADESs exhibit some degrees of toxicity to the cells, and the order of the cytotoxicity may be associated with the following factors:

1. pH. ChCl/malonic acid is the most toxic among all five NADESs, and one of the major reasons is the lowered pH induced by the presence of the organic acid.
2. Cellular metabolic pathways. The cells may have a high tolerance of those ingredients that are essentially required for cellular metabolism, such as choline, sugars, and glycerol. As a result, NADESs made from these components are less toxic. On the other hand, malonic acid is known to inhibit metabolic pathways such as the citric acid cycle, and this may account for the high toxicity exerted by NADESs made from this acid. Therefore, safe NADESs should be obtained by using biomaterials of cellular necessity.
3. Physical properties of the NADES, especially viscosity. It is believed that high viscosity is often related to an increased lethality. Among the five NADESs tested, ChCl/glycerol/H_2O (1:2:1) is the least toxic to all three cell lines, and coincidently it also has the lowest viscosity.
4. Interaction of NADES with the cellular membrane. The authors have tried to work out a mechanism for explaining the NADES cytotoxicity by employing the COSMO-RS (conductor-like screening model for real solvent) simulation. The model suggests a strong interaction between NADESs and some functional groups on the cellular membrane, leading to their penetration into the cells or accumulation and aggregation on the cell surface, thus possibly defining their cytotoxicity.

More recently, Hayyan's group [36] furthered their cytotoxicity studies by assessing two NADESs (namely ChCl/fructose (2:1) as NADES1 and ChCl/glucose (2:1) as NADES2) and one DES (*N,N*-diethyl ethanolammonium chloride/triethylene glycol (1:3) as DES1) using six human cancer cell lines. Cytotoxicity lies in the order of NADES2 < NADES1 < DES1, as can be compared from their EC_{50} values, thus providing evidence that NADESs are less toxic than DESs, and the results also reinforce the importance of the above-mentioned four factors in unraveling the cytotoxic mechanism. In particular, although glucose and fructose can both be taken as energy sources for cells, their high-level uptake may trigger an increase in the synthesis of advanced glycation end products and subsequently the synthesis of the reactive oxygen species (ROS). This situation is more severe for fructose and for cancer cells.

The membrane permeability assay has also verified that both DES and NADES are capable of perforating the cellular membranes, with DES being more destructive. This may be related to the cellular requirement. Entry of choline and choline-based moieties (such as those in NADES1 and NADES2) is naturally required in eukaryotic cells, although the DES1 species (i.e., *N,N*-diethyl ethanolammonium chloride and triethylene glycol) are not indigenous to the cell or perhaps are not required in similar amounts. Nevertheless, over a specified threshold concentration these ammonium cations may accumulate on the cellular membranes, disrupting the lipid bilayers and eventually leading to cellular necrosis. By comparing the EC_{50} values of the aqueous solutions of the two NADES and their raw materials (glucose, fructose, and choline), it can be ascertained that NADESs remain intact when crossing the cellular membrane.

The assessment of oxidative stress verifies that ROS were produced by the cells after the treatment with DESs or NADESs. Comparatively, the NADES-treated cells are less strenuous.

The in vivo toxicity assessment performed in mice presented a trend (NADES2 > NADES1 > DES1) that is opposite to the one obtained above after the in vitro assays. Here the higher toxicity of NADESs relative to DESs may be attributed to their high viscosity and the overall threshold concentrations, which are often lethal.

In a word, NADESs may be toxic to both bacterial and mammalian cells, implying that NADESs may possess antibacterial properties and can also be taken as potential anti-cancer agents. In terms of designing less toxic NADESs, the above-mentioned four factors have to be seriously considered.

4.2 *Biodegradability*

To date very few studies have been reported regarding the biodegradability of DESs/NADESs. The closed bottle test is the commonly used method to evaluate their aerobic biodegradability. According to the OECD guideline, all 20 DESs/NADESs tested by Zhao et al. [21] are readily biodegradable, with their biodegradation following the order of amine-based DESs ≈ sugar-based DESs > alcohol-based DESs > acid-based DESs. All 13 NADESs assessed by Huang et al. [26] can also be deemed as 'readily biodegradable.' Radošević et al. [31] have also reported that the three ChCl-based DESs they tested are all biodegradable, following the order ChCl/glycerol > ChCl/glucose > ChCl/oxalic acid, the same as found in [21].

However, among the eight DESs tested in our group [33], only two (i.e., ChCl/urea (1:1) and ChCl/acetamide (1:1)) could be considered as readily biodegradable. Comparatively, ChCl-based DESs were more biodegradable than ChAc-based DESs, and the DESs with urea and acetamide as the HBD were more susceptible than those with glycerol and ethylene glycol, which is in good agreement with the findings of Zhao et al. [21]. The fact that lower degradabilities were obtained in our study may be related to reasons such as reaction conditions, sources of the wastewater microorganisms obtained, and concentrations of the DESs and activated sludges used.

5 Applications of NADESs in Biotechnology

5.1 *Use of NADESs for Biocatalysis*

To date there are only three papers reporting the introduction of NADES into biocatalytic reaction systems: two for a free enzyme system [37, 38] and one for a whole-cell system [34].

5.1.1 Effect of NADESs on Enzyme Activity and Stability

Early in 2011 in their pioneering paper, Choi et al. [9] stated that laccase was completely dissolved but inactive in an NADES such as malic acid/ChCl (1:1), and that adding water to 50 vol% can turn the enzyme active again.

Khodaverdian et al. [38] examined the activity and stability of laccase in aqueous systems (pH 4.2) containing different betaine-based NADESs. Eight NADESs were prepared by mixing betaine with glycerol, urea, sorbitol, malic acid, oxalic acid, and citric acid. Upon addition of NADES, both activity and stability of the enzyme varied over a wide range, depending on the choice and content of the NADES added. Four of the NADESs could accelerate the enzymatic reaction. A maximal activity was obtained, which was three times as high as the activity obtained in the NADES-free system, when glycerol/betaine (2:1) was added (20 vol%). The effects of these NADESs on the kinetic parameters of the enzyme were further studied, and the results show that addition of these NADESs led to an increase in both K_m and k_{cat}, leading to a decrease in k_{cat}/K_m. In terms of stability, the half-life of the enzyme at 80°C was doubled when in the presence of sorbitol/betaine/H_2O (1:1:1). By comparing the activity and stability of the enzyme in the presence of the NADES and when the individual components of the NADES were added, the authors concluded that these activation and stabilization effects were caused by the NADES itself rather than by the presence of its separate components. These results are in good agreement with our previous studies regarding the impacts of DESs on the activity and stability of *Penicillium expansum* lipase [39] and horseradish peroxidase [40]. Interestingly, for the three betaine-based NADESs containing organic acids, malate-betaine-H_2O can enhance the laccase activity but the other two containing citric acid and oxalic acid seriously inhibit the enzyme. It is worth noting that a choline-based NADES (glycerol/ChCl (2:1)) was also tested in this study, and upon its addition both activity and stability of laccase dropped dramatically. Regarding the impact of the NADES on enzyme structure, the fluorescence spectroscopy of the enzyme in the presence of the NADES was monitored. Upon addition of three different betaine-based NADESs, the fluorescence intensity (I_{max}) was either enhanced, decreased, or unchanged, but the maximal emission wavelengths (λ_{max}) were all blue shifted, indicative of a more compact structure. In summary, this study appears to infer that laccase may be more compatible with betaine-based NADESs than with choline-based ones, and that the favorable betaine-based NADESs are capable of activating and stabilizing the enzyme.

Slightly earlier, the same research group [37] also demonstrated that the stability of chondroitinases ABCI (cABCI) can be significantly enhanced by addition of NADESs, varying in the order glycerol/ChCl (2:1) > glycerol/betaine (2:1) > no NADES. cABCI is an important clinical enzyme for treatment of spinal cord injuries, but its commercial application is seriously limited by the fact that this enzyme quickly loses its activity during storage. Therefore NADES addition provides a useful means of stabilizing the enzyme, which is beneficial to its clinical applications. The finding that the betaine-based NADES triggered more stable cABCI than the ChCl-based NADES is the opposite of what was observed in the above case of laccase. This suggests that the impact of NADESs on enzymatic performance may vary from enzyme to enzyme, depending on many factors, including protein structure and catalytic mechanism.

However, we have to admit that these are only very preliminary results so far, because there are still a lot of uncertainties and puzzles remaining unsolved. Typically, are there any correlations between activity, stability, and structure of the enzyme, and between the catalytic performance of the enzyme and the composition and combination of the NADES? Undoubtedly many more investigations have to be carried out to provide convincing insights into the mechanisms about how NADESs affect enzymatic performance.

5.1.2 Effect of NADESs on Whole-Cell Biocatalysis

In our very recent study [34], bioconversion of isoeugenol to vanillin catalyzed by *Lysinibacillus fusiformis* CGMCC1347 cells (Scheme 2) was taken as the model reaction to investigate whether whole-cell biocatalysis can be improved by adding DES/NADES. Twenty-one NADESs were prepared by mixing ChCl with a series of organic acids, polyols, and sugars. Twenty-four DESs were prepared by mixing two cholinium salts (ChCl and ChAc) and four H-bond donors (urea, acetamide, glycerol, and ethylene glycol) at a molar ratio of 1:2, 1:1, and 2:1, respectively.

As shown in Fig. 2, more than half of the NADESs tested (20 vol%) were capable of improving the production yield, and ChCl/lactic acid (4:1) and ChCl/raffinose (11:2) were the two that gave the highest conversions (132% and 131%, respectively relative to the yield obtained in the NADES-free solution). The yields seem to be sensitive to the choice of component selected for the preparation of the NADESs, varying in the order of organic acids < polyols < sugars. The low yields obtained in the presence of acid-containing NADESs are mainly caused by the significantly lowered pH they induced. The acceleration of the whole-cell catalyzed bioconversion was also observed with addition of all 24 DESs prepared.

To justify our assumption that one major impact of these additives (DESs/ NADESs) on the bioconversion might be through their interaction with the cells, we took DESs as the affecting agents to treat the cells, and used confocal laser scanning microscopy and flow cytometry tests to assess their membrane integrity

Scheme 2 Bioconversion of isoeugenol to vanillin, catalyzed by *Lysinibacillus fusiformis* CGMCC1347 cells

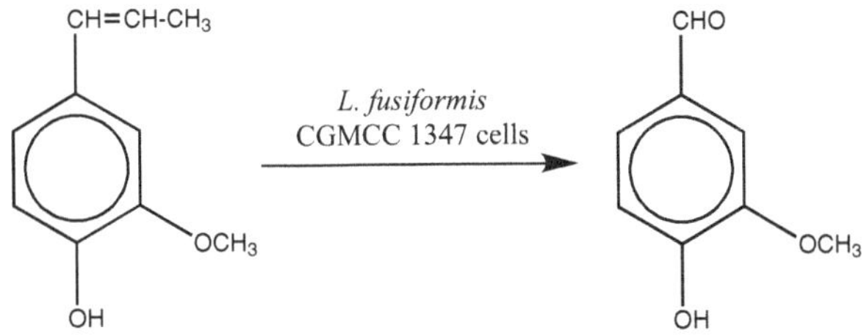

Isoeugenol Vanillin

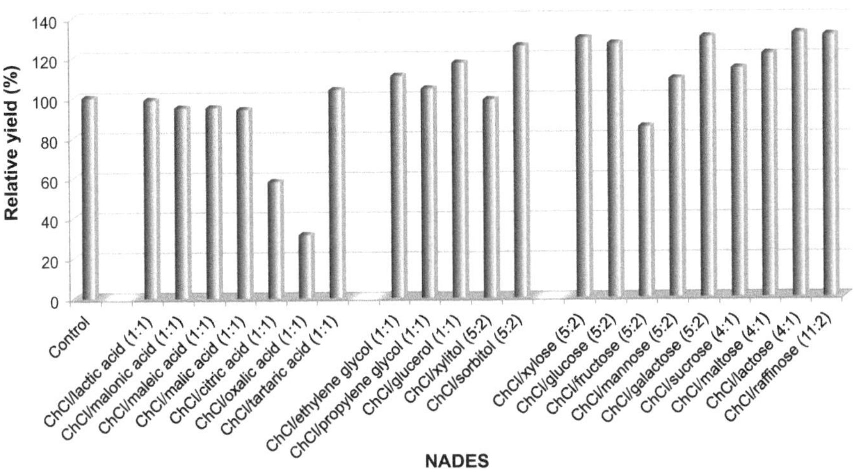

Fig. 2 Impact of NADES (1 vol% in aqueous solution) on the transformation of isoeugenol to vanillin catalyzed by *Lysinibacillus fusiformis* CGMCC1347 cells [34]

and cellular viability, respectively. Results from both tests have verified that addition of DES did to some degree disrupt the membrane of the cells, leading to their death, and the disruption became much more severe at a higher DES concentration. Determination of the change in OD260 and OD280 after removal of the DES-treated cells offers more evidence for the damage to the cell membranes, as an increase in both absorbances can be taken as a measure of the release of intracellular components (primarily nucleic acids and proteins) into the medium.

Results from all the above three tests agreed well with the bioconversion data, suggesting that addition of DES/NADES is beneficial to whole-cell biocatalysis, presumably by disrupting the cellular membranes and enhancing their permeability. The ruptured cellular membranes allow the substrate to get a better access to the enzyme responsible for the bioconversion by easily diffusing into the cells, also facilitating the product to pass more easily out of the cells. Additionally, because the enzyme may leak out through the disrupted/broken cellular membranes, it may have a direct contact with the substrate outside the cells for the reaction to occur.

It is to be noted that our conclusion above is somewhat in contrast to what was reported by Xu et al. [41]. In their study on ketone reduction catalyzed by *Acetobacter pasteurianus* G1.158 cells in DES/aqueous solutions, they found that, among the seven DESs tested, ChCl/ethylene glycol (1:2) gave the fastest initial reaction rate and highest production yield, also being biocompatible and triggering the cells to maintain relatively high cell membrane integrity.

We have to admit that the impact of (NA)DESs on whole-cell biocatalysis involves a complicated mechanism that cannot be simply explained by a single interpretation. In addition to the cellular membrane permeability, other important factors may also count, including viscosity and polarity of the reaction medium, toxicity of the additives to the cells, and direct interactions of these additives with the cells, with the enzymes responsible for the reaction, and with the substrates and/or the products. This reinforces the importance of carrying out more investigations to work out the mechanisms.

We also tried to immobilize the cells in PVA-alginate beads. The immobilized cells exerted both good activity and excellent operational stability. They offered an enhanced production yield in the presence of DES and NADES again, up to 181% of the yield obtained in a pure water system (Fig. 3), and their catalytic activity was well-maintained after being used for at least 13 cycles (each cycle lasting 72 h).

5.2 *Use of NADESs for Extraction*

Green and efficient extraction of natural products from biomass is considered an important field in the pharmaceutical and biochemical industries. Because of the unique merits of NADESs (such as sustainability, biodegradability, low toxicity, and adjustable solvent properties), especially their extraordinary solubilizing power for

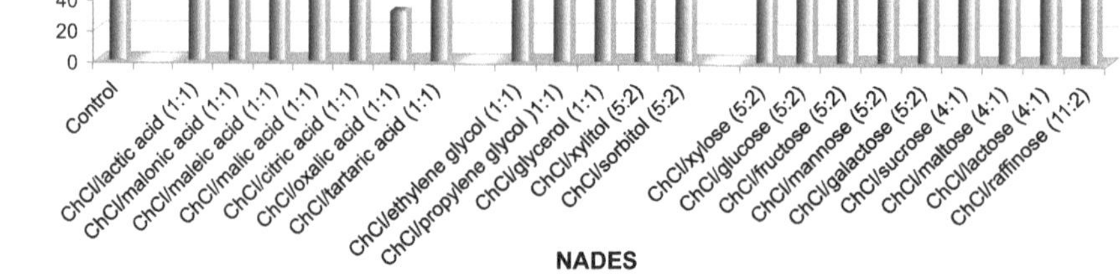

Fig. 3 Impact of NADES (20 vol% in aqueous solution) on the production of vanillin from isoeugenol, catalyzed by immobilized *Lysinibacillus fusiformis* CGMCC1347 cells [34]

natural products of diverse polarity, interest in using them for extraction of bioactive substances from natural sources has been increasing. Ruesgas-Ramón et al. [22] have written a review article extensively introducing the use of DES for extraction of phenolic compounds. In this chapter we focus on recent development in extraction with NADESs. This has recently been explored but is currently only limited to extraction of a few polyphenolic compounds such as rutin, anthocyanins, and other flavonoids/isoflavonoids [42].

A major obstacle to hinder the efficiency of NADESs as extraction solvents is their high viscosity. Fortunately this can be improved by addition of a certain amount of water and increasing the temperature to a certain degree, as has been mentioned above. Meanwhile, the polarity of NADESs also has to be considered as an important property affecting their extraction efficiency.

5.2.1 Extraction of Pigments from Flowers

The first report of applying NADESs to extraction was given by Choi's group [43]. They prepared seven NADESs composed of ChCl, organic acids, amino acids, sugars, and alcohols, and endeavored to: (1) compare their extractability of natural pigments of diverse polarity (e.g., hydroxysafflor yellow A, cartormin, carthamin) from safflower, (2) optimize the extraction parameters, and (3) recover the phenolic pigments from the NADESs extracts. Three of the NADESs, namely lactic acid/glucose (LGH), PMH, and sucrose/ChCl, were the most efficient ones in extraction, much better than ethanol, and major phenolic compounds were recovered from NADESs with a yield of 75–97%. This is the first example showing the potential of using NADESs as solvent for extraction and isolation of natural products from biomaterials.

In their next study [29], a green extraction method using NADESs was developed for extraction of anthocyanins from *Catharanthus roseus*, and their subsequent analysis was performed in combination with HPLC-DAD-based metabolic profiling. Among the seven NADESs tested, three (i.e., LGH, 1,2-propanediol/ChCl (PCH), and fructose/glucose/sucrose) presented an extractability similar to that of acidified methanol, the conventional solvent normally used for this extraction. The LGH extract was at least three times more stable than the pigments extracted from conventional organic solvents.

5.2.2 Extraction of Rutin

Inspired by the pioneering work of Choi's group indicating that rutin is 50–100 times more soluble in NADESs than in water, Zhao et al. [21] started to investigate the possibility of extracting rutin from *Sophora japonica* with DES and NADES. Among the 20 eutectic mixtures tested, 9 of them showed a better extractability, with ChCl/triethylene glycol the best, better than methanol and ethanol, the organic solvents traditionally used for rutin extraction. However, the five sugar-based NADESs did not give a high extraction yield.

Huang et al. [26] have also evaluated the extractability of rutin from tartary buckwheat hull with 13 NADESs. The solubility of rutin can be increased up to 1,577 times when placed in these NADESs, as compared to that in water. Among the 13 NADESs tested, 8 of them can extract more rutin than 80% ethanol. ChCl/glycerol (1:1) is the most efficient, producing a maximum extraction yield of 9.8 mg/g after optimization by adding water (optimal water content 20 vol%). Rutin can be recovered from this NADES with a recovery yield of 95% by using water as the most efficient antisolvent. After rutin recovery, the NADES can also be regenerated simply by vacuum evaporation. Recycling the NADES at least three times does not yield a significant loss in extraction efficiency. This extraction method is highlighted by several excellent features, such as the great solubilizing and extracting power of NADESs for rutin, simplicity and the 'green' nature of rutin recovery, and efficient NADES recycling, hence rendering this extraction process highly attractive.

5.2.3 Extraction of Other Phenolic Compounds

Table 1 lists some other examples of the use of NADESs for extracting bioactive phenolic compounds from natural sources. Among them, four have involved an evaluation of the bioactivity of the NADES extracts and are therefore discussed further below.

In the study carried out by Nam et al. [45], use of the NADES proline/glycerol (2:5) for flavonoid extraction from *Flos sophorae* is shown to be attractive, not only because it is more effective than methanol as an extraction solvent but also because the NADES-extract displays a higher anti-oxidant activity. This extra antioxidant activity derived from the NADES may be ascribed to its component, L-proline, which is known to have ROS scavenging activity.

Rajan et al. [46] have carried out an extraction of ginger (*Zingiber officinale* Roscoe) rhizome using 12 NADESs (prepared by different molar combinations of sucrose/citric acid, proline/lactic acid, proline/oxalic acid, and trehalose/citric acid). The bioactive constituents extracted from gingers were identified to be gingerols, shogaols, and other minor related compounds. The NADES extracts display not only a higher antioxidant activity relative to that of the organic solvent extracts but also a prominent antimicrobial activity against seven bacterial species: *Bacillus cereus*, *Escherichia coli*, *Klebsiella pneumonia*, *Salmonella typhi*, *Vibreo cholera*, *Staphylococcus aeureus*, *and Streptococcus viridans*.

The other study conducted by Radošević et al. [35] demonstrated that NADESs are beneficial solvents for extracting natural phenolic compounds from grapes. Among the five ChCl-based NADESs, the one containing malic acid showed the best performance in terms of extraction efficiency and antioxidant and antiproliferative activities. Some organic acids, including malic acid, which are good candidates for formation of NADESs are known to have various

Table 1 Examples of using NADESs for extracting bioactive phenolic products from natural sources

Ref.	Bioactive substances to be extracted	Natural source for the extraction	The best NADES used for the extraction
[44]	α-Mangostin	Mangosteen pericarp	ChCl/1,2-propanediol (1:3)
[45]	Flavonoids (quercetin, kaempferol, and isorhamnetin glycosides)	*Flos sophorae*	Proline/glycerol (2:5)
[46]	Gingerols, shogaols and other related compounds	Ginger	Proline/lactic acid (1:1) etc.
[47]	14 phenolics with diverse polarity	*Cajanus cajan* leaves	ChCl/maltose (1:2)
[48]	Total polyphenols	Dittany, fennel, marjoram, mint, sage	Lactic acid/glycine (3:1)
[49]	Total phenolic compounds	Extra virgin olive oil	LGH/H_2O (6:1:6)
[35]	Total phenolics Total anthocyanins	Grapes	ChCl/fructose (1.9:1)
[50]	Anthocyanins	Wine lees	ChCl/malic acid
[51]	Isoflavones (daidzin, genistin, genistein, daidzein)	Soy products	ChCl/citric acid (1:1)
[52]	Flavonoids (rutin, quercetin, kaempferol, and daidzein)	NADES-containing crude extracts	Glucose/ChCl/H_2O (2:5:5)
[53]	Phenolic compounds	Agro-food industrial by-products	LGH (5:1)

pharmacological effects including anti-oxidant and anti-cancer properties. A highly positive correlation was found between the phenolics content, anti-oxidant activity, and cytotoxicity toward cancer cells, indicating a potential anticancer property present in the grape extracts.

Bakirtzi et al. [48] tried to use five lactic acid-based NADESs as solvents for ultrasound-assisted extraction of antioxidant polyphenols from common native Greek medicinal plants such as dittany, fennel, marjoram, mint, and sage. Their data suggest that the NADES extracts with high polyphenol concentration may also possess high anti-radical activity and reducing power.

The above four studies offer good examples to show that NADESs are truly 'designer solvents', because we are able to fine-tune not only the physicochemical properties of the solvent but also the biological activities of the extracts. This may shed light on our further exploration on NADESs, leading to novel applications in the food and pharmaceutical industries.

5.2.4 Extraction of Gluten

Gluten is a composite of storage proteins that are stored together with starch in the endosperm of various cereal grains such as wheat, rye, barley, and oats. It is insoluble in water. In food processing, extraction of gluten from food is a critical step, and ethanol-water solution is usually utilized as the extractant. Lores et al. [54]

have employed an ultrasound-assisted extraction in combination with the use of diluted NADESs as the solvents for extracting gluten, which was then determined by a commercial immunoassay (ELISA). Their work demonstrated the feasibility of replacing the ethanol-water solution used for gluten extraction by some of the NADESs, especially the fructose/citric acid NADES. After extraction, the solubilized proteins maintained their structures well without significant changes. An inverse relationship was found between viscosity and solubilization ability of NADES, and a solvent with a viscosity close to that of water appeared to be most effective for gluten extraction. Additionally, the presence of citric acid in the NADES can help to prevent proteins from oxidation, thus eliminating the need for additional reducing agents such as β-mercaptoethanol, which is normally required for the ethanol-water extraction.

5.2.5 Deacidification of Crude Palm Oil

Crude palm oil contains a certain amount of free fatty acids which have to be removed by a process called deacidification; but the crude oil is also rich in natural antioxidant compounds such as carotenes and tocopherols, which are beneficial to human health and hence should be preserved during the refining process. In industry, free fatty acids are removed by steam stripping, a major disadvantage being a loss of the natural antioxidant compounds caused by the high temperatures used. In this sense, deacidification by liquid-liquid extraction with ethanol is recommended [55], but this solvent has a relatively lower extraction power for removing the free fatty acids.

By testing 13 NADESs based on betaine monohydrate, Zahrina et al. [56] found that betaine/glycerol (1:8) is highly selective as the solvent for liquid-liquid extraction of the crude palm oil because it has the highest distribution coefficient for palmitic acid and the lowest one for the antioxidants, thus facilitating a preferable selection. It has the ability to extract 34% of palmitic acid and still preserve 99% of the antioxidants in the refined palm oil. A comparison of the ^{1}H NMR spectra of the palm oil before and after extraction has revealed that the oil is stable during the whole extraction process. Infrared spectroscopy can confirm the interaction between palmitic acid and the NADES, illustrating the high affinity of the NADES for the fatty acid.

5.3 *Use of NADESs for Biomass Pretreatment*

Lignocellulosic biomass is the most abundantly available raw material on Earth for the production of biofuels, mainly bioethanol. It is composed of carbohydrate polymers (cellulose, hemicellulose) and a phenolic polymer (lignin). Pretreatment of lignocellulosic biomass is a prerequisite step, and separation of lignin from cellulose and hemicellulose fractions is critical for bioethanol production. However, these processes often involve the usage of organic solvents that are commonly used in laboratories and industry that pose important concerns over safety, health issues, and environmental pollution.

Francisco et al. [18] tried to test some NADESs as solvents for lignin, starch, and cellulose. Among the 26 NADESs they tested, most have high solubility for lignin and poor or negligible solubility for cellulose, showing a solubility for starch in between. Therefore, the separation of lignin/cellulose has a broad selection. The solubility can be tailored by carefully choosing the nature and ratio of constituents used for the NADES preparation. For example, the lactic acid/ChCl mixture does not dissolve cellulose at all but shows a good solubility for lignin, which increases with a higher acid/ChCl ratio. This is presumably because a higher acid proportion in the NADES allows stronger H-bonding to lignin. Preliminary tests on the solubility of wheat straw raw biomass samples in some selected NADESs have also shown promising results.

In another study, Kumar et al. [57] developed a recycling process in which the rice straw biomass was pretreated with NADES so that lignin and holocellulose were separated, NADES was recycled, and the residual holocellulose fraction was subjected to subsequent enzymatic hydrolysis toward fermentable sugar production. Five NADESs were prepared with lactic acid/betaine and lactic acid/ChCl combinations, all having good solubility for lignin but no solubility for both cellulose and semicellulose, making it practically feasible to proceed with delignification of rice straw by NADES pretreatment. Among the five NADESs tested, the lactic acid/ChCl (5:1) was the most efficient, achieving a maximum lignin extraction of 68 mg/g of the biomass. This lignin extraction can be further enhanced by addition of 5.0 vol% water during NADES pretreatment. The NADES can then be recovered and reused three times without any alterations. For the NADES-pretreated biomass sample, subtle structural alterations in the crystalline and amorphous regions of the cellulosic fractions were demonstrated by X-ray diffraction studies, FTIR, and TGA. The NADES-pretreated biomass can be further enzymatically hydrolyzed by cellulase with a maximum saccharification efficiency of 36.0% with a reducing sugar yield of 333 mg/g. Although this is relatively lower than obtained by using other two conventional pretreatment methods (621 mg/g and 356 mg/g for mild alkali pretreatment and dilute acid pretreatment, respectively), NADES pretreatment offers advantages such as no requirement for a post-process detoxification step because the degradation products such as furfural and hydroxymethyl furfural are not produced, and as a result a significant reduction in the overall process cost.

5.4 Use of NADESs for Clinical Therapy

Antibacterial photodynamic therapy (aPDT) is a form of phototherapy for bacterial infection, involving light and a photosensitizing chemical substance, used in conjunction with molecular oxygen to elicit cell death. Wilkene's group has demonstrated that photosensitizers (PS) such as curcumin (a hydrophobic PS) [58], THPP

(5,10,15,20-tetrakis(4-hydroxyphenyl)-porphyrin, a neutral porphyrin) [59], and TCPP (meso-tetra-(4-carboxyphenyl)-porphine, an anionic porphyrin) [60] can significantly enhance their phototoxicity against bacterial cells when formulated in NADES rather than in other aqueous preparations such as phosphate buffer saline. Their recent study [16] has further confirmed the antibacterial effect of NADES as solvent used in aPDT.

5.5 Use of NADESs for Preparation of Nutraceutical/ Pharmaceutical Products

Health-promoting products such as nutraceuticals, pharmaceuticals, and food supplements obtained by solvent extraction from plants are widespread. However, there are always two major issues to be concerned: safety of the solvent and bioavailability of the product. In terms of safety, NADESs come to be a good choice because of their well-known low toxicity.

Faggian et al. [27] selected rutin as the model compound to demonstrate the feasibility of introducing NADES into pharmaceutical and nutraceutical preparations. Rutin has a very poor solubility in water (0.12 mg/mL). Among the 30 NADESs they tested, 3 of them, proline/glutamic acid (2:1), proline/ChCl (1:2), and proline/ChCl (1:3), can dissolve more rutin than ethanol (solubility 2.4 mg/mL in ethanol), the solvent mostly used for rutin extraction. When proline/glutamic acid (2:1) was selected as the solvent for preparing the rutin formulation which was then administered orally to the Balb/c mice, a remarkable difference in the pharmacokinetic profile was observed, relative to that obtained by the rutin-water suspension. Rutin absorption was fast in both cases, but the rutin-NADES formulation exerted a nearly doubled maximal rutin concentration (C_{max}) and a much longer terminal half-life (t_{max}) in plasma, indicative of a much better bioavailability. Very recently, Sut et al. [61] also reported similar results after they tested NADES as solvent for the alkaloid berberine and had the NADES-berberine solutions orally administered to mice. These preliminary studies have thus showcased the potential use of NADES as solubilizing and formulating agent for increasing the absorption of poorly bioavailable natural products.

The great solubilizing power and intrinsic lack of toxicity have also enabled NADES to be introduced into pharmaceutical formulations. Choi's group [62] has demonstrated the feasibility and benefits of dissolving the poorly water-soluble salsalate in an NADES, 1,2-propanediol/ChCl/H_2O (1:1:1) as an alternative to DMSO for functional in vitro assays using brown cell adipocytes. Because of the increase in salsalate solubility, the pharmacological dose-response curve of salsalate activation can be greatly extended. In another study, Durand et al. [63] evaluated the ROS inhibiting activity of six polyphenolic compounds when formulated in the NADES 1,2-propanediol/ChCl/H_2O (1:1:1) by using the fibroblast cell line. All six antioxidants have their activity of ROS inhibition significantly improved, up to

360% relative to the NADES-free formulation. This NADES-induced improvement may result from an enhanced permeation capacity of the cellular membranes, or an optimization of activity within the intracellular medium by mechanisms that are yet to be found.

A recent trial showed that resveratrol, which is famous for its anti-oxidant and anti-inflammatory effects on human health, can enhance its hormetic mode of action by decreasing the matrix metalloprotease-9 (aMMP-9) activity, when formulated with an NADES, 1,2-propanediol/ChCl/H_2O (1:1:1) [64].

All the above studies have provided preliminary evidence of the feasibility of introducing NADESs into pharmaceutical/nutraceutical preparations. Whether they are really suitable to be used in vivo for drug administration still requires a lot more substantial investigations into their physicochemical, pharmacokinetic, and thermodynamic properties.

5.6 *Use of NADESs for Electrochemical Detection of Bioactive Materials*

Another application for NADES is to use it as an enhancing agent for electrochemical detection of bioactive materials such as polyphenols. Gomez et al. [65] developed a simple and sensitive electrochemical method for determining an important antioxidant, quercetin, using phosphate buffer with addition of an NADES, citric acid/glucose/H_2O (1:1:2), as the electrolyte in combination with carbon screen-printed electrodes. In the presence of 10 vol% of this NADES, the signal for quercetin was increased almost four-times relative to the signal obtained in the absence of the NADES. The proposed method exhibited a good analytical performance in terms of sensitivity, detection limit, and repeatability, highlighted by advantages such as sustainability, simplicity, fast speed, good stability, low cost, and complete portability.

The same group [66] further designed a novel electrochemical method for detection of oleuropein in complex plant matrices by a graphene oxide pencil graphite electrode in combination with a buffer modified with an NADES, 10 vol % of LGH/H_2O (5:1:3). As a result, an impressive signal enhancement (5.3 times) was obtained, relative to the signal obtained by the bare electrode with unmodified buffer.

6 Concluding Remarks

We have described above what is currently known about NADESs and their potential applications that have been explored in biotechnology and bioengineering. Although only in their infancy, NADESs have displayed many outstanding advantages and valuable potentials when employed in different biotechnology-related areas.

However, in order to take full advantage of the perceived benefits of NADESs, such as their cheapness, greenness, sustainability, excellent designability, and remarkable solubilizing power, it is necessary to carry out many more comprehensive investigations regarding their full physicochemical characterization, biosafety profiles, and correlations between the compositions and combinations of NADESs and their physicochemical properties and performance in various applications. This will undoubtedly give us a deeper understanding of this new solvent type at the molecular level, thus fostering the exploration of rational design of novel NADESs with specific properties and, in the meantime, unveiling their new applications.

References

1. MacFarlane DR, Seddon KR (2007) Ionic liquids—progress on the fundamental issues. Aust J Chem 60:3–5
2. Pham TPT, Cho C-W, Yun Y-S (2010) Environmental fate and toxicity of ionic liquids: a review. Water Res 44:352–372
3. Abbott AP, Capper G, Davies DL, Rasheed RK, Tambyrajah V (2003) Novel solvent properties of choline chloride/urea mixtures. Chem Commun 7:70–71
4. Yang Z, Wen Q (2015) Deep eutectic solvents as a new reaction medium for biotransformations. In: Paul BK, Moulik SP (eds) Ionic liquid based surfactant science: formulation, characterization and applications. Wiley, Hoboken, pp 517–531
5. Zhang Q, Vigier KDO, Royer S, Jérôme F (2012) Deep eutectic solvents: syntheses, properties and applications. Chem Soc Rev 41:7108–7146
6. Smith EL, Abbott AP, Ryder KS (2014) Deep eutectic solvents (DESs) and their applications. Chem. Rev. 114:11060–11082
7. Mbous YP, Hayyan M, Hayyan A, Wong WF, Hashim MA, Looi CY (2017) Applications of deep eutectic solvents in biotechnology and bioengineering—promises and challenges. Biotechnol. Adv. 35:105–134
8. Xu P, Zheng G-W, Zong M-H, Li N, Lou W-Y (2017) Recent progress on deep eutectic solvents in biocatalysis. Bioresour Bioprocess 4:34
9. Choi YH, van Spronsen J, Dai Y, Verberne M, Hollmann F, Arends IWCE, Witkamp G-J, Verpoorte R (2011) Are natural deep eutectic solvents the missing link in understanding cellular metabolism and physiology? Plant Physiol. 156:1701–1705
10. Paiva A, Craveiro R, Aroso I, Martins M, Reis RL, Duarte ARC (2014) Natural deep eutectic solvents – solvents for the 21st century. ACS Sustain Chem Eng 2:1063–1071
11. Durand E, Lecomte J, Villeneuve P (2016) From green chemistry to nature: the versatile role of low transition temperature mixtures. Biochimie 120:119–123
12. Dai Y, van Spronsen J, Witkamp GJ, Verpoorte R, Choi YH (2013) Natural deep eutectic solvents as new potential media for green technology. Anal. Chim. Acta 766:61–68
13. Gutiérrez MC, Ferrer ML, Mateo CR, del Monte F (2009) Freeze-drying of aqueous solutions of deep eutectic solvents: a suitable approach to deep eutectic suspensions of self-assembled structures. Langmuir 25:5509–5515
14. Dai Y, Witkamp G-J, Verpoorte R, Choi YH (2015) Tailoring properties of natural deep eutectic solvents with water to facilitate their applications. Food Chem. 187:14–19

15. Xin R, Qi S, Zeng C, Khan FI, Yang B, Wang Y (2017) A functional natural deep eutectic solvent based on trehalose: structural and physicochemical properties. Food Chem. 217:560–567
16. Wikene KO, Rukke HV, Bruzell E, Tønnesen HH (2017) Investigation of the antimicrobial effect of natural deep eutectic solvents (NADES) as solvents in antimicrobial photodynamic therapy. J Photochem Photobiol B 171:27–33
17. Gertrudes A, Craveiro R, Eltayari Z, Reis RL (2017) How do animals survive extreme temperature amplitudes? The role of natural deep eutectic solvents. ACS Sustain Chem Eng 5:9542–9553
18. Francisco M, van den Bruinhorst A, Kroon MC (2012) New natural and renewable low transition temperature mixtures (LTTMs): screening as solvents for lignocellulosic biomass processing. Green Chem. 14:2153–2157
19. Hayyan A, Mjalli FS, AlNashef IM, Al-Wahaibi T, Al-Wahaibi YM, Hashim MA (2012) Fruit sugar-based deep eutectic solvents and their physical properties. Thermochim Acta 541:70–75
20. Hayyan A, Mjalli FS, AlNashef IM, Al-Wahaibi YM, Al-Wahaibi T, Hashim MA (2013) Glucose-based deep eutectic solvents: physical properties. J Mol Liq 178:137–141
21. Zhao B-Y, Xu P, Yan F-X, Wu H, Zong M-H, Lou W-Y (2015) Biocompatible deep eutectic solvents based on choline chloride: characterization and application to the extraction of rutin from *Sophora japonica*. ACS Sustain Chem Eng 3:2746–2755
22. Ruesgas-Ramón M, Figueroa-Espinoza MC, Durand E (2017) Application of deep eutectic solvents (DES) for phenolic compounds extraction: overview, challenges, and opportunities. J. Agric. Food Chem. 65:3591–3601
23. Abbott AP, Harris RC, Ryder KS, D'Agostino C, Gladden LF, Mantle MD (2011) Glycerol eutectics as sustainable solvent systems. Green Chem. 13:82–90
24. Aroso IM, Paiva A, Reis RL, Duarte ARC (2017) Natural deep eutectic solvents from choline chloride and betaine – physicochemical properties. J. Mol. Liq. 241:654–661
25. Craveiro R, Aroso I, Flammia V, Carvalho T, Viciosa MT, Dionísio M, Barreiros S, Reis RL, Duarte ARC, Paiva A (2016) Properties and thermal behavior of natural deep eutectic solvents. J. Mol. Liq. 215:534–540
26. Huang Y, Feng F, Jiang J, Qiao Y, Wu T, Voglmeir J, Chen Z-G (2017) Green and efficient extraction of rutin from tartary buckwheat hull by using natural deep eutectic solvents. Food Chem. 221:1400–1405
27. Faggian M, Sut S, Perissutti B, Baldan V, Grabnar I, Dall'Acqua S (2016) Natural deep eutectic solvents (NADES) as a tool for bioavailability improvement: pharmacokinetics of rutin dissolved in proline/glycine after oral administration in rats: possible application in nutraceuticals. Molecules 21:1531
28. Dai Y, Verpoorte R, Choi YH (2014) Natural deep eutectic solvents providing enhanced stability of natural colorants from safflower (*Carthamus tinctorius*). Food Chem. 159:116–121
29. Dai Y, Rozema E, Verpoorte R, Choi YH (2016) Application of natural deep eutectic solvents to the extraction of anthocyanins from *Catharanthus roseus* with high extractability and stability replacing conventional organic solvents. J. Chromatogr. A 1434:50–56
30. Hayyan M, Hashim MA, Hayyan A, Al-Saadi MA, AlNashef IM, Mirghani MES, Saheed OK (2013) Are deep eutectic solvents benign or toxic? Chemosphere 90:2193–2195
31. Radošević K, Bubalo MC, Srček VG, Grgas D, Dragičević TL, Redovniković IR (2015) Evaluation of toxicity and biodegradability of choline chloride based deep eutectic solvents. Ecotoxicol Environ Saf 112:46–53
32. Hayyan M, Mbous YP, Looi CY, Wong WF, Hayyan A, Salleh Z, Mohd-Ali O (2016) Natural deep eutectic solvents: cytotoxic profile. Springerplus 5:913
33. Wen Q, Chen J-X, Tang Y-L, Wang J, Yang Z (2015) Assessing the toxicity and biodegradability of deep eutectic solvents. Chemosphere 132:63–69
34. Yang T-X, Zhao L-Q, Wang J, Song G-L, Liu H-M, Cheng H, Yang Z (2017) Improving whole-cell biocatalysis by addition of deep eutectic solvents and natural deep eutectic solvents. ACS Sustain Chem Eng 5:5713–5722

35. Radošević K, Curko N, Srček VG, Bubalo MC, Tomasevic M, Ganić KK, Redovniković IR (2016) Natural deep eutectic solvents as beneficial extractants for enhancement of plant extracts bioactivity. LWT Food Sci. Technol. 73:45–51
36. Mbous YP, Hayyan M, Wong WF, Looi CY, Hashim MA (2017) Unraveling the cytotoxicity and metabolic pathways of binary natural deep eutectic solvent systems. Sci. Rep. 7:41257
37. Daneshjou S, Khodaverdian S, Dabirmanesh B, Rahimi F, Daneshjoo S, Ghazi F, Khajeh K (2017) Improvement of chondroitinases ABCI stability in natural deep eutectic solvents. J. Mol. Liq. 227:21–25
38. Khodaverdiana S, Dabirmanesh B, Heydarib A, Dashtban-moghadama E, Khajeha K, Ghazi F (2018) Activity, stability and structure of laccase in betaine based natural deep eutectic solvents. Int J Biol Macromol 107(Pt B):2574–2579
39. Huang Z-L, Wu B-P, Wen Q, Yang T-X, Yang Z (2014) Deep eutectic solvents can be viable enzyme activators and stabilizers. J. Chem. Technol. Biotechnol. 89:1975–1981
40. Wu B-P, Wen Q, Xu H, Yang Z (2014) Insights into the impact of deep eutectic solvents on horseradish peroxidase: activity, stability and structure. J Mol Catal B Enzym 101:101–107
41. Xu P, Du P-X, Zong M-H, Li N, Lou W-Y (2016) Combination of deep eutectic solvent and ionic liquid to improve biocatalytic reduction of 2-octanone with *Acetobacter pasteurianus* G1M1.158 cell. Sci. Rep. 6:26158
42. Owczarek K, Szczepanska N, Plotka-Wasylka J, Rutkowska M, Shyshchak O, Bratychak M, Namiesnik J (2016) Natural deep eutectic solvents in extraction process. Ch&ChT 10:601–606
43. Dai Y, Witkamp G-J, Verpoorte R, Choi YH (2013) Natural deep eutectic solvents as a new extraction media for phenolic metabolites in *Carthamus tinctorius* L. Anal. Chem. 85:6272–6278
44. Mulia K, Krisanti E, Terahadi F, Putri S (2015) Selected natural deep eutectic solvents for the extraction of α-mangostin from mangosteen (*Garcinia mangostana* L.) pericarp. Int. J. Technol. 6(7):1211–1220
45. Nam MW, Zhao J, Lee MS, Jeong JH, Lee J (2015) Enhanced extraction of bioactive natural products using tailor-made deep eutectic solvents: application to flavonoid extraction from *Flos sophorae*. Green Chem. 17:1718–1727
46. Rajan M, Prabhavathy A, Ramesh U (2015) Natural deep eutectic solvent extraction media for *Zingiber officinale Roscoe*: the study of chemical compositions, antioxidants and antimicrobial activities. Nat. Prod. J. 5:3–13
47. Wei Z, Qi X, Li T, Luo M, Wang W, Zu Y, Fu Y (2015) Application of natural deep eutectic solvents for extraction and determination of phenolics in *Cajanus cajan* leaves by ultra performance liquid chromatography. Sep. Purif. Technol. 149:237–244
48. Bakirtzi C, Triantafyllidou K, Makris DP (2016) Novel lactic acid-based natural deep eutectic solvents: efficiency in the ultrasound-assisted extraction of antioxidant polyphenols from common native Greek medicinal plants. J. Appl. Res. Med. Aromat. Plants 3:120–127
49. Paradiso VM, Clemente A, Summo C, Pasqualone A, Caponio F (2016) Towards green analysis of virgin olive oil phenolic compounds: extraction by a natural deep eutectic solvent and direct spectrophotometric detection. Food Chem. 212:43–47
50. Bosiljkov T, Dujmic F, Bubalo MC, Hribar J, Vidrih R, Brncic M, Zlatic E, Redovnikovic IR, Jokic S (2017) Natural deep eutectic solvents and ultrasound-assisted extraction: green approaches for extraction of wine lees anthocyanins. Food Bioprod Process 102:195–203
51. Bajkacz S, Adamek J (2017) Evaluation of new natural deep eutectic solvents for the extraction of isoflavones from soy products. Talanta 168:329–335
52. Liu Y, Garzon J, Friesen JB, Zhang Y, McAlpine JB, Lankin DC, Chen S-N, Pauli GF (2016) Countercurrent assisted quantitative recovery of metabolites from plant-associated natural deep eutectic solvents. Fitoterapia 112:30–37
53. Fernández M, Espino M, Gomez FJV, Silva MF (2018) Novel approaches mediated by tailor-made green solvents for the extraction of phenolic compounds from agro-food industrial by-products. Food Chem. 239:671–678

54. Lores H, Romero V, Costas I, Bendicho C, Lavilla I (2017) Natural deep eutectic solvents in combination with ultrasonic energy as a green approach for solubilisation of proteins: application to gluten determination by immunoassay. Talanta 162:453–459
55. Gonçalves CB, Rodrigues CEC, Marcon EC, Meirelles AJA (2016) Deacidification of palm oil by solvent extraction. Sep Purif Technol 160:106–111
56. Zahrina I, Nasikin M, Krisanti E, Mulia K (2018) Deacidification of palm oil using betaine monohydrate-based natural deep eutectic solvents. Food Chem. 240:490–495
57. Kumar AK, Parikh BS, Pravakar M (2016) Natural deep eutectic solvent mediated pretreatment of rice straw: bioanalytical characterization of lignin extract and enzymatic hydrolysis of pretreated biomass residue. Environ. Sci. Pollut. Res. 23:9265–9275
58. Wikene KO, Bruzell E, Tønnesen HH (2015) Characterization and antimicrobial phototoxicity of curcumin dissolved in natural deep eutectic solvents. Eur. J. Pharm. Sci. 80:26–32
59. Wikene KO, Bruzell E, Tønnesen HH (2015) Improved antibacterial phototoxicity of a neutral porphyrin in natural deep eutectic solvents. J Photochem Photobiol B 148:188–196
60. Wikene KO, Rukke HV, Bruzell E, Tønnesen HH (2016) Physicochemical characterisation and antimicrobial phototoxicity of an anionic porphyrin in natural deep eutectic solvents. Eur. J. Pharm. Biopharm. 105:75–84
61. Sut S, Faggian M, Baldan V, Poloniato G, Castagliuolo I, Grabnar I, Perissutti B, Brun P, Maggi F, Voinovich D, Peron G, Dall'Acqua S (2017) Natural deep eutectic solvents (NADES) to enhance berberine absorption: an *in vivo* pharmacokinetic study. Molecules 22:1921
62. Rozema E, van Dam AD, Sips HCM, Verpoorte R, Meijer OC, Kooijman S, Choi YH (2015) Extending pharmacological dose-response curves for salsalate with natural deep eutectic solvents. RSC Adv. 5:61398–61401
63. Durand E, Lecomte J, Upasani R, Chabi B, Bayrasy C, Baréa B, Jublanc E, Clarke MJ, Moore DJ, Crowther J, Wrutniak-Cabello C, Villeneuve P (2017) Evaluation of the ROS inhibiting activity and mitochondrial targeting of phenolic compounds in fibroblast cells model system and enhancement of efficiency by natural deep eutectic solvent (NADES) formulation. Pharm. Res. 34:1134–1146
64. Shamseddin A, Crauste C, Durand E, Villeneuve P, Dubois G, Durand T, Vercauteren J, Veas F (2017) Resveratrol formulated with a natural deep eutectic solvent inhibits active matrix metalloprotease-9 in hormetic conditions. Eur. J. Lipid Sci. Technol. 119:1700171
65. Gomez FJV, Espino M, Fernández M, Raba J, Silva MF (2016) Enhanced electrochemical detection of quercetin by natural deep eutectic solvents. Anal. Chim. Acta 936:91–96
66. Gomez FJV, Spisso A, Silva MF (2017) Pencil graphite electrodes for improved electrochemical detection of oleuropein by the combination of natural deep eutectic solvents and graphene oxide. Electrophoresis 0:1–8

Adv Biochem Eng Biotechnol (2019) 168: 61–78
DOI: 10.1007/10_2018_64

Published online: 10 May 2018

Ionic Liquid Pretreatment of Lignocellulosic Biomass for Enhanced Enzymatic Delignification

Muhammad Moniruzzaman and Masahiro Goto

Contents

Abstract Ionic liquids (ILs), a potentially attractive "green," recyclable alternative to environmentally harmful volatile organic compounds, have been increasingly

M. Moniruzzaman
Chemical Engineering Department, Universiti Teknologi PETRONAS, Bandar Seri Iskandar, Perak, Malaysia

Center of Research in Ionic Liquids (CORIL), Universiti Teknologi PETRONAS, Bandar Seri Iskandar, Perak, Malaysia
e-mail: m.moniruzzaman@utp.edu.my

M. Goto (✉)
Department of Applied Chemistry, Graduate School of Engineering, Kyushu University, Fukuoka, Japan

Center for Future Chemistry, Kyushu University, Fukuoka, Japan
e-mail: m-goto@mail.cstm.kyushu-u.ac.jp

exploited as solvents and/or cosolvents and/or reagents in a wide range of applications, including pretreatment of lignocellulosic biomass for further processing. The enzymatic delignification of biomass to degrade lignin, a complex aromatic polymer, has received much attention as an environmentally friendly process for clean separation of biopolymers including cellulose and lignin. For this purpose, enzymes are generally isolated from naturally occurring fungi or genetically engineered fungi and used in an aqueous medium. However, enzymatic delignification has been found to be very slow in these conditions, sometimes taking several months for completion. In this chapter, we highlight an environmentally friendly and efficient approach for enzymatic delignification of lignocellulosic biomass using room temperature ionic liquids (ILs) as (co)solvents or/and pretreatment agents. The method comprises pretreatment of lignocellulosic biomass in IL-aqueous systems before enzymatic delignification, with the aim of overcoming the low delignification efficiency associated with low enzyme accessibility to the solid substrate and low substrate and product solubilities in aqueous systems. We believe the processes described here can play an important role in the conversion of lignocellulosic biomass—the most abundant renewable biomaterial in the world—to biomaterials, biopolymers, biofuels, bioplastics, and hydrocarbons.

Graphical Abstract

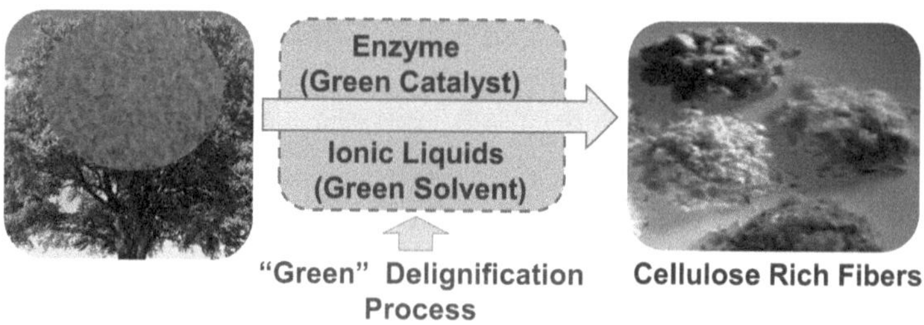

Keywords Cellulose fibers, Enzymatic delignification, Ionic liquids, Laccase, Lignin, Lignocellulosic biomass

1 Introduction

Plants are an abundant carbon-neutral renewable resource ("lignocellulosic biomass") for the production of bioenergy and biomaterials. In recent years the use of renewable resources, particularly lignocellulosic biomass-based raw materials, to replace synthetic materials/polymers for the manufacture of green materials has gained increased worldwide interest because of growing global environmental awareness, concepts of sustainability, and the absence of conflict between food and chemical/materials production. Most lignocellulosic biomasses (including wood) mainly consist of three subcomponents: the rigid semicrystalline polysaccharide cellulose, the amorphous multicomponent polysaccharide hemicellulose, and the amorphous aromatic polymer lignin. These three biopolymers are the primary constituents of plant cell walls, in which the cross-linked matrix of lignin and hemicelluloses embeds the cellulose fibers, forming a tight, compact structure. Among these biopolymers, celluloses—Earth's most abundant biopolymers—have been used extensively as a source of raw materials for biocompatible and biodegradable materials/biocomposites production, and lignin—Earth's second most abundant biopolymer—can be converted into value-added end products, especially fuel components, via depolymerization. However, structural heterogeneity and the complexity of cell wall microfibrils in most lignocellulosic biomass represent a big challenge in the clean separation of celluloses and lignin with minimal polymer degradation.

A number of techniques, including physical (e.g., pyrolysis and mechanical disruption) [1, 2], physicochemical (e.g., steam explosion and ammonia fiber explosion) [3, 4], chemical (e.g., acid hydrolysis, alkaline hydrolysis, and oxidative delignification) [5, 6], thermochemical [7], and enzymatic delignification methods [8–10], have been investigated for separation of cellulose and lignin at both laboratory scale and pilot plant levels. Most of these processes operate at high temperature and pressure and required highly concentrated chemicals. The conventional chemicals used for pulping processes, such as sulfate and sulfite, pose serious environmental hazards themselves to air and water. Furthermore, high temperature-based cooking processes result in the production of inhibitory chemicals of the products and degradation products.

Although enzymatic delignification by enzymes isolated from naturally occurring fungi, or using enzymes produced by genetically engineered fungi, can be performed in mild reaction conditions, this approach has been found to be very slow in aqueous systems, mainly because of difficulties in the enzymes accessing the solid substrate and the poor solubility of the products [11, 12]. In the last few years, ionic liquids (ILs)—a potentially environmentally benign, recyclable alternative to volatile organic compounds (VOCs)—have received tremendous academic and industrial attention as promising solvents for the pretreatment of lignocellulosic biomass before further processing [13–15]. ILs are composed entirely of ions (generally consisting of organic cations, namely derivatives of *N,N′*-substituted imidazolium, *N*-substituted pyridinium, tetraalkylated ammonium, and tetraalkylated phosphonium, and either organic or inorganic anions) and are liquids at ambient temperatures. Not only are ILs more environmentally attractive than VOCs but they also possess many attractive physicochemical properties, including high solvating power, excellent chemical and thermal stability, and so on. Recently, many ILs have been used to dissolve wood and other lignocellulosic biomass at high temperatures ($>100°C$), and cellulose-rich materials and lignin can then readily be separated by the addition of a variety of precipitating solvents [14, 16, 17]. It is reported that hydrophilic ILs can solubilize wood completely at $>100°C$, and cellulose-rich materials can readily be precipitated with an anti-solvent, such as water, acetone, or ethanol. The degree of polymerization of the regenerated cellulose was found to be reduced significantly compared to the cellulose in the original wood materials, which led to enhanced enzymatic cellulose hydrolysis, a desirable property for biomass for biofuels technologies [18]. However, the production of high-strength biomaterials and biocomposites requires strong cellulose nanofibers/microfibrils. Thus, research into the design and development of IL-based technology that can extract cellulose from lignocellulosic biomass with minimal structural alteration is needed. Other notable limitations of direct wood dissolution in ILs are the significant loss of cellulose during washing steps through the conversion of cellulose into sugars and only partial delignification of the wood biomass [19]. It is therefore desirable to develop an IL-assisted biomass pretreatment process that can extract strong cellulose fibers to be used to make high strength biomaterials, and 'pristine' lignin that can be converted into value-added products via depolymerization.

The enzymatic delignification of lignocellulosic biomass in IL or IL/water systems may be an efficient way to produce highly-purified cellulose nanofibers and lignin. The use of enzymes such as laccases (produced from different white-rot fungi) that can degrade lignin but leave the cellulose of the wood virtually untouched in ILs could have significant efficiency advantages over aqueous systems because the substrate and product solubility are expected to increase, notably in ILs. It should be noted that laccases from various white-rot fungi have been used extensively for lignin degradation from wood biomass [20]. It has been reported that laccases maintain their activity and stability in ILs only in the presence of >80% water in IL-water solutions [21], which is consistent with what was reported for other enzymes such as cellulase in IL-water reaction media [22].

In this chapter we discuss the effect of using ILs as pretreatment agents and/or solvents before enzymatic delignification of lignocellulosic biomass, including wood biomass and oil palm biomass. Physicochemical characterization of untreated and treated lignocellulosic materials is also discussed.

2 IL Pretreatment of Wood Biomass Followed by Enzymatic Delignification: A One-Step Process

Conventional delignification is generally carried out in aqueous medium, which is the best medium for enzymes. However, delignification in aqueous systems is very slow, mainly because of difficulties in the enzyme accessing the solid substrate and the poor solubility of the substrates and products, including lignin [12]. The poor solubility of substrate and products during wood delignification in aqueous systems can be overcome by using ILs as solvents/cosolvents. Unfortunately, the practical obstacle to using ILs for enzymatic delignification is that many ILs, particularly hydrophilic ones, have negative effects on enzyme structure, resulting in deactivation of the enzyme [13, 23]. However, such effects could be balanced by the increase in solubility of the substrates and products, leading to better overall process performance in terms of enhanced yield. It is also well-recognized that ILs can be used for effective pretreatment of lignocellulosic biomass before further enzymatic delignification and hydrolysis [13, 24].

2.1 IL Pretreatment of Wood Biomass for Enhanced Enzymatic Delignification

Enzymatic delignification efficiency can be improved by IL pretreatment of wood biomass (hinoki cypress, *Chamaecyparis obtusa*) before enzymatic delignification in aqueous systems [25]. A simplified overview of the experimental method is shown in Fig. 1. In this one-step process, pretreatment of wood biomass (hinoki cypress, 10 wt%) with IL [emim][OAc] (1-ethyl-3-methyl imidazolium acetate) was carried out in moderate conditions (80°C, 1 h) to swell the wood cells by partial dissolution. The selected IL [emim][OAc] can dissolve wood materials significantly [16, 19]. In general, temperatures of 80–130°C have been used to dissolve wood materials in ILs [19]. Although elevated temperatures (≥100°C) lead to complete dissolution of wood biomass, which favors delignification efficiency, the crystallinity of the regenerated cellulose-rich materials is decreased and loss of biopolymer is increased significantly [26, 27]. Thus, a moderate temperature was selected for this IL pretreatment to minimize the loss of the major extracts with minimal structural alteration. After completing pretreatment at 80°C with vigorous mechanical stirring, the mixture became dark and its viscosity increased, indicating that partial dissolution of the wood had occurred. The mixture was then diluted with acetate buffer and enzymatic delignification was carried out.

For delignification, aqueous solution containing enzyme (1,000 U/g commercial laccase Y120 [EC.1.10.3.2] from *Trametes* sp.) was added directly to the IL/wood mixture. The catalytic cycle of a laccase-mediator lignin degradation system is shown in Fig. 2. Wood delignification was performed in the presence of O_2 bubbles with a small stirrer bar at 50°C for 24 h when 1-hydroxybenzotriazole (1.5 wt% of wood biomass) was added as the mediator. The results clearly demonstrated that enzymatic delignification of wood biomass swollen by IL prior to the enzymatic delignification could be an efficient method for the removal of lignin to extract cellulose fibers (see Table 1). It was also found that delignification efficiency was highly dependent on the IL content in the IL-aqueous media (see Table 1, entries 3 and 4). Lower delignification at higher IL contents in IL-aqueous systems was

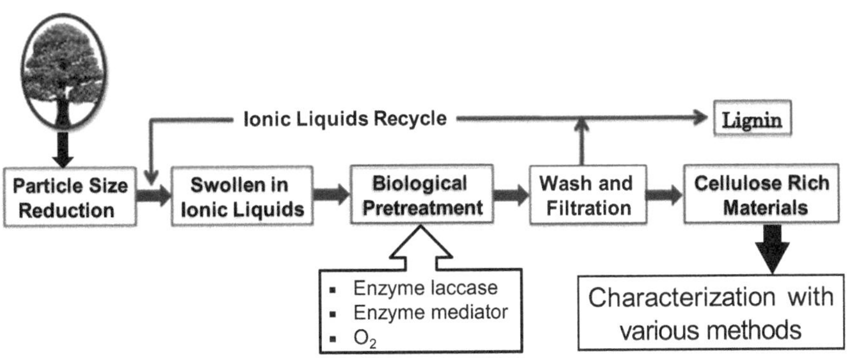

Fig. 1 Flowchart of ionic liquid (IL)-assisted enzymatic delignification of wood biomass

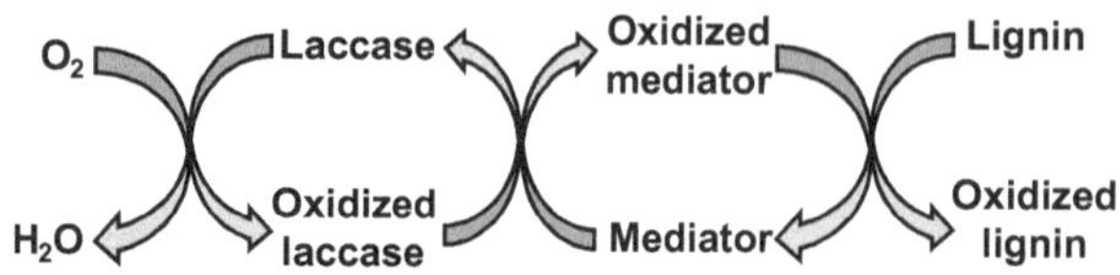

Fig. 2 Catalytic cycle of a laccase-mediator lignin degradation system

Table 1 Lignin extraction from wood biomass using various methods [25]

Entry	Delignification methods	Reaction media	ILs swollen prior to reaction[a]	Extracted lignin[b] (%)
1	Enzymatic[c]	Acetate buffer	No	10.85 ± 2.36
2	IL pretreated[d]	20 wt% IL in buffer	Yes	7.05 ± 1.84
3	Enzymatic[e]	20 wt% IL in buffer	Yes	28.93 ± 3.54
4	Enzymatic[f]	10 wt% IL in buffer	Yes	39.46 ± 4.60
5	Enzymatic[g]	5 wt% IL in buffer	Yes	50.10 ± 3.20

[a]Ground wood (200 mg) was incubated in 2 g IL at 80°C with vigorous magnetic stirring for 1 h
[b]Results are expressed as the percentage of extracted lignin relative to the lignin content in the original ground wood
[c]Reaction conditions: 200 mg wood chips, 10 mL of 100 mM sodium acetate buffer (pH 4.5), 50 U laccase, 50°C, 24 h, 500 rpm, 3 mg 1-hydroxybenzotriazole
[d]Wood (200 mg) swollen by 2 g IL, 8 mL acetate buffer, 50°C, 23 h and 500 rpm
[e]Wood (200 mg) swollen by 2 g IL, buffer 8 mL, 24 h, other reactions conditions the same as for entry 1
[f]All the conditions were the same as for entry 3 except for the addition of 18 mL of buffer instead of 8 mL
[g]All the conditions were the same as for entry 3 except for the addition of 38 mL of buffer instead of 8 mL

obtained because of the denaturing effect of IL on enzymes, which is a very common trend in the field of IL-based biocatalysis and biotransformations [23]. Compared to IL [bmim][Cl], IL [emim][OAc] was more suitable for pretreatment of wood biomass for enhanced enzymatic delignification [28], possibly because of its great ability to dissolve wood and its compatibility with enzymes. Another possible reason for the lower delignification efficiency when using IL [bmim][Cl] was the well-known adverse effect of Cl^- on enzyme performance [29].

2.2 *Effect of IL Pretreatment on Enzymatic Delignification*

As discussed in Sect. 2.1, delignification efficiency may be enhanced by a combination of factors. First, the swelling of wood biomass increases the surface area of the biomass available to the enzyme(s) [30]. A schematic diagram (Fig. 3) shows the

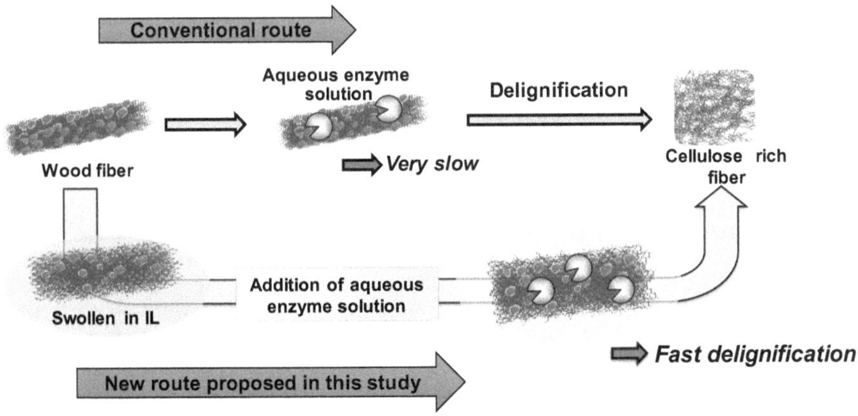

Fig. 3 Comparison of IL-assisted enzymatic delignification to conventional enzymatic delignification in aqueous systems [25]

advantages of IL-assisted enzymatic delignification over conventional enzymatic delignification in aqueous systems. In addition, during swelling at 80°C, ILs can dissolve some lignin, which can further increase enzyme accessibility [18, 31]. Second, the substrate and product solubility are expected to increase notably when using ILs, which is certainly beneficial for overcoming the low delignification efficiency associated with poor substrate and product solubility in aqueous systems.

To understand the correlation of wood biomass enzymatic delignification with IL pretreatment time, various samples of IL [emim][OAc]-pretreated wood biomass were prepared by changing the treatment time in the IL [28]. Generally, longer cooking time increases delignification efficiency [19], although use of energy becomes very important with long cooking times. The incubation time for pretreatment of wood in IL varied between 0.5 and 3 h at 80°C (Fig. 4). It was

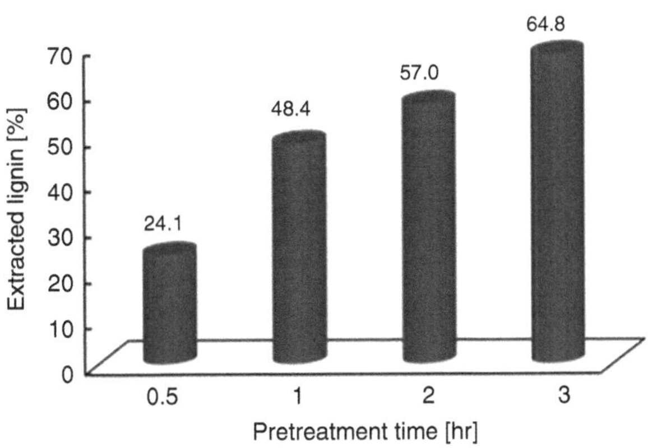

Fig. 4 Effect of cooking time in IL [emim][OAc] on delignification of wood biomass; 200 mg of ground wood were incubated in 2 g IL at 80°C with vigorous magnetic stirring [28]

found that the delignification efficiency increased with increasing pretreatment time. For example, on increasing the pretreatment time from 0.5 to 3 h, the delignification efficiency for wood biomass increased from 24.1 to 64.8%. This result is consistent with other reports in the literature [18].

2.3 *Characterization of Untreated and Treated Materials*

Extracted cellulose-rich wood fibers and untreated wood fibers were characterized by acid hydrolysis using Fourier-transform infrared spectroscopy (FTIR), scanning electron microscopy (SEM), and X-ray diffractometry (XRD) to reveal the compositional and structural impacts better [25]. As shown in Fig. 5, SEM images revealed that that, after pretreatment, wood cell networks composed of cellulose, hemicellulose, and lignin were broken down and cellulose fibers were partially separated into individual microsized fibers. Magnified images revealed that the surfaces of untreated fibers were very rough, possibly because of the coating of cellulose fibers with lignin, whereas those of treated fibers were smooth. This indicates the removal of lignin from the surface of the wood fibers during enzymatic treatment. FTIR results indicated that the cellulose-rich wood fibers produced from IL-pretreated wood fibers were richer in carbohydrates, consistent with the chemical composition study [25]. The XRD patterns of the untreated and treated wood fibers were examined to investigate the crystalline properties of the fibers [25]. There was a

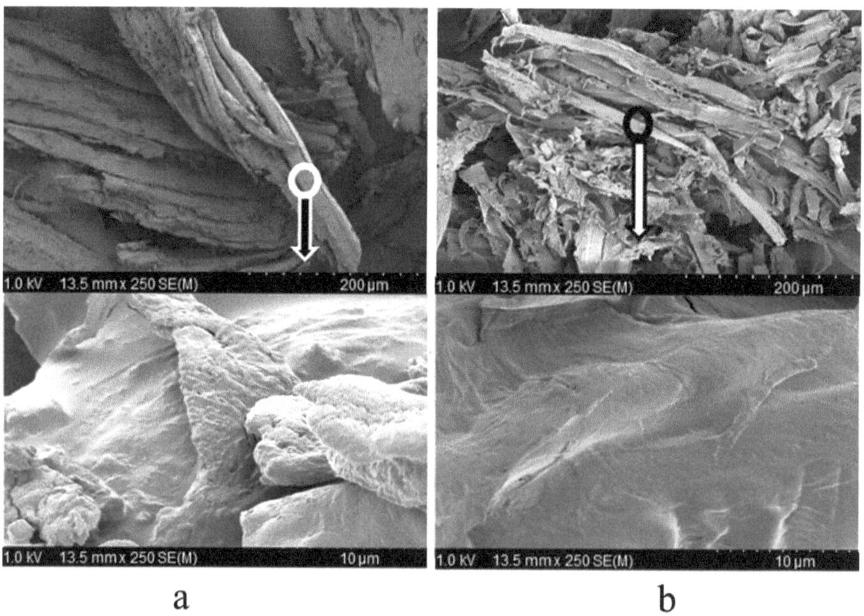

Fig. 5 SEM images of (**a**) untreated ground wood and (**b**) the corresponding enzymatic treated wood fibers [25]

higher degree of crystallinity in the structure of the treated fibers, which is undoubtedly caused by the partial removal of lignin and hemicellulose. This trend contrasts with that for cellulose fibers regenerated after dissolution in ILs at higher temperatures [18].

3 IL Pretreatment of Lignocellulosic Biomass Followed by Enzymatic Delignification of Recovered Biomass: A Two-Step Process

As discussed above, delignification efficiency can be improved by IL pretreatment of wood biomass before enzymatic reaction. However, in the one-step processes, IL remains in the wood cell wall, and thus enzyme performance was decreased by the well-known negative effects of IL on enzyme activity [22, 23]. To address this limitation, wood materials recovered after IL pretreatment can be used to accelerate enzymatic delignification [27, 32]. An effective pretreatment method for lignocellulosic biomass (e.g., wood and oil palm biomass) for isolating cellulose fibers via enzymatic delignification following IL pretreatment and a recovery step has been reported [33, 34].

3.1 IL Pretreatment of Wood Biomass Before Enzymatic Delignification: A Two-Step Process

3.1.1 Chemical Composition of Untreated and IL-Treated Wood Materials

A simplified flowchart of the IL pretreatment of wood biomass before enzymatic delignification is shown in Fig. 6. Pretreatment of wood materials with IL was carried out in moderate conditions (80°C for 1 h) to swell the wood cells by partial dissolution, and the product solids were recovered by addition of an anti-solvent [34]. The chemical compositions of untreated wood materials (UWMs) and recovered wood material (RWMs) after IL treatment are compared in Table 2. The yield of RWMs after IL treatment was about 87% of the untreated wood biomass weight. During IL treatment, swelling of wood cell walls occurs caused by partial breaking of the bonds between the major biopolymers in the wood matrix [18, 35]. Consequently, a small portion of the hemicellulose and lignin is solubilized, resulting in the decrease of their content in the wood material after the IL treatment. Thus, compared to the untreated sample, RWMs after IL treatment had a somewhat higher cellulose content caused by the removal of hemicellulose, lignin, and water/IL/acetone soluble extractives during the pretreatment/recovery processes.

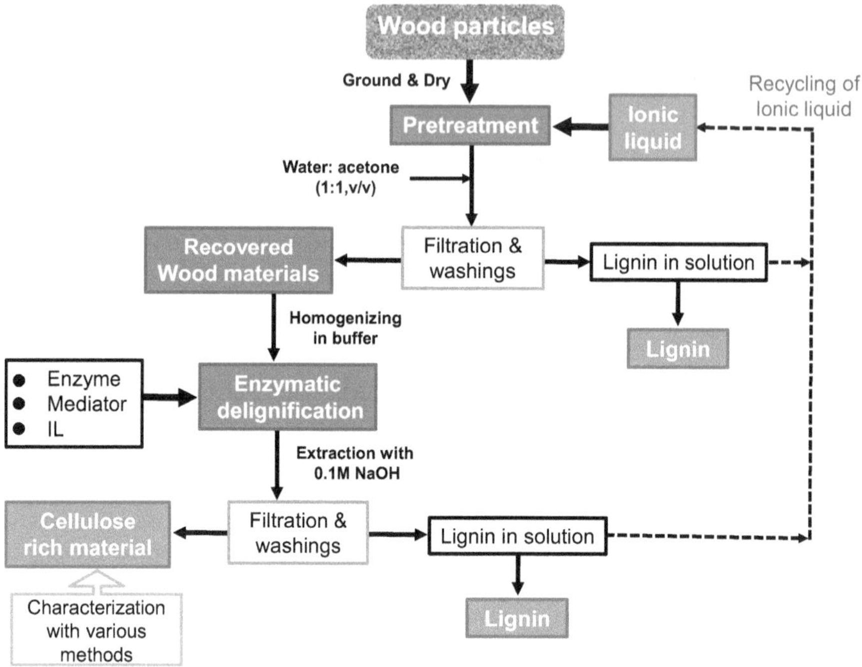

Fig. 6 Simplified schematic diagram for the delignifying process of wood biomass. Enzymatic delignification was performed in buffer with or without a small amount of IL and O_2 bubbles. The ionic liquid was [emim][OAc] [34]

Table 2 Chemical composition of untreated and treated wood materials[a] [34]

Materials	α-Cellulose (%)	Hemicelluloses (%)	Total lignin (%)[b]
Untreated wood materials	41.2 ± 1.3	25.2 ± 1.1	29.3 ± 1.7
RWMs after IL treatment	46.3 ± 2.1	15.3 ± 1.2	25.2 ± 3.2
Cellulose rich fibers (CRFs) (IL + enzymatic treatment)	73.1 ± 3.5	8.5 ± 1.0	9.8 ± 1.4

[a]The data are the average from three experiments and the standard deviation
[b]Total lignin = Klason lignin + acid soluble lignin

3.1.2 Enzymatic Delignification of RWMs After IL Treatment

To examine the effects of IL pretreatment on the subsequent enzymatic delignification, the RWMs were subjected to biological treatment (at 50°C for 24 h) using laccase as the catalyst. The reaction was carried out in acetate buffer containing 2.5 wt% IL [emim][OAc]. This small amount of IL was used during delignification because it was found that, although the initial activity of laccase decreased slightly in the presence of such IL concentrations, the stability of the enzyme at 50°C was significantly enhanced [34]. Another possible advantage of including the IL is to promote the dissolution of substrates and products, resulting in better overall process efficiency. To ensure these phenomena, the delignification efficiency of RWMs in aqueous media without IL was examined. The results indicated that the cellulose fibers produced only an aqueous medium contained 16.5 wt% lignin, whereas only 9.8 wt% lignin remained in the fibers obtained from aqueous medium containing 2.5% IL [34]. As Table 2 shows, after the enzymatic delignification of RWMs the α-cellulose content of the cellulose-rich materials obtained increased from 41.2 to 73.1%, whereas the hemicellulose and lignin contents were significantly decreased to 8.5 and 9.8%, respectively. This was as expected, because IL pretreatment swells the wood structure, allowing access of the enzyme deep into the wood biomass for improved delignification. Furthermore, the presence of IL may accelerate delignification efficiency by promoting the dissolution of substrate and products [36]. As shown in the process flow chart (Fig. 6), after the enzymatic delignification the sample was diluted with 0.1 M NaOH to extract the lignin. It was expected that a portion of the hemicellulose would also be leached in this step, which was reflected in the chemical composition of the final product cellulose-rich fibers (CRFs).

3.1.3 Characterization of Treated and Untreated Wood Fibers

Treated biomass mentioned in Sect. 3.1.1 and untreated wood materials were characterized by SEM, FTIR, XRD, TGA, and chemical methods [34]. SEM results indicated that RWMs after IL treatment have a relatively homogeneous macrostructure with more porosity compared to untreated materials, possibly caused by the fusing of wood fibers during the IL pretreatment. Interestingly, CRFs obtained after the enzymatic delignification had smooth, clean surfaces because most of the non-cellulosic materials (e.g., lignin) were removed during the IL and biological treatments. It was also found that wood cell networks RWMs composed of cellulose, hemicellulose, and lignin were broken down and cellulose fibers were partially separated into individual microsized fibers similar to the cellulose fibers obtained from wood by others chemical pretreatment [37]. As shown in Fig. 7, the lignin characteristic FTIR peaks at 1592/1503 cm^{-1} (C=C stretching vibration), 1,256 cm^{-1} (asymmetric bending in CH_3), and 1,251 cm^{-1} (C–O vibration in the syringyl ring) [38] remained (compared to untreated material) in the RWM sample, indicating that lignin in the wood biomass was not removed significantly during IL pretreatment. However, these peaks disappeared after enzymatic delignification because of the removal of most of the lignin.

The powder X-ray diffraction patterns of the UWMs, RWMs, and CRFs clearly showed that IL pretreatment did not significantly change the crystallinity of the wood fibers [34]. This is in contrast to other reports, where the degree of crystallinity of regenerated wood materials after dissolution in ILs at higher temperatures (100°C and higher) reduced significantly [26, 39]. Compared to UWMs and RWMs, CRFs showed increased intensity in the X-ray diffraction patterns. One possible explanation is that, because of the mild IL pretreatment conditions (80°C for 1 h), the crystalline regions of cellulose are unaffected but amorphous parts are partially

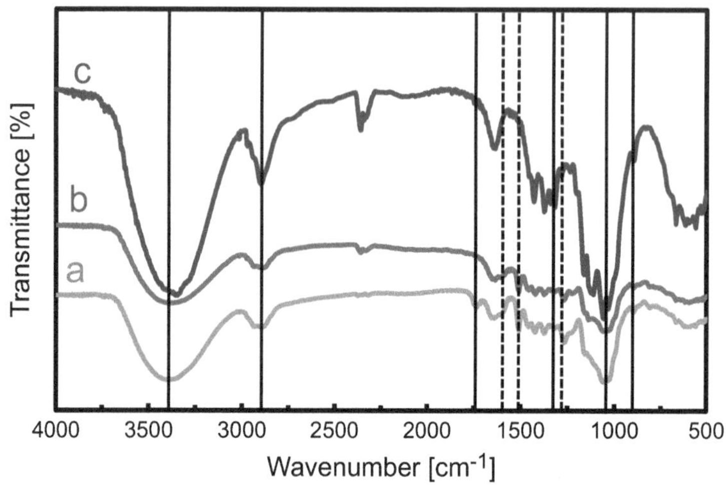

Fig. 7 Fourier-transform infrared spectra of (**a**) untreated wood materials, (**b**) recovered wood materials (RWMs) after IL pretreatment, and (**c**) cellulose-rich fibers after enzymatic delignification of RWMs. Vertical solid lines represent characteristic peaks of cellulose and hemicelluloses, vertical dashed lines mark characteristic peaks from lignin [34]

destroyed. If so, one would expect a higher degree of crystallinity in the structure of the CRFs, which is undoubtedly caused by the removal of most of the hemicellulose and lignin in the IL pretreatment followed by enzymatic delignification [37]. Investigation of the thermal properties of natural fibers is important to gauge their utility in biocomposite processing, in which the processing temperature for thermoplastic polymers rises above 200°C. The TGA results obtained for UWMs, RWMs, and CRFs clearly illustrated that the thermal stability of the wood fibers increased after IL treatment and further increased after enzymatic delignification [34].

3.2 IL Pretreatment of Oil Palm Biomass Before Enzymatic Delignification

As discussed in Sect. 3.1, the performance of an enzymatic delignification process can be improved significantly after swelling hinoki cypress wood in IL [emim][OAc]. However, pretreatment performances vary with feedstock type, IL type, and pretreatment conditions (e.g., temperature and time). Oil palm fronds biomass (OPFB), an abundant agricultural waste in Malaysia, can be converted to the value-added product CRF for biomaterial/biocomposite production. Financie et al. [33] reported a pretreatment method for OPFB for separation of CRFs via enzymatic delignification with prior IL pretreatment and recovery steps.

3.2.1 Effect of Treatment Temperature and Time on IL OPFB Pretreatment

Pretreatment of OPFB with IL [emim][dep] (1-ethyl-3-methylimidazolium-diethyl phosphate) was carried out in moderate conditions (70–100°C for 4 h) to swell the OPFB cells by partial dissolution, and the product solids were recovered by the addition of an anti-solvent [33]. IL [emim][dep] was suitable as a pretreatment agent because it has a greater ability to dissolve OPFB compared to other ILs, possibly because of its low viscosity and well-known high hydrogen bonding acceptor ability, which can facilitate dissolution of the biomass [40, 41]. The lignin and holocellulose (cellulose + hemicellulose) contents of untreated and IL-treated OPFB at various temperatures as well as different IL-pretreatment times (Fig. 8) were investigated. All recovered OPFB materials after IL treatment had a considerably lower lignin content than untreated OPFB. The lowest lignin content (about 7.5%) was found in OPFB treated with IL [emim][dep] at 100°C for 4 h, followed by 90°C (about 13.0%). However, recovered OPFB treated at 100°C contained less holocellulose than samples treated at 90°C, because the higher pretreatment temperature favored cellulose and hemicellulose degradation and the resulting products were easily washed out during separation steps [19]. To obtain materials with higher cellulose content, OPFB samples pretreated at 90°C for 4 h were used for subsequent enzymatic delignification processes.

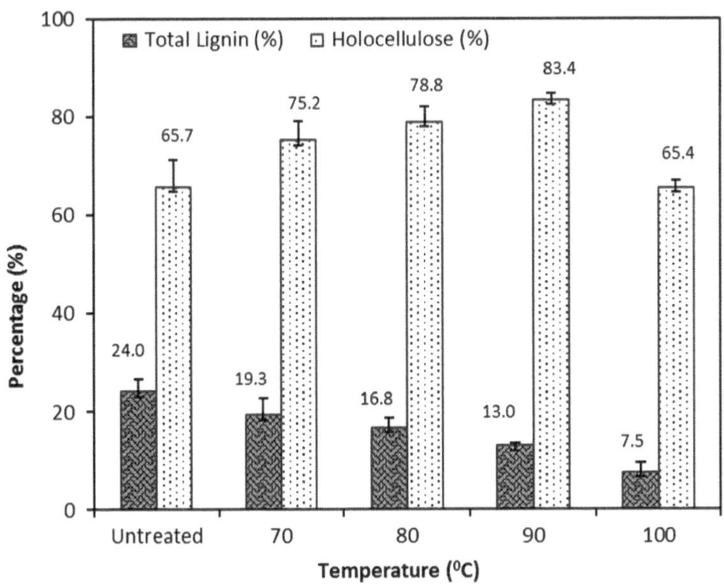

Fig. 8 Chemical composition of untreated and IL-treated oil palm fronds biomass (OPFB) with different pretreatment temperatures [33]

Table 3 Chemical composition of untreated and treated OPFB [33]

Entry	Pretreatment method	Total lignin (%)	α-Cellulose (%)	Hemicellulose (%)
1	Untreated	24.0 ± 2.5	45.7 ± 3.4	20.0 ± 2.9
2	Only IL pretreatment[a]	13.0 ± 0.5	65.1 ± 3.6	18.2 ± 3.6
3	Only enzymatic delignification[b]	19.6 ± 1.1	54.4 ± 2.0	15.1 ± 1.5
4	IL treatment followed by enzymatic delignification[c]	8.5 ± 1.7	68.5 ± 0.3	12.1 ± 1.7

[a]OPFB:IL = 1:10 (w/w) at 90°C for 4 h with vigorous magnetic stirring
[b]Enzymatic delignification of untreated OPFB with laccase in sodium acetate buffer, pH 4.5, at 50°C for 24 h
[c]OPFB pretreated with IL at 90°C for 4 h, followed by enzymatic delignification in sodium acetate buffer, pH 4.5, at 50°C, for 24 h

3.2.2 Enzymatic Delignification of IL-Treated and Untreated OPFB

After IL pretreatment, OPFB was subjected to enzymatic delignification at 50°C for 24 h using laccase in acetate buffer. Chemical compositions of various treated and untreated OPFBs are listed in Table 3. The results clearly indicated that combined pretreatment methods (IL treatment followed by enzymatic delignification) (Table 3, entry 4) were more effective than individual treatments (entries 2 and 3). As shown in entry 3, the conventional enzymatic approach did not have a significant effect on non-IL-pretreated OPFB. This trend is consistent with results reported for enzymatic delignification of wood biomass [12]. Pretreatment of OPFB with IL enhanced the enzymatic delignification efficiency significantly. After the enzymatic delignification of IL-treated OPFB, the α-cellulose content was increased from 45.7 to 68.5.1 wt%, whereas the lignin and hemicellulose content were reduced to 8.5 and 12.1 wt%, respectively. Indeed, improved enzymatic delignification of IL-treated OPFB was expected because IL pretreatment swells the OPFB structure to create a larger (more accessible) surface area for the enzyme, allowing deeper access into the OPFB for improved delignification [25]. Treated and untreated OPFB were characterized by SEM, FTIR, XRD, TGA, and chemical methods. The cellulose-rich OPFB fibers obtained from treated OPFB exhibited higher thermal stability than those from untreated OPFB, possibly because of the removal of lignin [33].

3.2.3 Recycling of ILs

Because ILs are considered expensive solvents, recycling of used ILs from the mixture after extraction of CRFs and lignin is very important for the development of an economic, energy-saving process [33]. After the separation of CRFs, the filtrate contains a solution of lignin, IL, acetone, and water. After evaporating acetone, lignin was precipitated, leaving IL in aqueous solution. Then water was evaporated from the solution under reduced pressure using a rotary evaporator to recover the IL. Recovered ILs were vacuum dried at 80°C. Over 90% of the IL was successfully recovered. ^1H NMR analysis was conducted to check for any changes in the

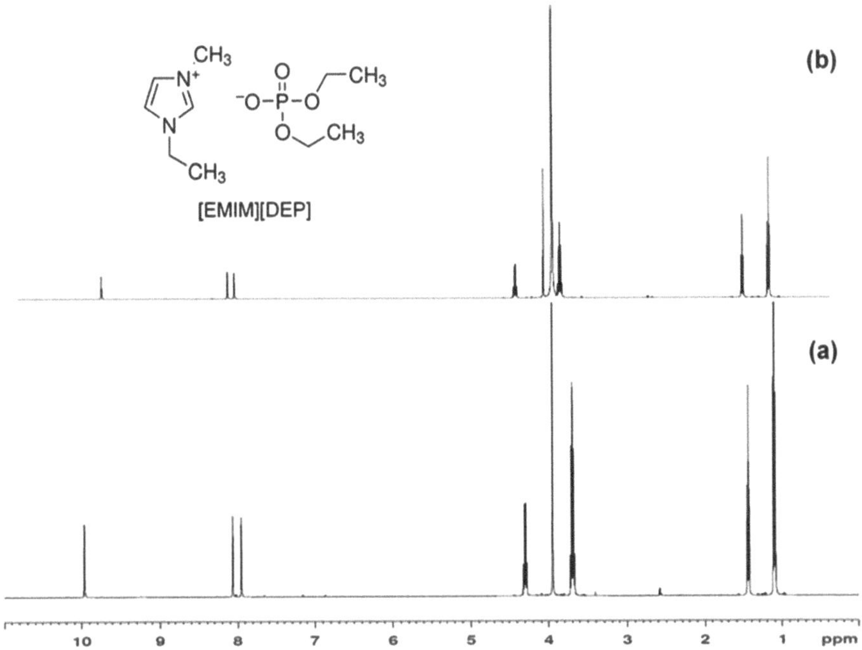

Fig. 9 ^{1}H NMR spectra of (**a**) original IL [emim][dep] and (**b**) IL [emim][dep] recovered after OPFB pretreatment [33]

recovered [emim][dep] structure compared with pure [emim][dep] but there was no structural change in the recovered [emim][dep] (Fig. 9). It can therefore be concluded that IL [emim][dep] can be reused for biomass pretreatment to reduce the process cost.

4 Conclusions

This chapter describes highly effective and clean pretreatment methods for isolation of cellulose fibers from wood biomass and oil palm biomass with minimum structural alteration. The delignification efficiency of IL-treated lignocellulosic biomass is notably improved in both one- and two-step processes. Characterization studies clearly indicate that, although IL pretreatment does not significantly change the cellulose chemical composition, crystal structure, or thermostability, it does change the structure to render a more accessible surface area, leading to enhanced enzymatic delignification. The cellulose fibers obtained have a higher degree of crystallinity and thermal stability than the native wood fibers.

Acknowledgment We thank Edanz Group (www.edanzediting.com/ac) for editing a draft of this chapter.

References

1. Mosier N, Wyman C, Dale B et al (2005) Features of promising technologies for pretreatment of lignocellulosic biomass. Bioresour Technol 96:673–686
2. Zhang M, Ju X, Song X et al (2015) Effects of cutting orientation in poplar wood biomass size reduction on enzymatic hydrolysis sugar yield. Bioresour Technol 194:407–410
3. Gabriele G, Cerchiara T, Salern G et al (2010) A new physical–chemical process for the efficient production of cellulose fibers from Spanish broom (*Spartium junceum* L.). Bioresour Technol 101:724–729
4. Xu H, Yu X, Mu X et al (2015) Effect and characterization of sodium lignosulfonate on alkali pretreatment for enhancing enzymatic saccharification of corn Stover. Ind Crop Prod 76:638–646
5. Morais AR, Pinto JV, Nunes D et al (2016) Imidazole: prospect solvent for lignocellulosic biomass fractionation and delignification. ACS Sustain Chem Eng 4:1643–1652
6. Zhao Y, Wang Y, Zhu JY et al (2008) Enhanced enzymatic hydrolysis of spruce by alkaline pretreatment at low temperature. Biotechnol Bioeng 99:1320–1328
7. Chen WH, Lin BJ, Huang MY et al (2015) Thermochemical conversion of microalgal biomass into biofuels: a review. Bioresour Technol 184:314–327
8. Bak JS, Ko JK, Choi IG et al (2009) Fungal pretreatment of lignocellulose by *Phanerochaete chrysosporium* to produce ethanol from rice straw. Biotechnol Bioeng 104:471–482
9. Balan V, da Costa Sousa L, Chundawat SP, Dale BE et al (2008) Mushroom spent straw: a potential substrate for an ethanol-based biorefinery. J Ind Microbiol Biotechnol 35:293–301
10. Castoldi R, Bracht A, Morais GRD et al (2014) Biological pretreatment of eucalyptus grandis sawdust with white-rot fungi: study of degradation patterns and saccharification kinetics. Chem Eng J 258:240–246
11. Mohanram S, Rajan K, Carrier DJ et al (2015) Insights into biological delignification of rice straw by trametes hirsuta and myrothecium roridum and comparison of saccharification yields with dilute acid pretreatment. Biomass Bioenergy 76:54–60
12. Sousa LD, Chundawat SPS, Balan V et al (2009) Cradle-to-grave assessment of existing lignocellulose pretreatment technologies. Curr Opin Biotechnol 20:339–347
13. Elgharbawy AA, Alam MM, Goto M et al (2016) Ionic liquid pretreatment as emerging approaches for enhanced enzymatic hydrolysis of lignocellulosic biomass. Biochem Eng J 109:252–267
14. Mora-Pale M, Meli L, Dordick JS (2011) Room temperature ionic liquids as emerging solvents for the pretreatment of lignocellulosic biomass. Biotechnol Bioeng 108:1405–1422
15. Sun N, Rodriguez H, Rogers RD et al (2011) Where are ionic liquid strategies most suited in the pursuit of chemicals and energy from lignocellulosic biomass? Chem Commun 47:1405–1421
16. Kilpelainen I, Xie H, Argyropoulos DS et al (2007) Dissolution of wood in ionic liquids. J Agric Food Chem 55:9142–9148
17. Mahmood H, Moniruzzaman M, Welton T (2017) Ionic liquids assisted processing of renewable resources for the fabrication of biodegradable composite materials. Green Chem 19: 2051–2075
18. Lee SH, Doherty TV, Dordick JS (2009) Ionic liquid-mediated selective extraction of lignin from wood leading to enhanced enzymatic cellulose hydrolysis. Biotechnol Bioeng 102:1368–1376
19. Sun N, Rahman M, Rogers RD (2009) Complete dissolution and partial delignification of wood in the ionic liquid 1-ethyl-3-methylimidazolium acetate. Green Chem 11:646–655
20. Blanchette R (1991) Delignification by wood decay fungi. Annu Rev Phytopathol 29:381–403
21. Shipovskove S, Gunaratne HQN, Seddon KR et al (2008) Catalytic activity of laccases in aqueous solutions of ionic liquids. Green Chem 10:806–810
22. Sivapragasam M, Moniruzzaman M, Goto M (2016) Recent advances in exploiting ionic liquids for biomolecules: solubility, stability, and applications. Biotechnol J 11:1000–1013

23. Moniruzzaman M, Nakashima K, Goto M et al (2010) Recent advances of enzymes in ionic liquids. Biochem Eng J 48:295–314
24. Qiu Z, Aita GM, Walker MS (2012) Effect of ionic liquid pretreatment on the chemical composition, structure and enzymatic hydrolysis of energy cane bagasse. Bioresour Technol 117: 251–256
25. Moniruzzaman M, Ono T (2012) Ionic liquid assisted enzymatic delignification of wood biomass: a new 'green' and efficient approach for isolating of cellulose fibers. Biochem Eng J 60:156–160
26. Labbe N, Kline LM, Moens L et al (2012) Activation of lignocellulosic biomass by ionic liquid for biorefinery fractionation. Bioresour Technol 104:701–707
27. Weerachanchai W, Leong SSJ, Chang MW et al (2012) Improvement of biomass properties by pretreatment with ionic liquids for bioconversion process. Bioresour Technol 111:453–459
28. Moniruzzaman M, Ono T, Uemura Y (2013) Improved biological delignification of wood biomass via ionic liquids pretreatment: a one step process. J Energy Technol Policy 3: 144–152
29. Moon YH, Lee SM, Koo YM et al (2006) Enzyme-catalyzed reactions in ionic liquids. Korean J Chem Eng 23:247–263
30. Singh S, Simmons BA, Vogel KP (2009) Visualization of biomass solubilization and cellulose regeneration during ionic liquid pretreatment of switch grass. Biotechnol Bioeng 104: 68–75
31. Fu D, Mazza G, Tamaki Y (2010) Lignin extraction from straw by ionic liquids and enzymatic hydrolysis of the cellulosic residues. J Agric Food Chem 58:2915–2922
32. Li C, Cheng G, Balan V et al (2011) Influence of physico-chemical changes on enzymatic digestibility of ionic liquid and AFEX pretreated corn Stover. Bioresour Technol 102: 6928–6936
33. Financie R, Moniruzzamana M, Uemura Y et al (2016) Enhanced enzymatic delignification of oil palm biomass with ionic liquid pretreatment. Biochem Eng J 110:1–7
34. Moniruzzaman M, Ono T (2013) Separation and characterization of cellulose fibers from cypress wood tretaed with ionic liquid prior to laccase treatment. Bioresour Technol 127: 132–137
35. Doherty TV, Mora-Pale M, Dordick JS et al (2010) Ionic liquid solvent properties as predictors of lignocelluloses pretreatment efficacy. Green Chem 12:1967–1975
36. Pu Y (2007) Ionic liquids as a green solvent for lignin. J Wood Chem Technol 27:23–35
37. Chen W, Yu H, Liu Y et al (2011) Individualization of cellulose nanofibers from wood using high-intensity ultrasonification combined with chemical pretreatments. Carbohydr Polym 83: 1804–1811
38. Labbe N, Rials TG, Kelley SS et al (2005) FT-IR imaging and pyrolysis-molecular beam mass spectrometry: new tools to investigate wood tissues. Wood Sci Technol 39:61–77
39. Lucas M, Wagner GL, Nishiyama Y et al (2011) Reversible swelling of the cell wall of poplar biomass by ionic liquid at room temperature. Bioresour Technol 102:4518–4523
40. Abe M, Fukaya Y, Ohno H (2010) Extraction of polysaccharides from bran with phosphonate or phosphinate-derived ionic liquids under short mixing time and low temperature. Green Chem 12:1274–1280
41. Zhao D, Li H, Zhang J et al (2012) Dissolution of cellulose in phosphate-based ionic liquids. Carbohydr Polym 87:1490–1494

Adv Biochem Eng Biotechnol (2019) 168: 79–104
DOI: 10.1007/10_2018_62

Published online: 10 May 2018

Activation of Lipase-Catalyzed Reactions Using Ionic Liquids for Organic Synthesis

Toshiyuki Itoh

Abstract The use of ionic liquids to replace organic or aqueous solvents in biocatalysis processes has recently received great attention, and much progress has been made in this area; the lipase-catalyzed reactions are the most successful. Recent developments in the application of ionic liquids as solvents in lipase-catalyzed reactions for organic synthesis are reviewed, focusing on the ionic liquid mediated activation method of lipase-catalyzed reactions.

Graphical Abstract

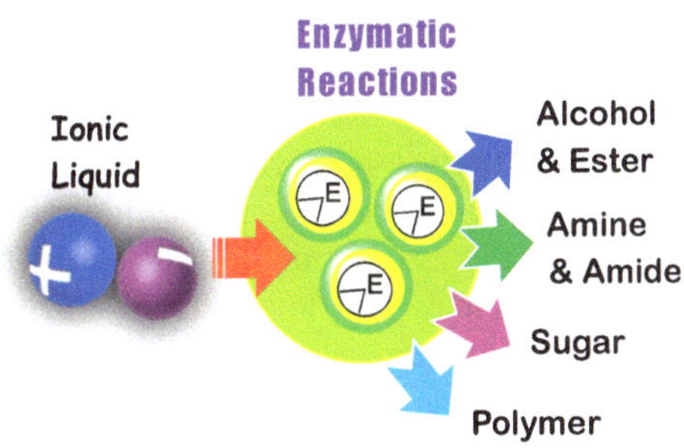

Keywords Activation, Enantioselectivity, Ionic liquids, Lipase-catalyzed reaction, Organic synthesis, Reaction rate, Stability of enzymes

T. Itoh (✉)
Department of Chemistry and Biotechnology, Graduate School of Engineering, Tottori University, Tottori, Japan

Center for Research on Green Sustainable Chemistry, Tottori University, Tottori, Japan
e-mail: titoh@chem.tottori-u.ac.jp; http://orcid.org/0000-0002-8056-6287

Contents

1 Introduction

The value of the enzymatic reaction in organic synthesis is highly respected from an environmentally-friendly aspect and has now reached an industrially-proven level in the pharmaceutical and food industries [1]. This chapter focuses on topics concerned with the improved performance of lipase-catalyzed reactions using ionic liquid engineering, because lipase (triglycerol acylhydrolases EC 3.1.1.3) is the most popular biocatalyst for organic chemists and the enzyme can catalyze a number of reactions, including hydrolysis, transesterification, alcoholysis, acidolysis, esterification, and aminolysis [1–3]. It is well known that both the reactivity and the enantioselectivity of lipase-catalyzed reactions are dependent on the solvent system [3]; the lipase-catalyzed reaction can proceed in many types of organic solvents or ionic liquids. Ionic liquids (ILs) have very good properties as reaction media for chemical reactions; they are non-volatile, non-flammable, have a low toxicity, and display a unique solubility for many organic and inorganic materials [4–8]. Based on these properties, numerous examples of biocatalysis-mediated reactions in ILs have been reported [9].

The publication history over the past 17 years for enzymatic reactions in ILs is shown in Fig. 1. There are 2,298 and 1,216 papers on the topics of "enzyme*" and "lipase*," respectively. Although it appears that the increase in the number of "lipase*" papers stopped in 2011, around 100 papers have still appeared year by year and almost of them are for "organic synthesis." Lipase-catalyzed reactions are still being widely used in the field of synthetic organic chemistry.

As shown in Fig. 1, lipase-catalyzed reactions using ILs as reaction media began in 2000/2001. The Sheldon group reported the first example of a lipase-catalyzed reaction in a pure IL in 2000 [10]. Itoh [11] and Kragl [12] independently reported

Fig. 1 Number of publications in the field of ILs from 2000 to 2017 (searched by "Web of Science," on December 6, 2017). A total of 82,407 references are listed based on the topic of "ionic liquid*" in the database of Web of Science Core Collections of Thomson Reuters, Ltd

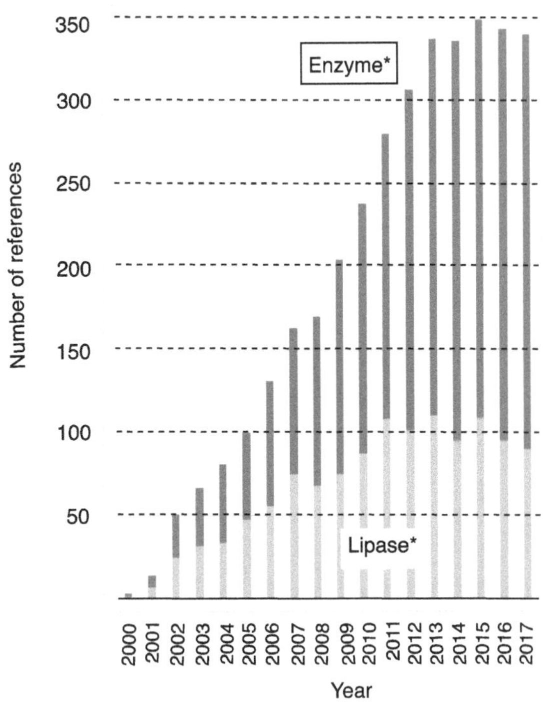

the first examples of enantioselective lipase-catalyzed reactions in ILs in early 2001. Husum et al. [13] Lozano et al. [14, 15], Kim et al. [16], and Park and Kazlauskas [17] reported lipase-catalyzed reactions in an IL solvent system later in the same year. Laszlo et al. also demonstrated the α-chymotrypsin mediated transesterification reaction of *N*-acetyl L-phenylalanine ethyl ester with 1-propanol in ILs or ILs with supercritical carbon dioxide ($ScCO_2$) [18]. Lozano et al. reported that the improved stability of lipases was significantly enhanced by the ILs [14, 19]. Itoh reported the remarkably increased stability of CAL in [C_4dmim][BF_4] [20]. Gubicza et al. reported that both the water contents in the ILs and IL species strongly influenced the enzyme activity [21–23]. Baldwin reported that Hofmeister ion interactions affected protein stability [24]. Zhao reported that the Hofmeister effect of ions of the solvent IL was related to the activity of the lipases [25] and, since that finding, numerous examples of lipase activity from this point of view have been reported [26–29]. It was known that the water contents in the ILs significantly influenced the activities of the lipases [30]. The introduction of the alkylether moiety in the ILs in the cationic or anionic parts generally provided good results [31]. Since then, numerous successful examples have been reported by many groups [9].

2 Typical Lipase-Catalyzed Reactions Using IL Solvent Systems

2.1 *Lipase-Catalyzed Enantioselective Transesterification of Secondary Alcohols Using an IL Solvent System*

The most important aspect of the biocatalysis reaction should be the enantioselective reaction. Typically, lipase-catalyzed enantioselective transesterification using an IL solvent system has been carried out as illustrated in Fig. 2 [11]. To a mixture of lipase in the ionic liquid were added a racemic alcohol ((±)-**1**) and vinyl acetate as the acyl donor and the resulting mixture was stirred at 35°C. After the reaction, ether or a mixed solvent of hexane and ether was added to the reaction mixture to form a biphasic layer, and the acetate product and unreacted alcohol were quantitatively extracted with the organic solvent layer. The enzyme remained in the ionic liquid phase and the addition of the next set of substrate (±)-**1** and acyl donor **3** caused the next cycle of the reaction to afford the chiral ester (*R*)-**2** and (*S*)-**1**. The lipase could therefore be used repeatedly anchored in the reaction media. This is a typical example, indicating a certain benefit of using ILs as the reaction media for enzymatic reactions. Selection of the appropriate ILs as a reaction medium is the key to realizing the desired reaction. The hydrophobic IL, [C_4mim][PF_6], was selected as a favorable solvent for the reaction because the system allowed a simple work-up process because of the insolubility of this IL in both water and extraction of the organic solvent. In contrast, a hydrophilic IL, such as [C_4mim][BF_4], was soluble in water and, therefore, it was difficult to remove a water-soluble by-product by a

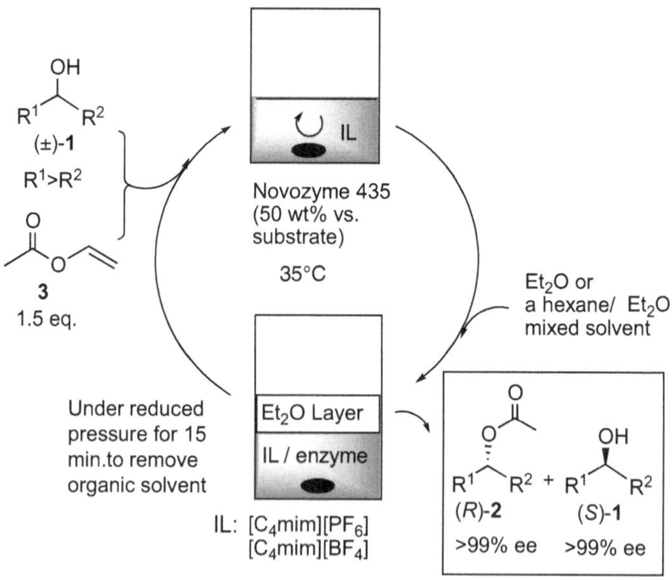

Fig. 2 Typical lipase-catalyzed enantioselective transesterification using the IL solvent system

simple work-up process, even though the enzyme exhibited a high activity in the solvent.

2.2 Lipase-Catalyzed Transesterification of Secondary Alcohols in an IL Under Reduced Pressure Conditions

It is well known that the usual methyl esters are not suitable for lipase-catalyzed transesterification as acyl donors because the reverse reaction with the produced methanol takes place. One of the important characteristics of an ionic liquid is its wide temperature range for the liquid phase and lack of vapor pressure. Using this property, the lipase-catalyzed reaction with a methyl ester as the acyl donor under reduced pressure conditions was reported by Itoh et al. [32]. Although it is essential to choose both substrate alcohols and an acyl donor ester, which have sufficiently higher boiling points compared to those of methanol or ethanol, the desired reaction proceeded smoothly under the conditions because the produced methanol or ethanol could be immediately removed from the reaction mixture and the reaction equilibrium occurs to produce the desired product. Phenylthioacetate **4a** was recommended as an acyl donor ester and thereby an efficient optical resolution was accomplished. This process allowed recycling of the enzyme five times with maintenance of the initial activity and succeeded in reducing the amount of the acyl donor ester (Fig. 3).

Lourenço et al. reported an efficient transesterification system for the lipase-catalyzed reaction under reduced pressure conditions using an IL as the acylating agent in an IL solvent (Fig. 4) [33, 34]. The authors designed an IL-type methyl ester **6**, which possessed the imidazolium salt moiety, and demonstrated that the ester was anchored in the IL ([C_4mim][PF_6]) solvent. After extraction of the 81% ee (*S*)-1-phenylethanol (**1a**), the IL layer that retained the imidazolium group substituted ester **7** was treated with ethanol to release 99% ee of the (*R*)-alcohol **1a**. The chromatography-free kinetic resolution has thus been accomplished through this

Fig. 3 Lipase-catalyzed transesterification using methyl ester as an acyl donor under reduced pressure conditions in the IL solvent system

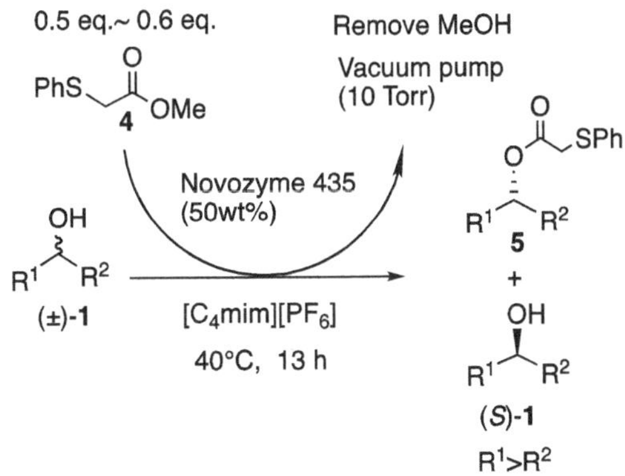

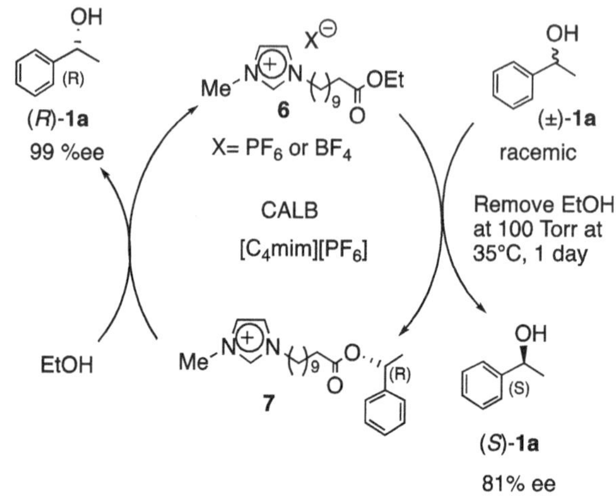

Fig. 4 Chromatographic separation free kinetic resolution of (±)-1-phenylethanol (**1a**) using the IL-type acyl donor in the IL solvent system under reduced pressure conditions

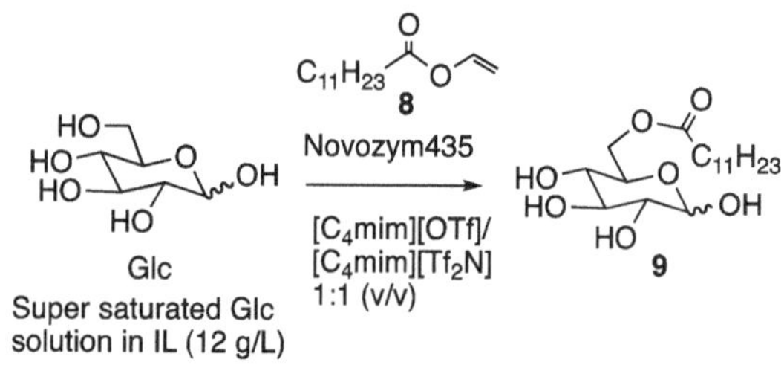

Fig. 5 The efficient lipase-catalyzed sugar ester synthesis in the IL solvent system using the super-saturated glucose solution in [C$_4$mim][OTf]

reaction system. Recently, Teixera et al. reported an extended version of this method [35].

2.3 *Use of Unique Solubility of ILs for Lipase-Catalyzed Regio-Selective Transesterification of Sugar Derivatives*

Another important property of the ILs is their unique solubility toward organic molecules. Using this property, Koo et al. reported the solubility-driven efficient lipase-catalyzed sugar fatty acid ester synthesis. They prepared a "super-saturated glucose solution in [C$_4$mim][OTf]" and subjected this to the lipase (Novozyme 435)-catalyzed transesterification in [C$_4$mim][Tf$_2$N] [36]. Because the method allowed a supply of a highly-concentrated substrate solution, the reaction proceeded very smoothly in the presence of the acyl donor **8** in the resulting mixed IL solvent of [C$_4$mim][OTf]:[C$_4$mim][Tf$_2$N] = 1:1 (v/v) to afford the desired fatty acid ester **9** in high yield (Fig. 5) [36].

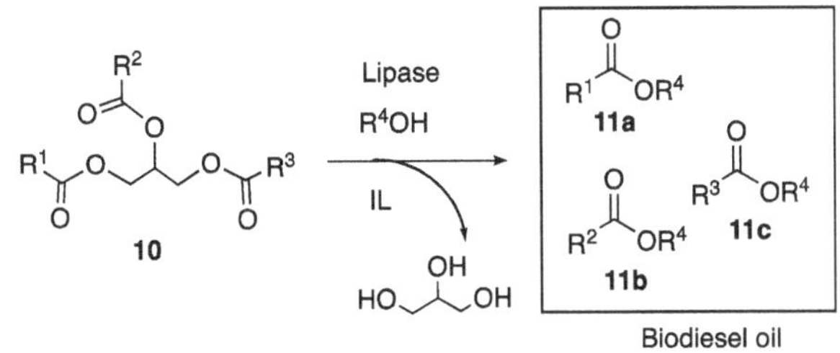

Lipase: *Candida antarctica, Burlholderia cepacia, etc.*

IL: $[C_2mim][OTf]$, $[C_4mim][Tf_2N]$,$[C_{18}tma][Tf_2N]$, $[Me(OCH_2CH_2)_3eim][OAc]$,$[OmPy][BF_4]$,$[Me(OCH_2CH_2)_3eim][Tf_2N]$,$[Me(OCH_2CH_2)_3\text{-Et-Pip}][Tf_2N]$, $[Me(OCH_2CH_2)_3\text{-}Et_3N][Tf_2N]$, or AMMOENG 102™

Fig. 6 Lipase-catalyzed biodiesel oil production in an IL solvent system

2.4 *Lipase-Catalyzed Reactions for Biodiesel Oil Production*

Enzymatic production of biodiesel oil has recently attracted strong interest from the standpoint of sustainable energy production. The diesel oils are hydrophobic compounds, and organic solvent-free separation from the IL reaction mixture has been easily achieved (Fig. 6) [37–42]. ILs could thus be used as the appropriate solvents for biodiesel oil **11** production from vegetable oil **10** through lipase-catalyzed transesterification. Ha et al. first demonstrated the lipase (from *Candida antarctica*)-catalyzed biodiesel production using soybean oil as the substrate in the presence of MeOH in the $[C_2mim][OTf]$ solvent system [42]. De Diego et al. prepared $[C_{16}tam][NTf_2]$, which was termed a "sponge-like IL(SLIL)." SLIL displayed a unique property as a reaction medium for the lipase-catalyzed reaction. The authors established an efficient protocol for biodiesel oil production employing the SLIL [43, 44]. Yang et al. demonstrated biodiesel oil production using an IL solvent system from various sources such as corn oil, microbial oil, and *Millettia pinnata* seed oil. They established that the yields were drastically improved compared to those in conventional organic solvents or solvent-free conditions [45–47].

2.5 *DKR Reaction Using a Combination of Transition Metal Catalyst in the Lipase-Catalyzed Reaction*

As mentioned before, lipases generally convert (*R*)-alcohol to the corresponding ester in the presence of an appropriate donor molecule in non-aqueous media, and lipase-catalyzed transesterification has been used for kinetic resolution (KR) of racemic substrates. This method allows a convenient use of both enantiomers. However, there is a serious limitation in the chemical yield of optically active products. ILs are suitable solvents for many types of transition metal catalysts. The use of an appropriate combination of racemization catalyst and lipase is the

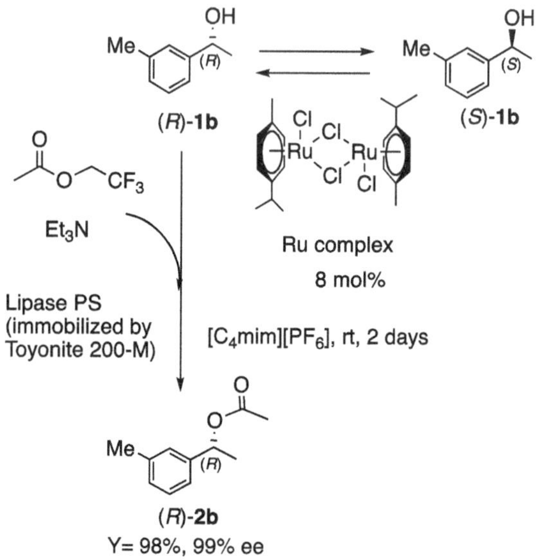

Fig. 7 Dynamic kinetic resolution (DKR) using a lipase-ruthenium combo catalyst system in an IL solvent system

key to realizing an efficient DKR system. Because preparation of chiral compounds is an important and challenging area of contemporary synthetic organic chemistry, extensive efforts have been devoted to developing an efficient DKR methodology, and numerous examples have been reported [48–50]. Kim et al. accomplished dynamic kinetic resolution (DKR) using a lipase-ruthenium combo catalyst system in an IL solvent system. The benzyl alcohol (±)-**1b** was quickly racemized by the Ru-complex during the lipase-catalyzed transesterification, and then the (*R*)-acetate **2b** was accumulated after the reaction (Fig. 7) [51].

Lozano et al. reported a transition metal-free DKR system using a combination of supercritical carbon dioxide ($ScCO_2$) and an IL in the presence of acidic regin as a Brønsted acid for the racemization catalyst of 1-phenylethanol [52, 53]. We also accomplished transition metal-free DKR of 1-phenyethanol and 2,3-dihydro-1*H*-inden-1-ol by a combination of *C. antarctica* lipase (CAL-B) and Zeolite using vinyl octanoate or *p*-chlorophenyl pentanoate as acyl donor in a mixed solvent system of ionic liquid and hexane in the presence of 1.0 eq. of water. Duplicated use of the catalysts has also been demonstrated in the DKR of 2,3-dihydro-1*H*-inden-1-ol using this reaction system [54]. However, because of the formation of an acid-catalyzed aryl ether by-product, the efficiency of DKR was inferior to that of transition metal-catalyzed DKR.

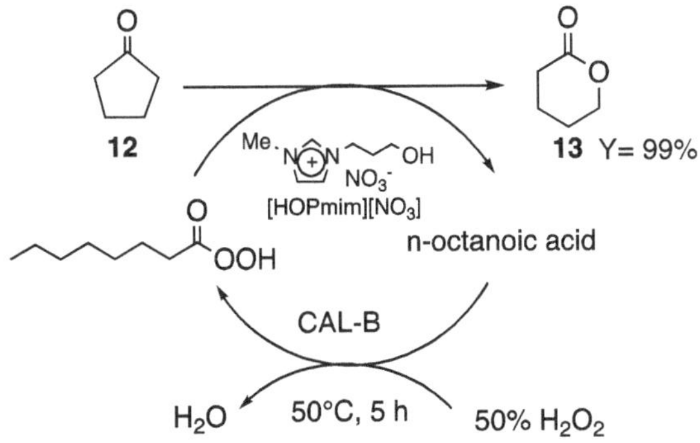

Fig. 8 Baeyer-Villiger oxidation using a combination of hydrogen peroxide and the lipase-catalyzed reaction

2.6 *Baeyer-Villiger Oxidation Using a Combination of Hydrogen Peroxide and Lipase-Catalyzed Reaction*

Lipase can catalyze the formation of peracid derivatives from esters simply by adding peroxide derivatives because hydrogen peroxide anions have a very strong nucleophilicity. Sheldon et al. reported the *C. antarctica* lipase (CAL-B)-catalyzed production of *n*-octanoic peracid by the reaction of octanoic acid with hydrogen peroxide in an IL solvent [10]. This is the first example of the lipase-catalyzed reaction in a pure ionic liquid solvent system. The resulting peracid was used for the epoxidation of olefins or Baeyer-Villiger oxidation against ketones. Using the reaction, Kotlewska et al. reported the Baeyer-Villiger oxidation of ketones to afford lactones in excellent yields (Fig. 8) [55]. The authors reported that [HOPmim][NO_3] generally gave the desired lactone in high yields; tetrahydro-2*H*-pyran-2-one (**13**) was thus obtained in 99% yield from cyclopentanone (**12**) when cyclopentanone was treated with 2 eq. of hydrogen peroxide in the presence of octanoic acid and CAL-B as a catalyst in [HOPmim][NO_3] (Fig. 8). Drozdz et al. also reported an efficient Baeyer-Villiger oxidation using a similar methodology [56].

As noted elsewhere in this chapter, lipases are generally tolerable to a wide number of substrates, and ILs allow various types of reactions that are impossible to realize using conventional molecular solvents or water as the reaction medium. However, very slow reactions or poor enantioselective reactions are also sometimes encountered [1–3]. To solve this problem, ionic liquid engineering for activating lipase-catalyzed reactions has progressed significantly in this decade. Ionic liquid engineering for activating lipase-catalyzed reactions is reviewed in the next section.

3 Improved Performance of Lipase-Catalyzed Reactions by IL-Coated Immobilization

Dordick et al. reported the salt-mediated activation of proteases (subtilisin and α-chymotrypsin) (Fig. 9) [57–59]. The authors reported that the alcoholysis reaction of the *N*-acetylphenylalanine ethyl ester **14** by these proteases afforded the corresponding acid **15** in good yield. The authors found that the "KCl-coating of the enzyme" by a lyophilization process led to modification of the activity of the original enzymes; the K_{cat} values for both enzymes were drastically improved by the KCl coating process. On the other hand, the K_m value of subtilisin decreased to 1/10 compared to the native enzyme, and no modification was observed for α-chymotrypsin. As the ILs are salts, it was anticipated that the lipases might be activated by a similar methodology.

Kim and Lee first reported the ionic liquid mediated activation of an enzyme focused on a modified enantioselectivity [60]. They prepared the ionic liquid coated lipase PS "IL-C-PCL" by mixing with the ionic liquid [3-Ph-C_3mim][PF_6] and reported that IL-C-PCL exhibited a slightly higher enantioselectivity than that of the commercial lipase PS-C for several alcohols, and no significant change was obtained in the reaction rate [60]. However, the reported E value increase was not significant; the E value of the reaction of 4-phenylbutan-2-ol ((±)-**16**) to the ester (*R*)-**17** was enhanced from 233 for the native PCL to 533 for the IL-C-PCL, as shown in Fig. 10. Because E values of over 200 should be very carefully examined [3] and no acceleration was recorded, the results remained at a doubtful level.

Itoh et al. reported the key solution to realize improved performance of the lipases using ILs [61]. The authors synthesized the 3-methyl-1-butylimidazolium cetyl-PEG (10)-sulfate ionic liquid (**IL1**) and prepared the "**IL1** coated *Burkholderia cepacia* lipase (**IL1**-PS)" [61, 62]. They reported that optimization of both the chain length of the alkyl group and PEG moiety of the anionic part of the IL was necessary, and the lyophilization process is essential for preparing the "**IL1**-coated lipase PS" (Fig. 11) [61–63]. The activation effect depended on the substrates, and the **IL1**-PS exhibited an excellent reactivity for many substrates. As shown in Fig. 11, 25-fold acceleration for the (*E*)-4-pehnylbut-3-en-2-ol (**18**), 1,000-fold for 1-naphthylethanol (**19**), and 18-fold for 1-(2-pyridyl)ethanol (**20**) were accomplished when these alcohols were subjected to the reaction, maintaining an excellent enantioselectivity [62]. It should be noted that both the increased reaction rate and enhanced enantioselectivity were also obtained when **18** was used as the substrate. Later, Lee and Kim prepared the

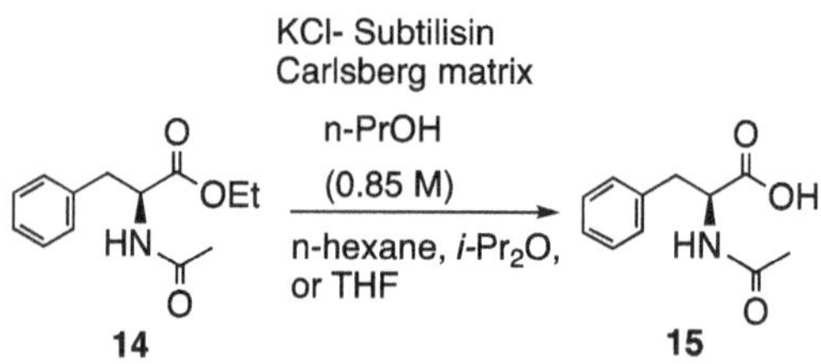

Fig. 9 Activation of Subtilisin Carlsberg-catalyzed alcoholysis by the KCl treatment

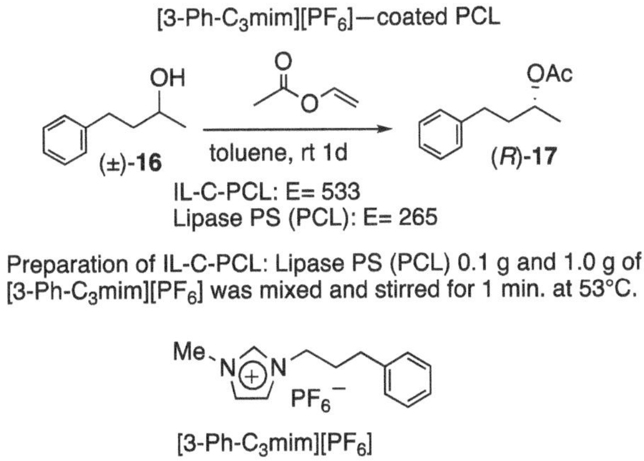

Fig. 10 Activation of lipase PS by the ionic liquid [3-Ph-C_3mim][PF_6]

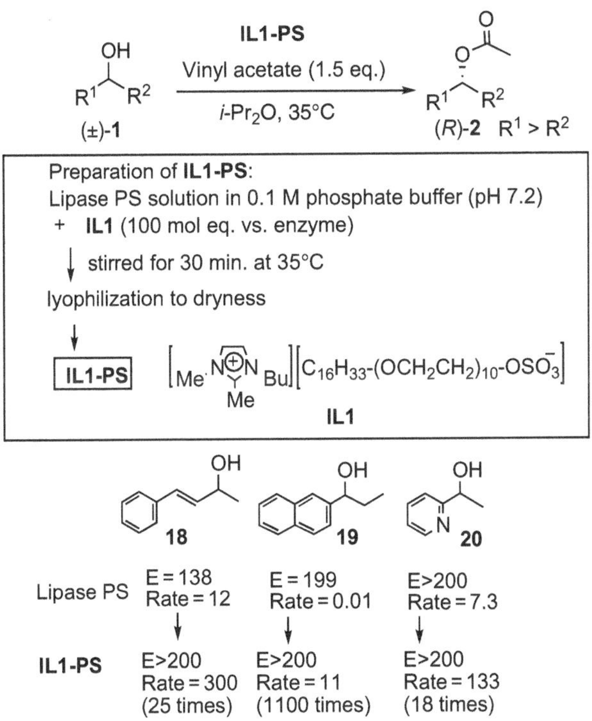

Fig. 11 Typical examples of IL1-PS-catalyzed enantioselective transesterification. Itoh et al. reported that the IL1-PS powder contains 3.3 wt% enzyme with 9.8 wt% IL1, the rest mainly consisting of very fine inorganic materials derived from celite

IL-coated lipase PS using dodecyl-imidazolium PF_6 as a coating material, and confirmed that lyophilization was indeed essential to achieve a significantly

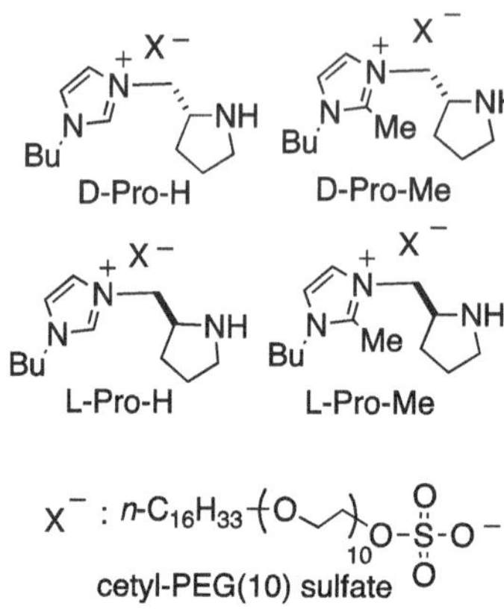

Fig. 12 Chiral imizadolium cetylPEG10 sulfate ILs as the coating materials for lipase PS

enhanced activity of the lipase as Itoh's group had established [64]. Lee et al. also reported the efficient stabilization of lipase through lyophilization of the ionic liquid type sol-gel silica [65].

Abe et al. synthesized chiral pyrrolidine-substituted imidazolium cetyl-PEG(10)-sulfate ionic liquids derived from D- and L-proline and used them as coating materials for lipase PS. The chirality of the pyrrolidine group strongly affected the reactivity of the lipase; the D-Pro-Me-coated PL displayed a 58-fold acceleration vs the commercial lipase PS and this was superior to that of the **IL1**-PS for the transesterification of 1-phenylethanol as a model substrate (Fig. 12) [66].

It is known that several amino acids have been used as enzyme stabilizers. For example, commercial lipase PS involves ca. 20 wt% glycine as an essential stabilizer. Inspired by the results, Yoshiyama et al. prepared the coated lipase PS combining the amino acid with **IL1** and found an interesting synergetic effect of the amino acids and **IL1**; the combination of L-proline and **IL1** effectively improved the activity of lipase PS such that the reaction rate of the enzyme was superior to that of **IL1**-PS when 1-phenylethanol was subjected to the reaction (Fig. 13) [67]. Synergetic activation of the lipase using a combination of amino acids with **IL1** provided a very simple way to activate a lipase.

Rahman et al. recently prepared amino acid ILs, $[N_{2222}]$[His] and $[N_{2222}]$[Asn], and used them as coating materials for *Candida rugose* lipase; the enzymes exhibited high reactivities during the esterification of oleyl alcohol with carboxylic acid **21** (Fig. 14) [68]. The reactivity depended on the amino acid moiety, and the $[N_{2222}]$[His]-coated lipase showed a higher reactivity than $[N_{2222}]$[Asn]. The authors speculated that the reaction of the substrate **21** with the imizadole moiety of $[N_{2222}]$[His] formed the acylinium salt, which displayed high activity as an acylating agent, and thus smooth acylation of **21** took place and efficiently released ester **22**.

Kim et al. reported activation of the lipase by ionic surfactant compounds that have an alkyl PEG moiety [69, 70]; potassium 3,5-bis(2-(2-(2-ethoxyethoxy)ethoxy)ethoxy)benzoate (**ISCB1**) was used as a coating material for *B. cepacia* lipase

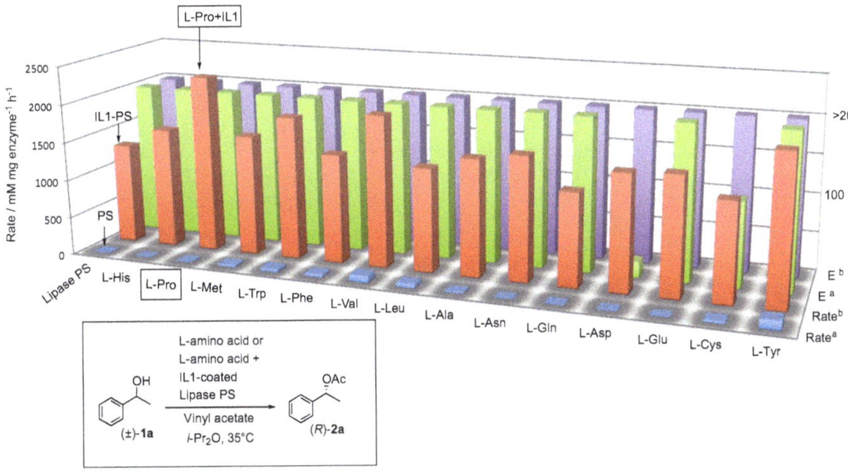

Fig. 13 Co-operative activation of lipase PS with IL1 and amino acid. Reproduced with permission from [67]. Copyright 2018, The Chemical Society of Japan

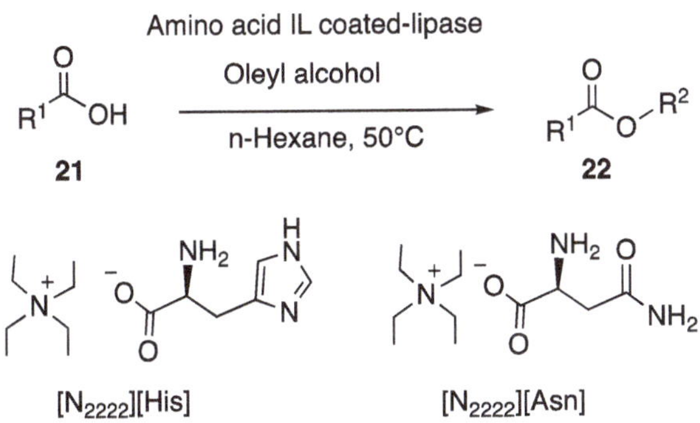

Fig. 14 Activation of *C rugose* lipase by the coating with amino acid ILs

through the lyophilization and the resulting lipase (**ISCB1**-PS) exhibited an excellent reactivity to various alcohols. The important finding was that the enzyme was applicable for the Ru complex mediated DKR of 24 types of alcohols with 96~>99% ee in 90~97% yields (Fig. 15) [69]. Lee et al. further reported the synergetic effect of sugars and an ionic surfactant for activation of the lipoprotein lipase (LPL). A remarkable acceleration of the reaction rate was recorded when 4-nitrophenylacetate (**23**) was subjected to an alcoholysis reaction using the **ISC-glcN**-coated LPL; the K_{cat}/K_m value was 4,000-fold higher than that of the native LPL (Fig. 16) [71].

As already noted, chirality of the imidazolium cation of the coating material strongly influenced the lipase reactivity [66]. The work clearly indicated that the cationic part of the coating ILs influenced the lipase reactivity. Matsubara et al. synthesized four types of phosphonium alkyl PEG sulfate ionic liquids and used

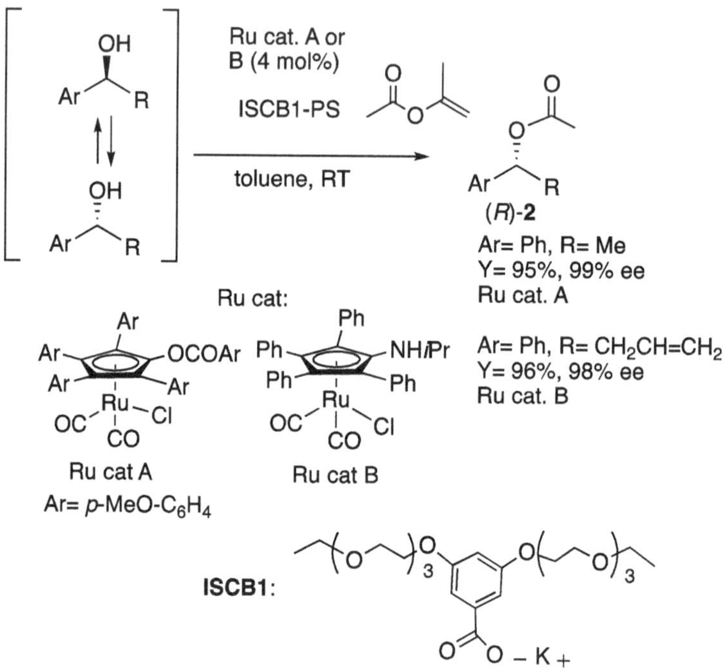

Fig. 15 DKR using IL surfactant coated lipase PS (ISCB1-PS) in the presence of Ru complex

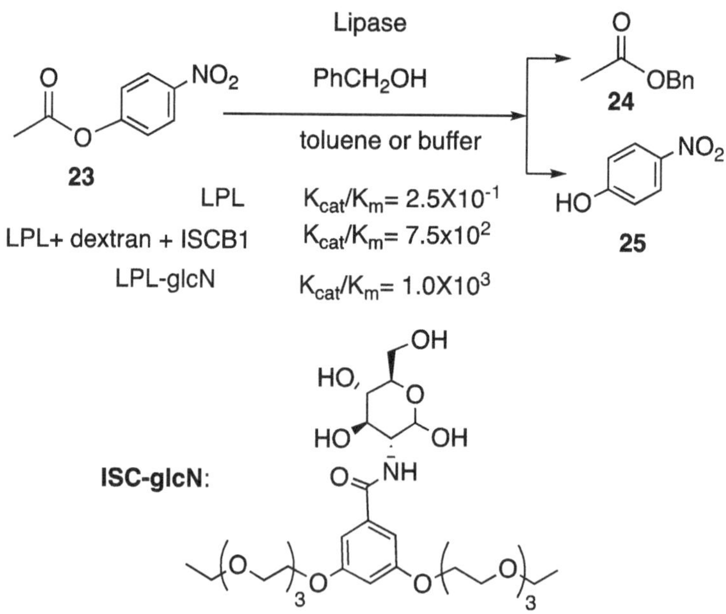

Fig. 16 Activation of lipoprotein lipase by coating using a combination of sugar with ISCB1

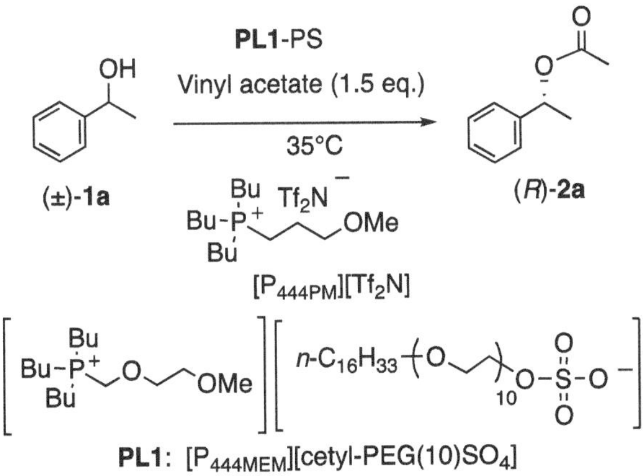

Fig. 17 PL1-PS-catalyzed acylation of 1-phenylethanol in an IL solvent system

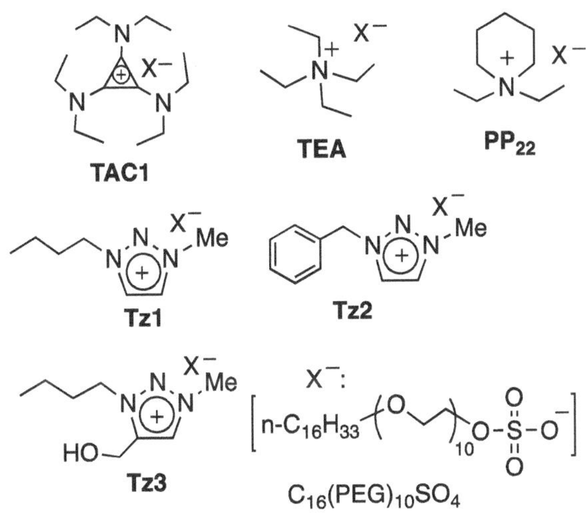

Fig. 18 IL type coating materials for *Bukholderia cepacia* lipase (lipase PS) activation

them as coating materials for the lipase PS. Among them, tributyl(2-methoxyethyl) phosphonium cetyl-PEG10 sulfate (**PL1**) was found to be the best coating material (Fig. 17) [72]. The **PL1**-coated lipase PS displayed a high reactivity in the transesterification of broad types of secondary alcohols **1** using vinyl acetate as an acyl donor to form esters (*R*)-**2** with excellent enantioselectivities. The authors demonstrated the recyclable use of **PL1**-PS in an IL, [P_{444PM}][Tf_2N], and repeated the enantioselective transesterification reaction of (±)-**1a** 10 times with a rate superior to those in the *i*-Pr_2O solvent.

Inspired by the results, Itoh's group further developed novel IL-type activating agents for the lipase PS (Fig. 18) [73, 74]. Kadotani prepared three quaternary ammonium salts, **TAC1**, **TEA**, and **PP$_{22}$**, and investigated their activation properties. The resulting quaternary ammonium salt-coated lipase PSs exhibited different substrate specificities compared to those of the native PS or **IL1**-PS; the **TAC1**-

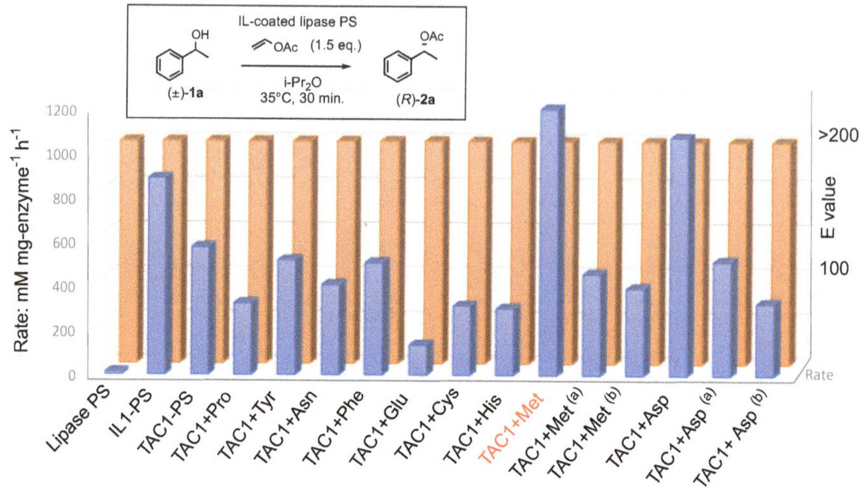

Fig. 19 Effect of coating on lipase PS only with TAC1 or with both an amino acid and TAC1 on the transesterification of (±)-1-phenylethanol (**1a**) using vinyl acetate as acyl donor. TAC1 and amino acid were used as 100 eq. vs lipase protein unless otherwise cited. (a) The amino acid was added at 0.5 eq. vs **TAC1**. (b) The amino acid was added at 2.0 eq. vs **TAC1**.Reproduced with permission from [73]. Copyright 2018, The American Chemical Society

coated lipase PS (**TAC1**-PS) was especially suitable for the transesterification of 1-(pyridin-2-yl)ethanol, 1-(pyridin-3-yl)ethanol, 1-(pyridin-4-yl)ethanol, and (*E*)-4-phenylbut-3-en-2-ol [73].

The authors also reported the cooperative activation of amino acids with TAC1. Interestingly, L-methionine (L-Met) was found to be the best partner with **TAC1** [73], though the best reactant of **IL1** was L-proline [67]. As shown in Fig. 19, the 1:1 mixture of **TAC1** with L-Met provided the highest activation, and the reaction rate reached 1,200 mM/h/mg enzyme, which was superior to that of **IL1-PS**. On the other hand, the increased or decreased amount of the amino acids (both L-Met and L-Asp) caused a significant drop in the reaction rate (Fig. 19) [73]. The combination of the amino acids contributed to increasing the K_{cat} value that agrees with the previously reported results [67].

Nishihara prepared three types of triazolium cetyl-PEG10 sulfate ionic liquids, **Tz1**, **Tz2**, and **Tz3**, and investigated their properies for activation of lipase PS, and they revealed that both the reaction rate and enantioselectivity were strongly modified by the triazolium type IL coating [74].

Itoh's group reported the results of designing IL solvents suitable for recycling use of their IL-coated enzyme [75–77]. Abe et al. found that $[P_{444MEM}][Tf_2N]$ was suitable for the reaction of 1-phenylethanol (**1a**). On the other hand, $[N_{221MEM}]$ $[Tf_2N]$ or $[P_{444MEM}][Tf_2N]$ gave similar results when (±)-4-phenylbut-1-en-3-ol (**26**) was subjected to the reaction (Fig. 20) [77]. Using these ILs, they accomplished the recycling use of **IL1**-PS with higher reaction rates compared to those in *i*-Pr_2O. The authors observed increased K_{cat} values for these ILs compared to those in *i*-Pr_2O, although the viscosities of these ILs were ca. 200~300 times higher than *i*-

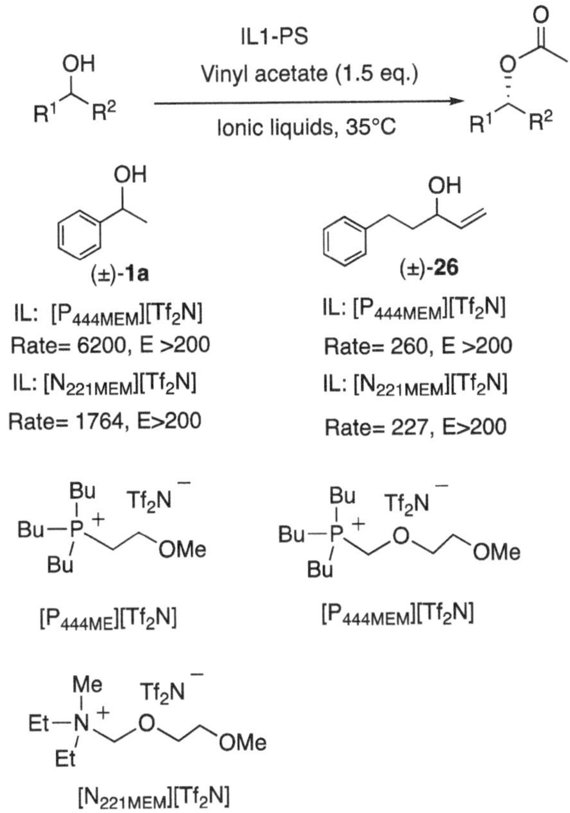

Fig. 20 IL1-PS- catalyzed acylation in different IL systems

Pr_2O [77]. These results indicated that the catalytic activity of the lipase is independent of the mass transfer rate of the substrate in the solvent system [77].

Nishihara et al. recently reported amazingly successful results of the recycling use of their IL-coated enzyme in in the $[N_{221MEM}][Tf_2N]$ solvent system. They prepared triazolium cetyl-PEG10 sulfate ILs, **Tz1**, **Tz2**, and **TZ3** (Fig. 18), and found that **Tz1-PS** exhibited excellent results when the enzyme was used as a catalyst for the transesterification of 1-phenylethanol (**1a**) in $[N_{221MEM}][Tf_2N]$, almost the same activity as that of the freshly prepared one even after storing for 2 years in $[N_{221MEM}][Tf_2N]$ (Fig. 21) [74]. These results indicated the possibility of IL engineering for enzyme stabilization.

What is the origin of the increased activity of a lipase by the ILs? It is known that the lipase PS protein has a lid part consisting of hydrophobic amino acid chains that covers the entrance portion of the enzyme. Kazlauskas et al. reported that treatment by *i*-PrOH of the *C. rugose* lipase through lyophilization altered the lid part to an open form with resulting activation of the enzymatic reaction [78]. Since then, numerous examples have been reported that the reactivity of the lipases depended on the lid part [79–83]. Quilles et al. also reported that the activation phenomenon was observed when the lipase from *Thermomyces lanuginose* or *C. antarctica* was treated with a surfactant [84]. Interaction of the IL against this lid might therefore

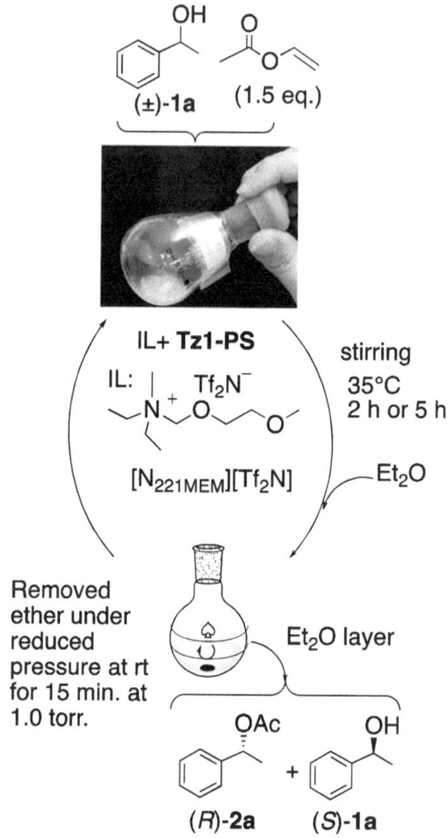

Fig. 21 Recyclable system of **Tz1**-PS using $[N_{221MEM}]$ $[Tf_2N]$ as a solvent. Picture in this figure reproduced with permission from [74]. Copyright 2018 Royal Society of Chemistry

play an important role in activation of the enzyme. Because **IL1** has an amphiphilic property, this might contribute to concentrating the hydrophobic substrate on the enzyme proteins that are located in the lid, so that the initial acceleration of the rate may be realized. MD simulation of the structural and dynamic features of a lipase from *C. rugose* in various reaction media indicated that the ILs influence both the enzyme structure and the dynamics of the protein [85]. Kurata et al. reported that the enzyme stability depended on the ratio of the α-helix and β-sheet of the enzyme protein; the former decreased and the latter increased in the ILs and caused deactivation [86]. Byrne et al. reported that the pH of the reaction mixture had a significant impact on the stability of the *T. lanuginosus* lipase. The authors also reported that ILs, which have a high hydrogen bonding basicity (β-value on Kamlet-Taft parameter over 0.8), caused a significant denaturation of the enzyme protein [87].

There exist numerous water molecules on the surface of the enzyme, and the water molecules play an important role in maintaining the motion flexibility of the enzyme protein and displaying the enzyme activity [88]. It was reported that enhancing the rigidity of the flexible segment within the active site was important for improving the enzyme kinetic stability [89–91]. As already described, it has been established that ILs, which have alkyl PEG sulfate anions, work as excellent activating agents of the *B. cepacia* lipase. Because the alkyl PEG moiety is essential to increase the reaction rate, it is assumed that the alkyl PEG moiety in the IL1 binds with the bottom portion and prevents removal of these surface water molecules, thus contributing to the improved flexibility of the enzyme. It had therefore been speculated that these two factors might be simultaneously improved by an appropriate design of the coating materials and reaction medium.

Kim et al. investigated the relationship between the enhanced activity of lipase B from *C. antarctica* and its molecular structure, and the authors proposed the following role of the cations of the ILs in the enzyme activity [92]: (1) the size of the cation and anion are comparable and show a high ion coordination number, so that the anion can promote the interaction between the gating residues of the enzyme and, consequently, the closing of the catalytic cavity which might provide a poor enzymatic activity, and (2) in the case of the IL which has a hydrophobic substituent, the IL has a strong interaction with the hydrophobic amino acid residue which results in the disruption of an α-helix and causes a very large cavity opening where anions can diffuse in and destabilize the catalytic trial [92]. For more detailed discussions, see recent review [9].

4 Future Perspective of Using ILs for Enzymatic Reactions

The use of ILs as reaction media changes the kinetics of the chemical reactions compared to those in aqueous or classical molecular solvents. Replacement of the reaction media from classical organic solvents or water by ILs in industry has unfortunately still not been realized. In order for ILs to become more popular, the key is how to reduce the amount of the ILs used, at the same time obtaining a certain benefit. The fields we should focus our investigation for utilizing the ILs in biotechnology in the future are discussed. Two properties of the ILs might become very important as research projects in this field.

4.1 Stabilizing Ability of ILs Toward Enzymes

As already mentioned, there is a possibility of an enhanced stability of a protein by an IL-coating; in fact, the ionic liquid-coated lipase is very stable in an IL or conventional organic solvent in the absence of substrates [62, 74]. Gutiérrez et al. reported interesting results from the application of ILs with microbes; the authors using a freeze-drying process for the incorporation of bacteria in DES in its pure state. This process allowed the outstanding preservation of the bacteria integrity and viability [93]. Brogan and Hallett demonstrated that engineering the surface of a protein to yield protein-polymer surfactant nano-constructs allows for dissolution of the dry protein into dry ionic liquids [94]. The authors have shown that this method can deliver protein molecules with a nearly native structure into both hydrophilic and hydrophobic anhydrous ionic liquids and the protein stability significantly increases in the ionic liquid.

4.2 *Importance of ILs as Enzyme Activating Agent*

The viscosities of many ILs are at least several hundred times higher than that of conventional organic liquids, such as *i*-Pr_2O, toluene, and hexane. However, the reaction rates of an ionic liquid coated IL (IL1-PS)-catalyzed transesterification of many secondary alcohols in the IL reaction media were superior to those of organic liquids. These results clearly indicated that the catalytic activity of the lipase is independent of the mass transfer rate of the substrate in the solvent system [83].

ILs obviously offer a high potential to replace classic flammable and toxic organic solvents. However, replacement of the reaction media from classical organic solvents or water to ILs in industry has not been realized to date. It is thought that one serious barrier might be the high viscosity of the ILs, because complete replacement of the reaction process is needed to use these viscous solvents. It is therefore believed that the importance of ILs should be highlighted not as solvents but as the controlling agents of enzymes. In such cases, we need only small quantities of the ILs and no modification of the present reaction process. Although successful examples have to date been limited only to lipases, it is believed that this methodology might become more important in the future. In the chemical industry field, the same trend is observed; investigations of using ILs as a solvent of chemical reactions have decreased. Instead, supported ionic liquid (SILP) catalysis [95] has currently attracted strong interest and has reached a practical level [96].

5 Conclusion

The activating method of the lipase-catalyzed transesterification using ionic liquid technology was chosen as the main topic of this review. An ionic liquid has a certain advantage over conventional organic solvents because the solvent makes it possible to use the enzyme repeatedly and has lower volatility and lower flammability. As already mentioned, the phosphonium ionic liquid and ammonium ionic liquids, which have alkylether moieties, worked as excellent reaction media for lipase-catalyzed transesterification, especially for ionic liquid-coated lipase. We can recover the ILs and repeatedly use them following a simple purification after the reaction. In fact, we always recycle our ionic liquids after the reaction and have not wasted any in the past. We are still using several ionic liquids that have more than an 18-year history. Innovation of the reaction medium used in a chemical reaction can provide a breakthrough and this is true, even in enzymatic reactions. It is hoped that this chapter may provide encouragement for the reader's research.

Acknowledgment Some work leading to results introduced in this chapter has been conducted in our laboratory. The financial support for a Grant-in-Aid for Scientific Research from the Japan Society for the Promotion of Science (Kakenhi Kiban) is acknowledged.

Glossary

$[C_2mim]^+$ 1-Ethyl-3-methylimidazolium
$[C_4mim]^+$ 1-Butyl-3-methylimidazolium
$[C_4dmim]^+$ 1-Butyl-2,3-dimethylimidazolium
$[HOPmim]^+$ 1-Methyl-3-(3-hydroxy)propylimidazolium
$[3\text{-}Ph\text{-}C_3mim]^+$ 1-Methyl-3(3-phenylpropyl)imidazolium
$[C_{16}tam]^+$ *N*-Cetyl-*N*,*N*,*N*-trimethylammonium
$[Me(OCH_2CH_2)_3\text{-}Et\text{-}Pip]^+$ Ethyl (2-(2-methoxyethoxy)ethoxy)ethylpiperidin-1-ium
$[Me(OCH_2CH_2)_3\text{-}Et_3N]^+$ Triethyl (2-(2-methoxyethoxy)ethoxy)ethylammonium
$[Me(OCH_2CH_2)_3eim]^+$ 1-Ethyl-3-(2-(2-methoxyethoxy)ethoxy)ethylimidazolium
L-Pro-Me (*S*)-1-Butyl-2-methyl-3-(pyrrolidin-2-ylmethyl)-1*H*-imidazol-3-ium
L-Pro-H (*S*)-1-Butyl-3-(pyrrolidin-2-ylmethyl)-1*H*-imidazol-3-ium
D-Pro-H (*R*)-1-Butyl-3-(pyrrolidin-2-ylmethyl)-1*H*-imidazol-3-ium
D-Pro-Me (*R*)-1-Butyl-2-methyl-3-(pyrrolidin-2-ylmethyl)-1*H*-imidazol-3-ium
$[N_{221MEM}]^+$ *N*,*N*-Diethyl-*N*-methyl-*N*-(2-methoxyethoxymethyl)ammonium
$[N_{2222}]^+$ *N*,*N*,*N*-Tetraethylammonium
$[P_{444ME}]^+$ Tributyl(2-methoxyethyl)phosphonium
$[P_{444MEM}]^+$ Tributyl(2-methoxyethoxyethyl)phosphonium
$[TAC1]^+$ Tris(diethylamino)cyclopropenium
$[TEA]^+$ *N*,*N*,*N*,*N*-Tetraethylammonium
$[PP_{22}]^+$ *N*,*N*-Diethylpiperidinium
$[Tz1]^+$ 1-Butyl-3-methyltriazolium
$[Tz2]^+$ 1-Benzyl-3-methyltriazolium
$[Tz3]^+$ 1-Butyl-3-methyl-5-(hydroxymethyl)triazolium
$[OAc]^-$ Acetate
$[BF_4]^-$ Tetrafluoroborate
$[NO_3]^-$ Nitrate
$[OTf]^-$ Trifluoromethanesulfonate
$[Tf_2N]^-$ Bis(trifluoromethylsulfonyl)amide. $[Tf_2N]$ is widely known as "bis (trifluoromethylsulfonyl)imide." However, "imide" means "an amido compound which is connected with two carbonyl groups," so $[Tf_2N]$ should be labelled as "bis(trifluoromethylsulfonyl)amide" according to the IUPAC rule
ISCB1 Potassium 3,5-bis(2-(2-(2-ethoxyethoxy)ethoxy)ethoxy)benzoate (ISCB1)
IL1 1-Butyl-2,3-dimethylimidazolium 3,6,9,12,15,18,21,24,27,30-decaoxacetyltriacontyl sulfate (cetyl-PEG10 sulfate)
DES Deep eutectic solvent
AMMOENG 102™ See Fig. 22

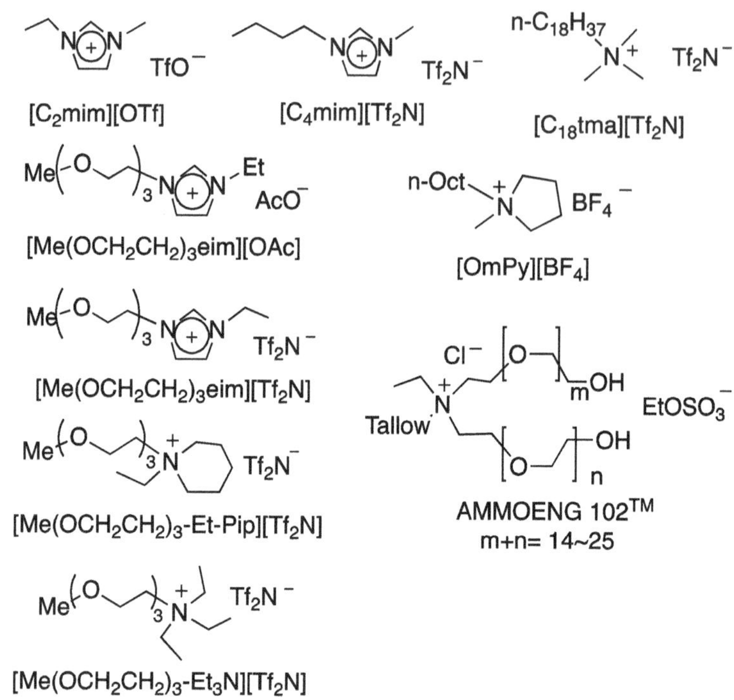

Fig. 22 Chemical structures of typical ILs appearing in the text

References

1. Wong C-H, Whitesides GM (1994) Enzymes in synthetic organic chemistry. Pergamon, Oxford
2. Bornscheuer UT, Kazlauskas RJ (1999) Hydrolases in organic synthesis: regio-and stereo-selective biotransformations. Wiley, Chichester
3. Faber K (2011) Biotransformations in organic chemistry, a textbook.6th edn. Springer, Heidelberg
4. Wasserscheid P, Welton T (eds) (2002) Ionic liquids in synthesis. Wiley, Weinheim
5. Chowdhury S, Mohan RS, Scott JL (2007) Reactivity of ionic liquids. Tetrahedron 63: 2363–2389
6. Ranke J, Stolte S, Störmann R, Arming J, Jastorff B (2007) Design of sustainable chemical products. The example of ionic liquids. Chem Rev 107:2183–2206
7. Plechkova NV, Seddon KR (2008) Applications of ionic liquids in the chemical industry. Chem Soc Rev 37:123–150
8. Hallett JP, Welton T (2011) Room-temperature ionic liquids: solvents for synthesis and catalysis. 2. Chem Rev 111:3508–3576
9. Itoh T (2017) Ionic liquids as tool to improve enzymatic organic synthesis. Chem Rev 117: 10567–10607
10. Lau RM, van Rantwijk F, Seddon KR, Sheldon RA (2000) Lipase-catalyzed reactions in ionic liquids. Org Lett 2:4189–4191
11. Itoh T, Akasaki E, Kudo K, Shirakami S (2001) Lipase-catalyzed enantioselective acylation in the ionic liquid solvent system: reaction of enzyme anchored to the solvent. Chem Lett 30: 262–263

12. Schöfer SH, Kaftzik N, Wasserscheid P, Kragl U (2001) Enzyme catalysis in ionic liquids: lipase catalysed kinetic resolution of L-phenylethanol with improved enantioselectivity. Chem Commun 5:425–426
13. Husum TL, Jørgensen CT, Christensen MW, Kirk O (2001) Enzyme catalysed synthesis in ambient temperature ionic liquids. Biocatal Biotransformation 19:331–338
14. Lozano P, De Diego T, Guegan JP, Vaultier M, Ibora JL (2001) Stabilization of α-chymotrypsin by ionic liquids in transesterification reactions. Biotechnol Bioeng 75:563–569
15. Lozano P, De Diego T, Carrie D, Vaultier M, Iborra JL (2001) Over-stabilization of Candida antarctica lipase B by ionic liquids in ester synthesis. Biotechnol Lett 23:1529–1533
16. Kim K-W, Song B, Choi M-Y, Kim M-J (2001) Biocatalysis in ionic liquids: markedly enhanced enantioselectivity of lipase. Org Lett 3:1507–1509
17. Park S, Kazlauskas RJ (2001) Improved preparation and use of room-temperature ionic liquids in lipase-catalyzed enantio- and regioselective acylations. J Org Chem 66:8395–8401
18. Laszlo JA, Compton DL (2001) α-Chymotrypsin catalysis in immobilized-based ionic liquids. Biotechnol Bioeng 75:181–186
19. Lozano P, De Diego T, Carrié D, Vaultier M, Iborra JL (2003) Enzymatic ester synthesis in ionic liquids. J Mol Catal B 21:9–13
20. Itoh T, Nishimura Y, Ouchi N, Hayase S (2003) 1-Butyl-2,3-dimethylimidazolium tetrafluoroborate; the most desirable ionic liquid solvent for recycling use of enzyme in lipase-catalyzed transesterification using vinyl acetate as acyl donor. J Mol Catal B 26:41–45
21. Gubicza L, Nemestóthy N, Fráter T, Bélafi-Bakó K (2003) Enzymatic esterification in ionic liquids integrated with pervaporation for water removal. Green Chem 5:236–239
22. Ulbert O, Fráter T, Bélafi-Bakó K, Gubicza L (2004) Enhanced enantioselectivity of *Candida rugosa* lipase in ionic liquids as compared to organic solvents. J Mol Catal B 31:39–45
23. Fehér E, Ileová V, Kelemen-Horváth I, Bélafi-Bakó K, Polakovic M, Gubicza L (2008) Enzymatic production of isoamyl acetate in an ionic liquid–alcohol biphasic system. J Mol Catal B 50:28–32
24. Baldwin RL (1996) How Hofmeister ion interactions affect protein stability. Biophys J 71: 2065–2063
25. Zhao H, Cambell SM, Jackson L, Song Z, Olubajo O (2006) Hofmeister series of ionic liquids: kosmotropic effect of ionic liquids on the enzymatic hydrolysis of enantiomeric phenylalanine methyl ester. Tetrahedron 17:377–383
26. Yang Z (2009) Hofmeister effects: an explanation for the impact of ionic liquids on Biocatalysis. J Biotechnol 144:12–22
27. Zhao H (2010) Methods for stabilizing and activating enzymes in ionic liquids—a review. J Chem Technol Biotechnol 85:891–907
28. Kumar A, Venkatesu P (2014) Does the stability of proteins in ionic liquids obey the Hofmeister series? Int J Biol Macromol 63:244–253
29. Zhao H (2016) Protein stabilization and enzyme activation in ionic liquids: specific ion effects. J Chem Technol Biotechnol 91:25–50
30. Barahona D, Pfromm P, Rezac ME (2006) Effect of water activity on the lipase catalyzed esterification of geraniol in ionic liquid [bmim]PF6. Bioetechnol Bioeng 93:318–324
31. Itoh T (2015) Chapter 5: ionic liquid mediated activation of lipase-catalyzed reaction. In: Handy S (ed) Ionic liquids-current state of the art. InTech, Croatia, pp 121–137
32. Itoh T, Akasaki E, Nishimura Y (2002) Efficient lipase-catalyzed enantioselective acylation under reduced pressure conditions in an ionic liquid solvent system. Chem Lett 31:154–155
33. Lourenço NMT, Afonso CAM (2007) One-pot enzymatic resolution and separation of sec-alcohols based on ionic acylating agents. Angew Chem Int Ed 46:8178–8181
34. Lourenco NMT, Monteiro CM, Afonso CAM (2010) Ionic acylating agents for the enzymatic resolution of sec-alcohols in ionic liquids. Eur J Org Chem 2010:6938–6943
35. Teixeira R, Lourenco NMT (2014) Enzymatic kinetic resolution of sec-alcohols using an ionic liquid anhydride as acylating agent. Tetrahedron 25:944–948

36. Lee SH, Ha SH, Nguyen MH, Chang W-J, Koo Y-M (2008) Lipase-catalyzed synthesis of glucose fatty acid ester using ionic liquids mixture. J Biotechnol 133:486–489
37. Sheldon RA (2016) Biocatalysis and biomass conversion in alternative reaction media. Chem Eur J 22:12984–12999
38. Nawshad MN, Elsheikh YA, Mutalib MIA, Bazmi AA, Khan RA, Khan H, Rafiq S, Man Z, Khan I (2015) An overview of the role of ionic liquids in biodiesel reactions. J Ind Eng Chem 21:1–10
39. Liu C-Z, Wang F, Stiles AR, Guo C (2012) Ionic liquids for biofuel production: opportunities and challenges. Appl Energy 92:406–414
40. Fauzi AHM, Amin NAS (2012) An overview of ionic liquids as solvents in biodiesel synthesis. Renew Sust Energ Rev 16:5770–5786
41. Andreani L, Rocha JD (2012) Use of ionic liquids in biodiesel production: a review. Braz J Chem Eng 29:1–13
42. Ha SH, Lan MN, Lee SH, Hwang SM, Koo Y-M (2007) Lipase-catalyzed biodiesel production from soybean oil in ionic liquids. Enzyme Microb Technol 41:480–483
43. De Diego TD, Manjón A, Lozano P, Vaultier M, Iborra JL (2011) An efficient activity ionic liquid-enzyme system for biodiesel production. Green Chem 13:444–451
44. De Diego TD, Manjón A, Lozano P, Iborra JL (2011) A recyclable enzymatic biodiesel production process in ionic liquids. Bioresource Technol 102:6336–6339
45. Lai J-Q, Hu Z-L, Wang P-W, Yang Z (2012) Enzymatic production of microalgal biodiesel in ionic liquid [BMIm][PF_6]. Fuel 95:329–333
46. Laia J-Q, Hub Z-L, Sheldon RA, Yang Z (2012) Catalytic performance of cross-linked enzyme aggregates of Penicillium expansum lipase and their use as catalyst for biodiesel production. Process Biochem 47:2058–2063
47. Huang Z-L, Yang T-X, Huang J-Z, Yang Z (2014) Enzymatic production of biodiesel from *Millettia Pinnata* seed oil in ionic liquids. Bioenergy Res 7:1519–1528
48. Pámies O, Bäckvall JE (2004) Chemoenzymatic dynamic kinetic resolution. Trends Biotechnol 22:130–135
49. Pellissier H (2008) Recent developments in dynamic kinetic resolution. Tetrahedron 64: 1563–1601
50. Cheng G, Xia B, Wu Q, Lin X (2013) Chemoenzymatic dynamic kinetic resolution of α-trifluoromethylated amines: influence of substitutions on the reversed stereoselectivity. RSC Adv 3:9820–9828
51. Kim M-J, Kim HM, Kim D, Ahn Y, Park J (2004) Dynamic kinetic resolution of secondary alcohols by enzyme–metal combinations in ionic liquid. Green Chem 6:471–474
52. Lozano P, De Diego T, Larnicol M, Vaultier M, Iborra JL (2006) Chemoenzymatic dynamic kinetic resolution of *rac*-1-phenylethanol in ionic liquids and ionic liquids/supercritical carbon dioxide systems. Biotechnol Lett 28:1559–1565
53. Lozano P, De Diego T, Vaultier M, Ibora JL (2009) Dynamic kinetic resolution of sec-alcohols in ionic liquids/supercritical carbon dioxide biphasic systems. Int J Chem React Eng 7:A79
54. Shimomura K, Harami H, Matsubara Y, Nokami T, Katada N, Itoh T (2015) Lipase-mediated dynamic kinetic resolution (DKR) of secondary alcohols in the presence of zeolite using an ionic liquid solvent system. Catal Today 255:41–48
55. Kotlewska AJ, van Rantwijk F, Sheldon RA, Arends IWCE (2011) Epoxidation and Baeyer–Villiger oxidation using hydrogen peroxide and a lipase dissolved in ionic liquids. Green Chem 13:2154–2160
56. Drozdz A, Chrobok A, Baj S, Szymanska K, Mrowiec-Bialon J, Jarzebski AB (2013) The chemo-enzymatic Baeyer–Villiger oxidation of cyclic ketones with an efficient silica-supported lipase as a biocatalyst. Appl Catal A 467:163–170
57. Khmelnitsky YL, Welch SH, Clark DS, Dordick JS (1994) Salts dramatically enhance activity of enzymes suspended in organic solvents. J Am Chem Soc 116:2647–2648

58. Ru MT, Hirokane SY, Lo AS, Dordick JS, Reimer JA, Clark SS (2000) On the salt-induced activation of lyophilized enzymes in organic solvents: effect of salt kosmotropicity on enzyme activity. J Am Chem Soc 122:1565–1571
59. Lindsay JP, Clark DS, Dordick JS (2004) Combinatorial formulation of biocatalyst preparations for increased activity in organic solvents: salt activation of penicillin amidase. Biotechnol Bioeng 85:553–560
60. Lee JK, Kim M-J (2002) Ionic liquid-coated enzyme for biocatalysis in organic solvent. J Org Chem 67:6845–6847
61. Itoh T, Han S-H, Matsushita Y, Hayase S (2004) Enhanced enantioselectivity and remarkable acceleration on the lipase-catalyzed transesterification using novel ionic liquids. Green Chem 6:437–439
62. Itoh T, Matsushita Y, Abe Y, Han S-H, Wada S, Hayase S, Kawatsura M, Takai S, Morimoto M, Hirose Y (2006) Enhanced enantioselectivity and remarkable acceleration of lipase-catalyzed transesterification using an imizadolium PEG-alkyl sulfate ionic liquid. Chem Eur J 12:9228–9237
63. Han S-H, Hirakawa T, Fukuba T, Hayase S, Kawatsura M, Itoh T (2007) Synthesis of enantiomerically pure cycloalkenols via combination strategy of enzyme-catalyzed reaction and RCM reaction. Tetrahedron 18:2484–2490
64. Lee JK, Kim M-J (2011) Ionic liquid co-lyophilized enzyme for biocatalysis in organic solvent: remarkably enhanced activity and enantioselectivity. J Mol Catal B 68:275–278
65. Lee SH, Dong DT, Ha S-H, Chang W-J, Koo Y-M (2007) Using ionic liquids to stabilize lipase within sol–gel derived silica. J Mol Catal B 45:57–61
66. Abe Y, Hirakawa T, Nakajima S, Okano N, Hayase S, Kawatsura M, Hirose Y, Itoh T (2008) Remarkable activation of an enzyme by (R)-pyrrolidine-substituted imidazolium alkyl PEG sulfate. Adv Synth Catal 350:1954–1958
67. Yoshiyama K, Abe Y, Hayse S, Nokami T, Itoh T (2013) Synergetic activation of lipase by an amino acid with alkyl-PEG-sulfate ionic liquid. Chem Lett 42:663–665
68. Rahman MBA, Jumbri K, Hanafiah NAMA, Abdulmalek E, Tejo BA, Basri M, Salleh AB (2012) Enzymatic esterification of fatty acid esters by tetraethylammonium amino acid ionic liquids-coated Candida rugose lipase. J Mol Catal B 79:61–65
69. Kim HJ, Choi YK, Lee J, Lee E, Park J, Kim M-J (2011) Ionic-surfactant-coated Burkholderia cepacia lipase as a highly active and enantioselective catalyst for the dynamic kinetic resolution of secondary alcohols. Angew Chem Int Ed 50:10944–10948
70. Kim C, Lee J, Choi J, Oh Y, Choi YK, Choi E, Park J, Kim M-J (2013) Kinetic and dynamic kinetic resolution of secondary alcohols with ionic-surfactant-coated Burkholderia cepacia lipase: substrate scope and enantioselectivity. J Org Chem 78:2571–2578
71. Lee E, Oh Y, Choi YK, Kim M-J (2014) Aqueous-level turnover frequency of lipase in organic solvent. ACS Catal 4:3590–3592
72. Matsubara Y, Kadotani S, Nishihara T, Fukaya Y, Nokami T, Itoh T (2015) Design of phosphonium alkyl PEG sulfate ionic liquids as a coating materials for activation of Burkholderia cepacia lipase. Biotechnol J 10:1944–1951
73. Kadotani S, Inagaki R, Nishihara T, Nokami T, Itoh T (2017) Enhanced activity of a lipase by the coating with tris(diethylamino)cyclopropenium alkyl-PEG sulfate ionic liquid and cooperative activation with an amino acid. ACS Sustain Chem Eng 5:8541–8545
74. Nishihara T, Shiomi A, Kadotani S, Nokami T, Itoh T (2017) Enhanced activity and remarkable improved stability of a Burkholderia cepacia lipase by the coating with a triazolium alkyl-PEG sulfate ionic liquid. Green Chem 19:5250–5256
75. Abe Y, Kude K, Hayase S, Kawatsura M, Tsunashima K, Itoh T (2008) Design of phosphonium ionic liquids for lipase-catalyzed transesterification. J Mol Catal B 51:81–85
76. Abe Y, Yoshiyama K, Yagi Y, Hayase S, Kawatsura M, Itoh T (2010) A rational design of phosphonium salt type ionic liquids for ionic liquid coated-lipase catalyzed reaction. Green Chem 12:1976–1980

77. Abe Y, Yagi Y, Hayase S, Kawatsura M, Itoh T (2012) Ionic liquid engineering for lipase-mediated optical resolution of secondary alcohols: design of ionic liquids applicable to ionic liquid coated-lipase catalyzed reaction. Ind Eng Chem Res 51:9952–9958
78. Colton IJ, Sharmin NA, Kazlauskas RJ (1995) A 2-propanol treatment increases the enantioselectivity of Candida rugosa lipase toward esters of chiral carboxylic acids. J Org Chem 60: 212–217
79. Overbeeke PLA, Govardhan C, Khalaf N, Jongejan JA, Heijnen JJ (2002) Influence of lid conformation on lipase enantioselectivity. J Mol Catal B 10:385–393
80. Halimi H, De Caro J, Carrière F, De Caro A (2005) Closed and open conformations of the lid domain induce different patterns of human pancreatic lipase antigenicity and immunogenicity. Biochim Biophys Acta 1753:247–256
81. Mathesh M, Luan B, Akanbi TO, Weber JK, Liu J, Barrow CJ, Zhou R, Yang W (2016) Opening lids: modulation of lipase immobilization by graphene oxides. ACS Catal 6: 4760–4768
82. Skjold-Jørgensen J, Vind J, Moroz OV, Blagova E, Bhatia VK, Svendsen A, Wilson KS, Bjerruma MJ (2017) Controlled lid-opening in thermomyces lanuginosus lipase– an engineered switch for studying lipase function. Biochim Biophys Acta 1865:20–27
83. Riccardi L, Arencibia JM, Bono L, Armirotti A, Girotto S, De Vivo M (2017) Lid domain plasticity and lipid flexibility modulate enzyme specificity in human monoacylglycerol lipase. Biochim Biophys Acta 1862:441–451
84. Quilles JCJ, Brito RR, Borges JP, Aragon CC, Fernandez-Lorente G, Bocchini-Martins DA, Gomes E, Da Silva R, Boscolo M, Guisane JM (2015) Modulation of the activity and selectivity of the immobilized lipases by surfactants and solvents. Biochem Eng J 93:274–280
85. Eisenmenger MJ, Reyes-De-Corcuera JI (2010) Enhanced synthesis of isoamyl acetate using an ionic liquid-alcohol biphasic system at high hydrostatic pressure. J Mol Catal B 67:36–40
86. Kurata A, Senoo H, Ikeda Y, Kaida H, Matsuhara C, Kishimoto N (2016) Properties of an ionic liquid-tolerant Bacillus amyloliquefaciens CMW1 and its extracellular protease. Extremophiles 20:415–424
87. Han Q, Wang X, Byrne N (2016) Understanding the influence of key ionic liquid properties on the hydrolytic activity of Thermomyces lanuginosus lipase. ChemCatChem 8:1551–1556
88. Xin JY, Zhao Y-J, Shi Y-G, Xia C-G, Li S-B (2005) Lipase-catalyzed naproxen methyl ester hydrolysis in water-saturated ionic liquid: significantly enhanced enantioselectivity and stability. World J Microbiol Biotechnol 21:193–199
89. Lee SH, Koo Y-M, Ha SH (2008) Influence of ionic liquids under controlled water activity and low halide content on lipase activity. Korean J Chem Eng 25:1456–1462
90. Micaêlo N, Soares C (2008) Protein structure and dynamics in ionic liquids. Insights from molecular dynamics simulation studies. J Phys Chem B 112:2566–2572
91. Noritomi H, Chiba H, Kikuta M, Kato S (2013) How can aprotic ionic liquids affect enzymatic enantioselectivity? J Biomed Sci Eng 6:954–959
92. Kim HS, Ha SH, Sethaphong L, Koo Y-M, Yingling YG (2014) The relationship between enhanced enzyme activity and structural dynamics in ionic liquids: a combined computational and experimental study. Phys Chem Chem Phys 16:2944–2953
93. Gutiérrez M, Ferrer ML, Yuste L, Rojo F, del Monte F (2010) Bacteria incorporation in deep-eutectic solvents through freeze drying. Angew Chem Int Ed 49:2158–2162
94. Brogan APS, Hallett JP (2016) Solubilizing and stabilizing proteins in anhydrous ionic liquids through formation of protein-polymer surfactant Nanoconstructs. J Am Chem Soc 138: 4494–4501
95. Riisager A, Fehrmann R, Haumann M, Wasserscheid P (2006) Supported ionic liquid phase (SILP) catalysis: an innovative concept for homogeneous catalysis in continuous fixed-bed reactors. Eur J Inorg Chem 2006:695–706
96. Abai M, Atkins MP, Hassan A, Holbrey JD, Kuah Y, Nockemann P, Oliferenko AA, Piechkova NV, Rafeen S, Rahman AA et al (2015) An ionic liquid process for mercury removal from natural gas. Dalton Trans 44:8617–8624

Adv Biochem Eng Biotechnol (2019) 168: 105–132
DOI: 10.1007/10_2018_77

Published online: 29 November 2018

Whole-Cell Biocatalysis in Ionic Liquids

Ngoc Lan Mai and Yoon-Mo Koo

Contents

Abstract The use of whole-cell biocatalysis in ionic liquid (IL)-containing systems has attracted increasing attention in recent years. Compared to bioreactions catalyzed by isolated enzymes, the major advantage of using whole cells in biocatalytic processes is that the cells provide a natural intracellular environment for the enzymes to function with in situ cofactor regeneration. To date, the applications of whole-cell biocatalysis in IL-containing systems have focused on the production of valuable compounds, mainly through reduction, oxidation, hydrolysis, and transesterification reactions. The interaction mechanisms between the ILs and biocatalysts in whole-cell biocatalysis offer the possibility to effectively integrate ILs with biotransformation. This chapter discusses these interaction mechanisms between ILs and whole-cell catalysts. In addition, examples of whole-cell catalyzed reactions with ILs will also be discussed.

N. L. Mai and Y.-M. Koo (✉)
Department of Biological Engineering, Inha University, Incheon, South Korea
e-mail: ymkoo@inha.ac.kr

Graphical Abstract

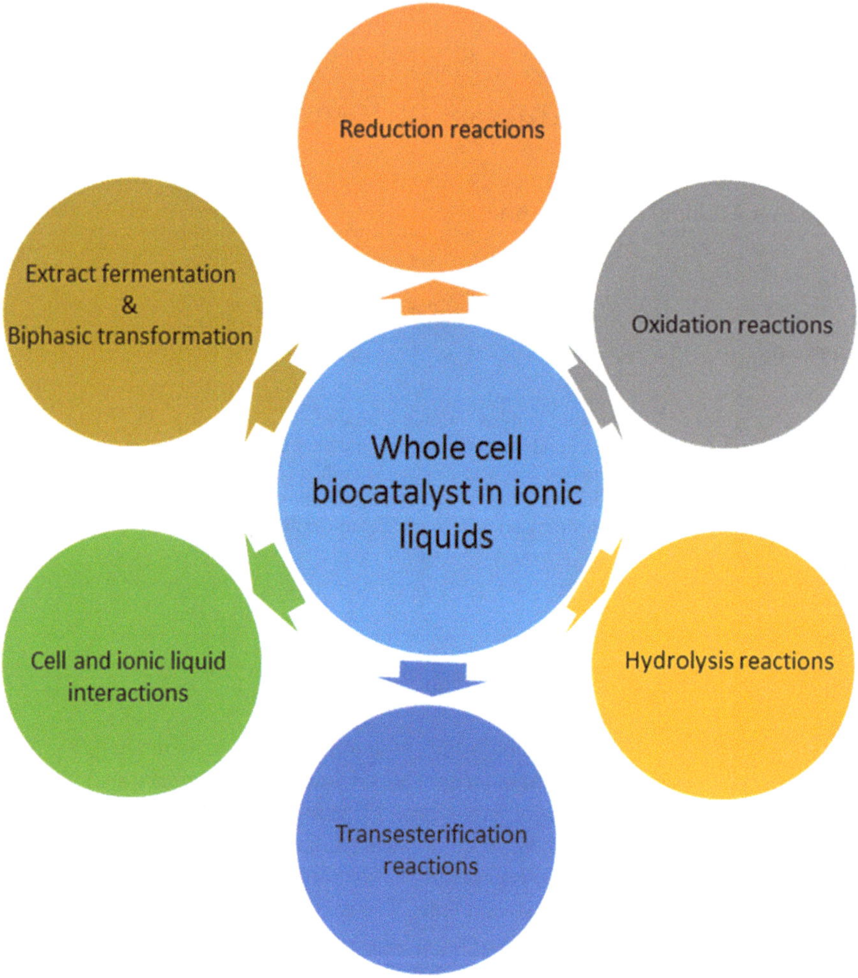

Keywords Biocatalysis, Biocompatibility, Ionic liquid, Toxicity, Whole-cell

Abbreviations

Cations

[Bmim]	1-Butyl-3-methylimidazolium
[BP]	1-Butyl-1-pyrrolidinium
[BPy]	*N*-butyl-pyridinium

[C_2OHmim]	1-(2′-hydroxyl)ethyl-3-methylimidazolium
[CABHEM]	PEG-5 Cocomonium
[Emim]	1-Ethyl-3-methylimidazolium
[EtOHNMe$_3$]	2-Hydroxy ethyl trimethyl-ammonium
[Hmim]	1-Hexyl-3-methylimidazolium
[HMPL]	1-Hexyl-1-methylpyrrolidinium
[NMeOct$_3$]	Methyltrioctylammonium
[Omim]	1-Methyl-3-octylimidazolium
[$P_{6,6,6,14}$]	Trihexyltetradecylphosphonium
[Pmim]	1-Propyl-3-methylimidazolium

Anions

[BF_4]	Tetrafluoroborate
[E_3FAP]	Tris(pentafluoroethyl)trifluorophosphate
[ES]	Ethylsulfate
[MDEGSO$_4$]	Ethylenglycolmonomethyl-ethersulfate
[Me_2PO_4]	Dimethyl phosphate
[MS]	Methylsulfate
[NO_3]	Nitrate
[OS]	Octylsulfate
[PF_6]	Hexafluorophosphate
[Tf_2N]	bis[(Trifluoromethyl)sulfonyl]mide
[TfO]$^-$	Trifluoromethanesulfonate
[TOS]	Tosylate

1 Introduction

Ionic liquids (ILs) are organic salts that consist of ions and are liquids at room temperature. ILs are generally considered to be green alternatives to traditional volatile organic solvents due to their unique physicochemical properties, such as negligible vapor pressures, low melting points, varying viscosities, larger electrochemical windows, thermal stability, and ability to dissolve a variety of materials [1]. In addition, the physical properties of ILs, such as polarity, hydrophobicity, and solvent miscibility behaviors, can be finely tuned through appropriate modification of the cations and anions; thus, they have been referred to as "designer solvents" for various applications [2]. ILs have been widely used in analytical chemistry [3–6], electrochemistry [7–10], gas separation and purification [11–19], and biotechnology [20–28].

Applications of ILs in biotcatalysis have been extensively studied during the past decade, with a strong focus on catalysis using isolated enzymes, which remain active in ILs and exert good performance. There are several role models in biocatalytic process with ILs. Thus far, ILs have been used as pure solvents, co-solvents in the aqueous phase, biphasic systems together with other solvents, and even with combinations of polymers for the enhancement of both chemical and enzymatic reactions. The beneficial effects of ILs on enzymatic activity and selectivity are partly attributed to the high polarity of these solvents, with the ability to increase the solubility of polar substrates while the biocatalyst exhibits high activity. The influence of ILs on the structure, activity, stability, and selectivity of isolated enzymes, as well as the applications of enzymes in such media, have been extensively reviewed.

Much attention has been paid to the whole-cell-based biocatalytic process in ILs in recent years. However, compared to enzymatic catalysis, studies of whole-cell biocatalysis in the presence of ILs have been less well explored. To date, the use of ILs in whole-cell processes has been mainly limited to "extraction fermentation" or "biphasic transformation" processes, in which water-immiscible ILs act as a substrate reservoir and in situ extractant for the removal of toxic products to avoid inhibition of the cells. In such applications, the IL phase can generally be recycled without loss of productivity of the process, thus making the use of an IL economically feasible. Moreover, another one of the major advantages of using whole cells rather than isolated enzymes biocatalysts is that they are able to regenerate efficiently expensive cofactors in redox reactions. In addition, the cells provide a natural environment for the enzymes, preventing denaturation and conformational changes in the protein structure that may lead to loss of activity in nonconventional reaction media, such as ILs. Furthermore, there is no need for costly enzyme purification in whole-cell processes.

Whole-cell biocatalysts exhibit an increased stability and additional benefits compared with isolated enzymes. Unfortunately, ILs are reported to have toxic effects on bacteria, yeasts, and fungi, which are limitations of the biotransformation applications. Therefore, an understanding of the mechanisms of the effects of ILs on microorganism cells is necessary to reveal possible approaches to enhance the biocatalysis yield; this research area has thus attracted much attention and interest in recent years [29–31]. The applications of ILs in whole-cell biocatalysis are summarized in this chapter. The characteristics of ILs in whole-cell catalysis are also discussed, including the effects of IL properties on the microbial cells, substrates, and products involved in synthesis reactions; interactions between ILs and microbial cells; and typical whole-cell biotransformation reactions in ILs.

2 Characteristics of Whole-Cell Catalysts in Ionic Liquids

Various kinds of anions and cations available for use as ILs have been studied so far. Newly designed ILs are being developed into green solvents, with the range of their applications expanding enormously. There are approximately one trillion (10^{18}) accessible ILs. However, only a few ILs can be used for biotransformation. The structures of some cations and anions commonly used in whole-cell biocatalysis are summarized in Fig. 1.

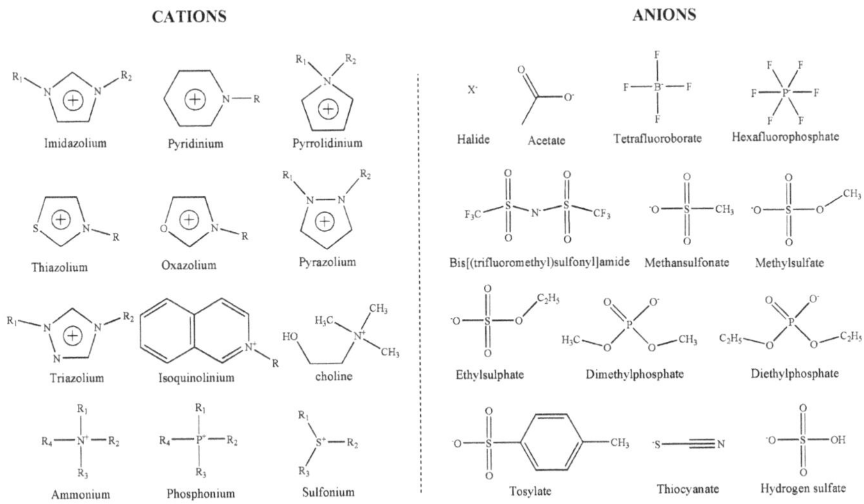

Fig. 1 Structure of common ionic liquid cations and anions used in whole-cell biotransformation

Composition and structure type are the key factors affecting the physicochemical properties of ILs, together with biological performances. Moreover, the tunable properties of ILs vary with modified structures. Compared with conventional organic solvents, ILs are beneficial and indispensable alternatives to organic solvents. The properties of ILs have been described in a number of publications [1, 32]. The most important of these properties for whole-cell biotransformation are the melting point, viscosity, density, and polarity.

ILs have a low melting point, which is determined by the types of cations and anions. The melting point of ILs is much lower with weaker molecular interactions, lower structural symmetry, and uniformly distributed cation/anion charges. In addition, ILs have a broad temperature window at or below 300°C, in which condition they remain as liquids [33]. In light of the importance of temperature to microorganism growth conditions, ILs with an equal melting point must be developed and employed.

Most ILs are much more viscous than organic solvents. The viscosity of ILs is usually affected by the strong intermolecular forces between solvent molecules, the alkyl chain length, organic co-solvents, and water. For ILs, the strong forces are their inherent charge–charge interactions; van der Waals forces also exist between molecules. The reducing van der Waals interactions can cause slightly lower viscosity by reducing the surface area of the molecules [34]. ILs are less viscous in the presence of water or organic co-solvents because the hydrogen bonding and strength of their van der Waals interactions are lower. However, ILs with longer alkyl chains exhibit much higher viscosity than those with shorter chains on the cation [20]. Because the use of a viscous IL-containing medium causes some operations (including mixing, filtrating, and other mass transfer) to be more difficult and inhibits the metabolism of cells, less viscous ILs need to be designed for wide applications.

The density of ILs (ranging from 1.1 to 1.6 g cm^{-3}) is generally larger than that of typical solvents or water. The higher density allows ILs to be separated from aqueous cell-containing phases [26]. Very slight differences of structure can result in changes of density. In many cases, the density decreases by the reduced volume of the anion or the increasing alkyl chain length of the cation. This is related to mass transfer in reaction systems, and ILs with designed structure and density are beneficial for cell biotransformation.

Polarity is one of the most important properties of ILs. In general, solvents with high polarity can affect the enzymes and microbial cells, resulting in enzyme activity reduction or deactivation. On the contrary, ILs with high polarity can make enzymes stable and selective, leading to an increased reaction rate. Because ILs have the ability to dissolve many polar or nonpolar substances, they can be used in both hydrophilic substrate reactions and hydrophobic substrate reactions [35, 36]. However, it is difficult to define the polarity of one kind of ILs, because different results can be observed when it is determined based on the shift of the charge-transfer absorption band of a solvatochromic probe, such as Reichardt's dye and Nile red dye [37, 38]. Methods with partitions and fluorescence probes have also been used to determine polarity [39]. All the results revealed that ILs are polar, with a polarity between that of water and some alcohols. However, the polarity varies with the structure changes of cations/anions and decreases with an increase of the alkyl chain length. In cell biotransformation, the polarity is essential to mass transfer. An IL with shorter alkyl chains on the cation has higher polarity and lower viscosity at the same time. Such an IL helps to accelerate the catalysis reactions. However, it is uncertain whether the viscosity and the polarity changes affect the changing reaction rate together or separately. The viscosity change may be the key factor because the viscosity is affected by the alkyl chain length more. In addition, the polarity of ILs is sometimes susceptible to water and temperature [40]. Furthermore, the presence of water can affect the properties of ILs. However, the solubility of water in ILs varies depending on the types of anions that are present. This effect should be taken into consideration when ILs are used for biotransformations in culture broths with cells [30].

In whole-cell biotransformation, ILs used in the media have inevitable effects on the microorganism cells. In general, ILs directly affect cell membrane permeability and hence can inhibit or activate the growth of cells. In addition, ILs can affect the environment surrounding the cells or interact with the substrates involved in the biotransformation, as depicted in Fig. 2 [30].

2.1 Interactions of Ionic Liquids with Cells in Whole-Cell Biocatalysis

2.1.1 Toxicity Toward Microorganism Cells

Although ILs are generally regarded as "green" solvents, their toxicity to microorganisms has been reported as a key drawback for whole-cell applications. A

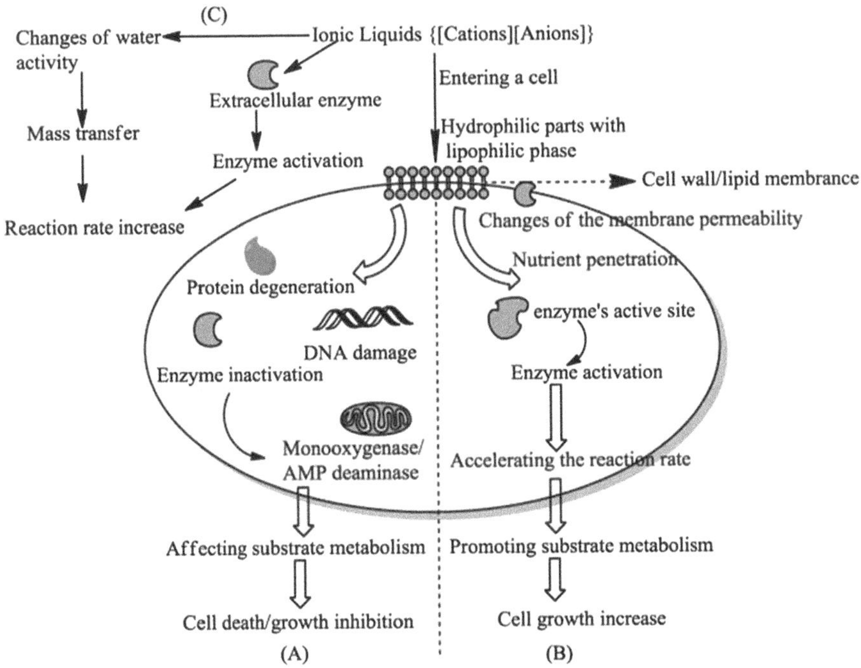

Fig. 2 Possible interactions of ILs with cells in whole-cell biotransformation. Some ILs may affect the cell microstructure and lead to cell growth inhibition/cell death (**A**) or increase of cell growth (**B**). Some ILs interact with the surrounding environment and accelerate the mass transfer and reaction rate (**C**). Figure was adapted from ref. [30]

number of studies have evaluated the toxicity of ILs towards microorganism cells. For example, Wood et al. studied the toxicity of more than 90 ILs from diverse structural classes toward *Escherichia coli* using high-throughput methods, including agar diffusion tests and growth inhibition tests in liquid media. A wide range of ILs containing imidazolium, pyridinium, quaternary ammonium, alkanolammonium, and quaternary phosphonium cations, combined with a diverse range of anions, were screened. In general, imidazolium salts with short alkyl chains were relatively non-toxic, especially when paired with alkyl sulfate anions. Methylpyrrolidinium salts were also very promising, whereas water-miscible quaternary ammonium salts were generally toxic. However, there was evidence that increasing the alkyl chain length increased their toxicity [41]. Similar relationships between increasing lengths of the alkyl chain/lipophilicity of the imidazolium cation and the increasing toxicity of the ILs have been widely reported elsewhere [42, 43]. The hydrophobicity, which corresponds to the increasing alkyl chain length on the IL cation, is found to induce rising toxicity. In comparison to some commonly used industrial solvents such as phenol, toluene, and benzene, the octyl- and hexylsubstituted ILs are more toxic. Apart from the alkyl chain length of the cation in ILs, the anion was also found to be crucial for the toxicity against different microorganisms. For example, Santos et al.

investigated the toxicity of various ILs to nine microorganisms of interest to the food industry. The authors found that ILs containing choline as the cationic moiety were more biocompatible because they allowed the growth of all the studied microorganisms, whereas the [Tf_2N] anion-based ILs showed serious toxicity [44].

2.1.2 Effects on Cell Membrane Permeability

Cell membrane permeability is an important factor in the efficiency of whole-cell biocatalytic processes [29]. The interaction of ILs with the phospholipid bilayer and the increase in membrane permeability as cells are exposed to ILs have been confirmed [45, 46]. ILs may be able to increase cell membrane permeability and thus allow substrates and products to diffuse more readily into and out of the cells, resulting in acceleration of biocatalysis [47]. Conversely, a loss of membrane integrity could reduce cell viability, compromise cell metabolism, and diminish biocatalytic reaction yields [29, 48].

Evan et al. used a supported phospholipid bilayer formed by 1,2-dielaidoylphosphocholine (DEPC) and 1,2-dimyristoylphosphoglycerol on different supports and large unilamellar vesicles of DEPC to investigate the interactions between ILs and cell membranes [49–51]. The study's results demonstrated that the instability of a fluid bilayer increased as the length of the alkyl group of the cation increased. The large unilamellar vesicles were mostly stable in the presence of the anions, as evidenced by continued retention of encapsulated materials. In the case of a 1-butyl-3-methylimidazolium cation (100–1,500 mM), size measurements revealed that encapsulated material was released through small holes in the bilayer. However, total vesicle disruption was the release mechanism in the case of 1-hexyl or octyl-3-methylimidazolium above 200 mM. The cations of some ILs clearly act as short-chain surfactants, potentially disintegrating a phospholipid bilayer, such as a cell membrane. The effects of the different anions commonly found in ILs were also tested on the bilayer. [BF_4] and [PF_6] did not induce any leakage, suggesting that these anions do not interact with the lipid bilayer. However, [Tf_2N] showed small and unstable defects in the lipid bilayer at concentrations of 500 mM [49].

Bräutigam et al. investigated the effects of imidazolium-based ILs on the membrane integrity of recombinant *E. coli* in a continuously stirred reactor. The results showed that [PF_6] anions affected the cell membranes of *E. coli* marginally, which decreased the membrane integrity only to 70% compared with 95% in a pure aqueous system after 5 h of incubation. In contrast, [Tf_2N] anions seemed to be more toxic to the cell membrane. The most negative effect was measured for ILs with [E3FAP] anions [45].

Xiao et al. studied the effects of various ILs on the cell membrane permeability of *Acetobacter* sp. CCTCC M209061 [52]. The changes in the A260 and A280 values – indicators of the release of intracellular components (presumably mostly nucleic acids and proteins) into the medium after removal of the cells – were taken as a measure of the ILs' effects on the cell membrane permeability. Different types of ILs had distinctly different effects on the permeability of the cell membrane. With [BF_4]-

based ILs, A260 and A280 values due to the release of cellular components were relatively high, suggesting that these ILs greatly increased cell membrane permeability. However, in the presence of these [BF_4]-based ILs, the cells had very poor catalytic activity. Thus, it appeared that the [BF_4]-based ILs damaged the cell membrane too seriously to maintain effective catalytic activity of the cells. Among all the ILs investigated, the greatest cell membrane integrity and the lowest A260 and A280 values were recorded for [EOHmim][NO_3], which also gave the best catalytic performance of the cells. Thus, a moderate increase in cell membrane permeability appeared to enhance the reaction efficiency whereas a large increase in permeability would put an irreversibly activity-lowering effect on the cell.

2.2 *Effect of Ionic Liquids on Enzymes Within Microbial Cells*

Because ILs can permeate through the cell membrane, they can they can interact with enzymes within the cell. However, the effects of ILs on intracellular enzymes are inevitable. For example, Lou et al. used fluorescence microscopy analysis and showed that ILs could enter *Rhodotorula* sp. AS2.2241 cells and accumulate inside the cell membranes, where the redox enzymes distributed, indicating that ILs might influence the catalytic performances of the intracellular enzymes [47]. Subsequently, several researchers further demonstrated that ILs significantly affected the enzyme systems present in *Aspergillus nidulans* cells and consequently led to a change in the metabolism of intracellular chemicals [53–56]. For example, the exposure of *A. nidulans* to [Emim][Cl] not only led to the accumulation of proteins involved in stress response and autolysis, but it also altered the expression of proteins involved in carbohydrate and energy metabolism [56]. Similarly, it was also observed that the IL [Amim][BF_4] affected the expression of enzyme genes present in *Vibrio qinghaiensis* sp.-Q67 cells, where several genes expressing luciferase, superoxide dismutase, and catalase were upregulated in a time-dependent hormesis manner [57]. These observations clearly indicate that ILs could affect the enzymes present in microbial cells at the gene level. However, the specific effects and detailed mechanisms of ILs on the intracellular enzymes involved in biocatalysis remains largely unknown.

2.3 *Effect of Ionic Liquids on Substrate and Product Partition Coefficients*

Pfruender et al. have proposed that the equilibrium distribution coefficients of substrates and products are important, together with data about the toxicity of ILs, in predicting the process efficiency of whole-cell biocatalysts in IL/aqueous systems

[58]. Investigating the reduction of 4-chloroacetophenone to (*R*)-1-(4-chlorophenyl) ethanol (1-4-Cl-PE) catalyzed by *Lactobacillus kefir* as a model reaction, they calculated that systems containing organic solvents (e.g., an aqueous:decane biphasic system) would have diminished yield because of the low distribution coefficient (log D) value for the product 1-4-Cl-PE (1.54) between the aqueous phase and decane – which meant that in a 20% solvent setup, approximately 11.5% of the product was lost because it remained in the aqueous phase. With [Bmim][Tf_2N] as the cosolvent, the log D value was substantially greater (2.03); this loss would account for only 3.7%.

In addition, greater partitioning of toxic substrates and products into the IL phase could reduce the effect of such toxic compounds on the cells, as well as alleviate any substrate or product inhibition observed in monophasic aqueous systems [59, 60]. Partition coefficients in biphasic systems containing ILs are a function both of the IL and the chemical nature of the substrates and products. Larger partition coefficients, in favor of the increased partitioning of the substrate and product into the IL phase, correlated with greater biocompatibility of the IL with the cells and overall biocatalytic process efficiency [61].

3 Whole-Cell Catalyzed Transformations in Ionic Liquid-Containing Systems

To date, most of the studies on the use of whole-cell biocatalysts in ILs have focused on the production of chiral compounds, obtaining high yields and enantiomeric excess (ee) values in such reaction media, especially high substrate loading benefits from the introduction of hydrophobic ILs to an IL/buffer biphasic system. In addition, most of the work has involved biphasic reaction systems formed by hydrophobic ILs with an aqueous phase. In fact, biphasic reaction systems have certain advantages over monophasic reaction modes, such as facilitated product isolation and recovery of the product and the IL, which make their use more attractive. Nevertheless, some applications also involve hydrophilic ILs as co-solvents in the aqueous phase. Most of the reactions studied in whole-cell biocatalysis involving ILs are asymmetric reductions of ketones by oxidoreductases for the production of chiral alcohols. This class of enzymes takes particular advantage of the whole-cell approach due to its need for cofactors to perform the reduction. Nevertheless, applications involving ILs are becoming increasingly diverse, and various reactions have been investigated in ILs using whole-cell catalysts. In addition, the cells used as catalysts cover a wide range of microorganisms, including bacteria, yeasts, and filamentous fungi. Here, whole-cell catalyzed reduction, oxidation, hydrolysis, and transesterification reactions in ILs are discussed.

3.1 Reduction Reactions

The first study of whole-cell biocatalysis with ILs was reported by Cull et al. in 2000. In their study, the reduction of 1,3-dicyanobenze (1,3-DCB) to 3-cyanobenzamide (3-CB) using the bacteria *Rhodococcus* R312 was investigated in a biphasic system containing an aqueous buffer and water-immiscible ILs (Scheme 1). The production of 3-CB in buffer/IL was optimized and compared with a standard buffer/organic solvent system (toluene). The biotransformation of 1,3-DCB in both buffer/toluene and buffer/[Bmim][PF_6] systems showed similar profiles for the conversion of 1,3-DCB initially to 3-CB and then 3-cyanobenzoic acid. However, the initial rate of 3-CB production in the buffer/[Bmim][PF_6] system was somewhat lower due to the reduced rate of 1,3-DCB mass transfer from the more viscous [Bmim] [PF_6] phase. It was also shown that the specific activity of the biocatalyst in the buffer/ [Bmim] [PF_6] system was almost an order of magnitude greater than in the buffer/ toluene system, which suggests that the rate of 3-CB production was limited by substrate mass transfer rather than the activity of the biocatalyst. Moreover, in the buffer/toluene system, cells were found to aggregate near the interface, leading to the formation of stable emulsions, which were not found with the solvent system containing IL. Hence, the use of an IL in place of an organic solvent could simplify downstream separations [62].

Howarth et al. reported preliminary results from the asymmetric reduction of a range of aliphatic and cyclic ketones to corresponding alcohols (Scheme 2) with immobilized baker's yeast in a biphasic medium with [Bmim][PF_6]. Yeast cells were immobilized in calcium alginate beads for quicker separation of the cells by filtration. A high volume ratio of ILs to water (10:1, v/v) was used. The yields and enantiomeric excess value from the biotransformation in aqueous-IL media were compared to the results with different organic solvents (hexane, benzene, toluene, petroleum ether, and carbon tetrachloride), as published in the literature. For some of the systems, higher yields were obtained with the IL than with organic solvents [63].

Pfruender et al. demonstrated the application of ILs as a substrate reservoir and in situ extracting agent in an efficient multiphase process for the whole-cell-catalyzed synthesis of fine chemicals exemplified by the asymmetric reduction of 4-chloroacetophenone to (*R*)-1-(4-chlorophenyl)ethanol with *Lactobacillus kefir* (Scheme 3). Among the tested ILs, [Bmim][Tf_2N] gave higher product yields (92.8%); the yield was 88.2% with the other ILs. These yields were two times higher than those obtained in the monophasic aqueous system, which suffers from both substrate and product toxicity (88–93% vs. 46%). Yields using the

Scheme 1 Reduction of 1,3-dicyanobenze

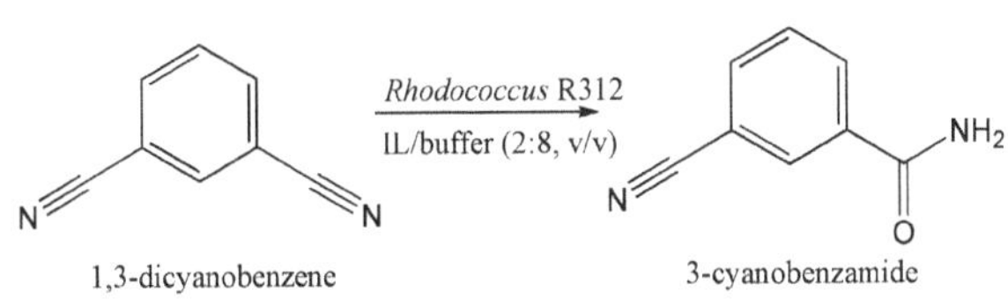

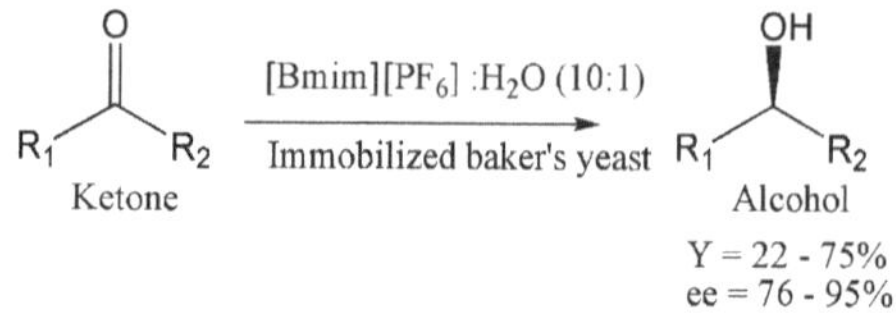

Scheme 2 Reduction of ketones

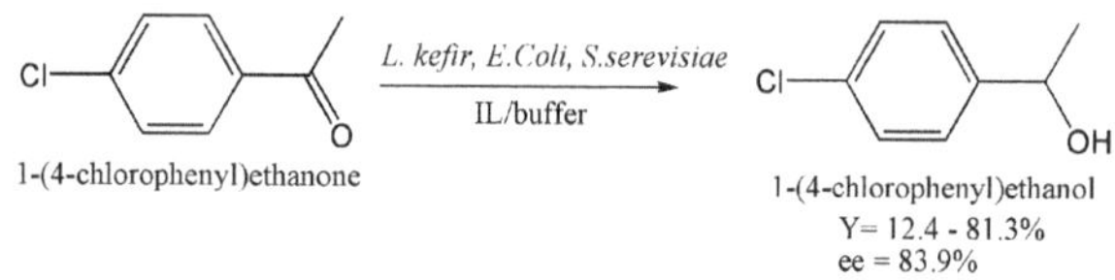

Scheme 3 Reduction of 4-chloroacetophenone

organic solvent methyl-*tert*-butyl-ether (MTBE) were very low at 4%, which indicates that cell viability with MTBE was very low. Moreover, the reduction of 4-chloroacetophenone to (*R*)-1-(4-chlorophenyl)ethanol with high product concentration (82 g L^{-1}) and purity (99.6% ee) were achieved in an IL-containing system without the addition of cofactors and without damaging the biocatalyst [58]. In their extended study, [Bmim][PF_6], [Bmim][Tf_2N], and [$NMeOct_3$][Tf_2N] were tested for their biocompatibility towards *E. coli* and *Saccharomyces cerevisiae* for the reduction of ketones to their corresponding enantiopure alcohols. Together with previously published results, it is shown that these water-immiscible ILs do not damage microbial cells. Thus, they can be used as a substrate reservoir and in situ product extracting agent for biphasic whole-cell biocatalytic processes, replacing organic solvents and thereby increasing process efficiency [64].

Matsuda et al. investigated the asymmetric reduction of *o*-fluoroacetophenone to (*S*)-1-(*o*-fluorophenyl)ethanol in both hydrophilic ([Emim][BF_4]) and hydrophobic ([Bmim][PF_6]) ILs using immobilized dried cells of *Geotrichum candidum* IFO 5767 (Scheme 4) [65]. They demonstrated the important role of minimal quantities of water in ILs for whole-cell catalysts. Neither of the reaction systems showed conversion of the substrate after 16 h when the dried cells were added into the pure ILs. To improve the activity, water was added to the reaction mixture, then the yield increased drastically to give (*S*)-1-(*o*-fluorophenyl)ethanol in [Bmim][PF_6], while the reaction in [Emim][BF_4] did not proceed. The difference in yield between the two reactions can be attributed to the difference in the water miscibility: Water is not miscible with [Bmim][PF_6]; thus, the water and ILs form two separate layers (a two-layer reaction), while [Emim][BF_4] is completely miscible with water. The results imply that the water layer around the cell is necessary for the reaction to proceed. Therefore, to create the water layer in [Emim][BF_4], a water-absorbing polymer was added to the mixture of water and cells to immobilize the cell around the polymer physically. This guaranteed the presence of a minimal quantity of water

Scheme 4 Reduction of o-fluoroacetophenone

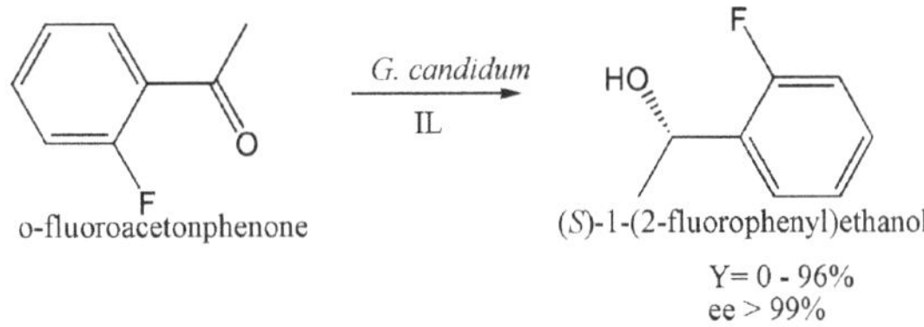

near the biocatalyst. Indeed, using this setup, the reaction system with [Emim][BF_4] reached a very satisfying 96% yield. In addition, even for the reaction system with water-immiscible ILs, the yield increased from 84% to 92% when using the cells immobilized on the water-absorbing polymer. This observation was related to the increased surface area between the ILs and the water layers. Finally, the polymer might also stabilize the pH and thereby create more favorable reaction conditions. Selectivity was constantly very satisfying, with enantiomeric excesses >99% whenever the reduction was observed.

To confirm the success of this reaction setup, other ILs and substrates were tested. The yield was strongly affected by the anions: [PF_6] and [BF_4] anions gave the highest yield (88–96%), the [Tf_2N] anion gave moderate yield (49–61%), whereas sulfur functionality anions (i.e., [TfO], [MS], and [ES]) resulted in a reaction that hardly proceeded. However, the kind of imidazolium cation did not affect the reaction outcome. The hydrophobic property of ILs also did not relate to the yield when the cell was immobilized on a water-absorbing polymer, although it strongly affected the yield of the reaction that did not use a water-absorbing polymer. In addition, the reduction of acetophenone derivatives (e.g., benzyl acetone, 2-hexanone, β-keto ester, fluorinated epoxy ketone) showed that *G. candidum* immobilized on a water-absorbing polymer accepts a large variety of substrates with moderate to excellent yield and enantioselectivity.

Lou et al. investigated the effects of several reaction-engineering parameters on the reduction of acetyltrimethylsilane to enantiopure (*S*)-1-trimethylsilylethanol catalyzed by immobilized *S. cerevisiae* cells in IL-containing media [66]. The results demonstrated that [Bmim][PF_6] and [Bmim][BF_4] can markedly boost the activity and the stability of the immobilized cells. The reaction temperature, buffer pH, and substrate concentration were optimized. Under optimal conditions, the initial reaction rate, maximum yield, and product enantiomeric excess were 63.4 mM h^{-1}, 99.9%, and >99.9% with the [Bmim][PF_6]/buffer system (1/6, v/v). The values obtained with a [Bmim][BF_4]/buffer system (10%, v/v) were 74.5 mM h^{-1}, 99.2%, and >99.9%, respectively, which were much higher than those achieved with either an *n*-hexane/buffer (2/1, v/v), biphasic system, or aqueous buffer.

Hussain et al. investigated the reduction of 6-bromo-β-tetralone to (*S*)-6-bromo-β-tetralol by yeast *Trichoderma capitatum* MY 1890 and baterium *Rhodococcus erythropolis* MA7213 in monophasic and biphasic systems (Scheme 5) [67]. The authors evaluated some of the physical properties of the solvents thought to influence the outcome of the biotransformation, such as water miscibility, density,

Scheme 5 Reduction of 6-bromo-β-tetralone

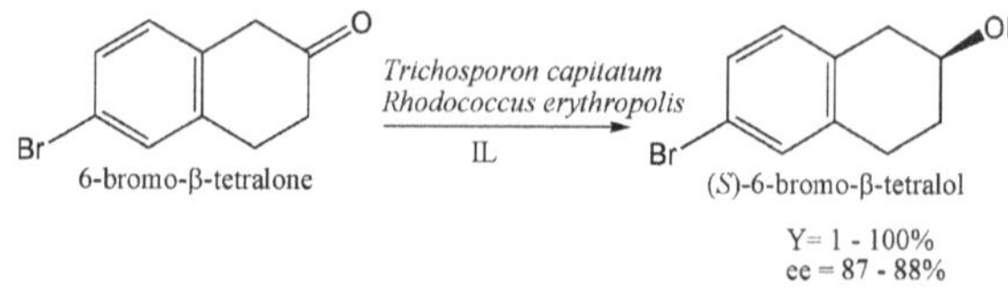

and viscosity, as well as biocompatibility and substrate solubility. The aim was to possibly correlate the observed conversions and reaction rates to these properties. Several reaction setups were compared, including water-miscible and water-immiscible ILs (20% v/v), as well as the reference system composed of 10% (v/v) ethanol in the aqueous phase. The following ILs were investigated as the solvent phase or as an additive in the biotransformation: [Emim][TOS], [$NMeOct_3$][Tf_2N], [Bmim][BF_4], [Bmim][PF_6], [Bmim][$MDEGSO_4$], [Bmim][OS], and [CABHEM] [MS]. Only [$NMeOct_3$][Tf_2N] and [BMIm][PF_6] formed a second liquid phase, whereas the rest were used as additives or co-solvents in the process. The choice of a monophasic (co-solvent) or biphasic system can affect the rate of reaction, downstream processing, and product separation. The ILs were evaluated by measuring the cell viabilities, initial rate of product formation, and the final substrate conversion; these were compared with organic solvents. The bacterial cell viabilities at a 1:1 (v/v) ratio of ILs to aqueous phase were measured by a plate count method. The hydrophobic ILs ([Bmim][PF_6] and [$NMeOct_3$][Tf_2N]) and the hydrophilic ILs ([Emim][TOS] and [Bmim][BF_4]) showed >80% cell viability, compared with 10% cell viability using 10% by volume of ethanol solution; no cell viability was observed with toluene and the other ILs, such as [Bmim][$OcSO_4$], [Bmim] [$MDEGSO_4$], and [CABHEM] [$MeSO_4$]. Hence, by changing the anion group from [$OcSO_4$] to [BF_4] with a [Bmim] cation, cell viability was increased. Overall, the yeast cells seemed to have better reaction performance than bacteria. With yeast, 100% conversion was obtained with 20% volume concentration of the IL [Emim] [TOS]. Enantiomeric excess values with ILs and 10% ethanol were similar (88%). Substrate conversion with the water-immiscible ILs ([Bmim][PF_6] and [$NMeOct_3$] [Tf_2N]) and water-miscible [Bmim][BF_4] were between 35 and 60% with low initial rate of product formation, although the cell viabilities were >80%. The lower conversion was attributed to the slower mass transfer rate of the substrate from the IL phase to the aqueous phase (for the biphasic systems) and also attachment of ILs to the cell membrane preventing effective transfer of substrate into cells. With the bacterium *Rhodococcus erthyropolis* MA1392, substrate conversion was around 28% in an aqueous system with dissolved [Emim][TOS], which was higher than in the 10% vol. of ethanol/water system. Cells were observed to form aggregates in 10% ethanol solutions; however, in [Emim] [TOS], the cells were dispersed in the medium. Although higher cell viabilities were obtained with [Bmim][PF_6], [Bmim] [BF_4], and [$NMeOct_3$][Tf_2N] compared with 10% ethanol (14%), the final conversion rates were similar (7–17%). The observations made correlated only partially with the viability measurements, indicating that the toxicity of the different solvents tested is probably one of the reasons for variations in biocatalytic efficacy, but not the only one. No predictable relationship between the physical properties evaluated and the activity data could be established. Cell viability data and substrate solubility are thus apparently not sufficient to predict the efficiency of the biotransformation. The lack of knowledge concerning the interaction of ILs with the cells and the resulting effects on the reaction rate does not permit complete understanding of the mechanisms taking place. The authors concluded that screening of ILs was currently still a more viable method for applications in industry than rational selection.

Bräutigam et al. reported a systematic procedure based on physical properties to identify more commercially available ILs exhibiting the potential to improve the efficiency of whole-cell biocatalysis [45]. These ILs were rated by their biocompatibility, their substrate- and product-specific distribution coefficients, and the performed asymmetric reductions of several prochiral ketones, among others. With the use of recombinant *E. coli* as a biocatalyst, overproducing *Lactobacillus brevis* alcohol dehydrogenase and *Mycobacterium vaccae* N10 formate dehydrogenase for cofactor regeneration, the great potential of asymmetric whole-cell biocatalysis in biphasic IL/water systems was demonstrated in simple batch processes.

To identify the ILs that are suitable for biphasic biocatalysis, the total number of commercially available ILs was narrowed down by physical properties. Obviously, important preconditions for the application as a second liquid phase in a biphasic process are immiscibility with water and a melting point below 30°C. An IL density above 1.2 g/cm^3 is necessary for simple and efficient phase separation after biotransformation. The viscosity of ILs saturated with water should be below 400 mm^2/s due to their major impact on dispersion quality and mass transfer limitations. As a result, nine ILs were selected for further evaluation. The results for the biocompatibility of ILs and their distribution coefficients indicated that [PF_6]-based ILs possessed the best qualities for use in biphasic whole-cell biocatalysis compared with the other ILs. However, in exemplarily performed conversions of 600-mM 4-chloroacetophenone on a 1.4-mL scale, [Tf_2N]-based ILs showed similarly high yields to [PF_6]-based ILs. As this example indicates, the interactions between ILs and biocatalysts and between ILs and substrates/products (log D) aere not sufficient for an estimation of the best-suited IL for an entire reaction system.

Bräutigam et al. applied their systematic approach to choose suitable ILs for whole-cell biocatalysis from a larger pool of ILs (21 ILs with seven cation classes and three anion classes) in 2009 [60]. The best results were achieved with [PF_6]- and [Tf_2N]-based ILs, whereas [FAP]-based ILs showed minor qualifications. For example, the use of [HMPL][Tf_2N] as second liquid phase for asymmetric synthesis of (*R*)-2-octanol resulted in a space–time-yield of 180 g L^{-1} d^{-1}, a chemical yield of 95%, and an enantiomeric excess of 99.7% in a simple batch process.

Schroer et al. investigated the reduction of 2,5-hexanedione to the corresponding (2*R*,5*R*)-hexanediol catalyzed by recombinant *E. coli* cells expressing an alcohol dehydrogenase from *Lactobacillus brevisi*. Because the conversion of 2,5-hexanedione to 2,5-hexanediol is a two-step reaction, an excess of the co-substrate 2-propanol is necessary to achieve sufficient regeneration of cofactors. As a result, acetone is formed and accumulated in the system with a need to remove

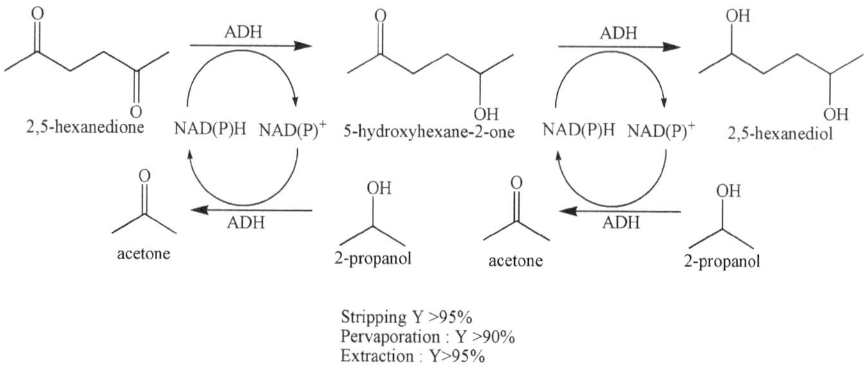

Scheme 6 Reduction of 2,5-hexanedione with substrate-coupled cofactor regeneration

to overcome the reaction thermodynamic equilibrium (Scheme 6). The water-immiscible ILs were used to extract acetone from the reaction system and compared with two other in situ removal techniques. In contrast to a process without acetone removal, where 54% yield could be reached, the yield was increased to >90% when a pervaporation system was applied or when acetone was removed by stripping (air sparging). The partition coefficient of the acetone in the [Bmim][Tf_2N]/buffer medium was higher than in a methyl *tert*-butyl ether (MTBE)/buffer medium. However, the partition coefficient of 2-propanol in the [Bmim][Tf_2N]/buffer medium was slightly lower than in MTBE/buffer medium. Using the [Bmim][Tf_2N]/buffer medium also improved the substrate availability to the cells. Accordingly, higher yields >95% were obtained with ILs compared to only 24% in MTBE. In addition, the stability test showed that stripping seemed to harm the recombinant *E. coli* cells more, leading to only 25% relative activity. However, recombinant *E. coli* in the biphasic system with ILs showed intermediate stability, with 63% relative activity left [68].

The asymmetric reduction of ethyl 2-oxo-4-phenylbutyrate (EOPB) to synthesize optical active ethyl 2-hydroxy-4-phenylbutyrate (EHPB) (Scheme 7) catalyzed by *S. cerevisiae* in nonaqueous solvents including ethyl ether, benzene, and [Bmim][PF_6] was investigated by Shi et al. [69]. The addition of small amounts of water (1–3% (v/v)) to nonaqueous media resulted in the formation of (*S*)-enantiomer in the IL medium (28%), whereas higher enantiomeric excess of (*R*)-enantiomer was observed in benzene (80%) and ethyl ether (70%). However, in an IL–water (10:1, v/v) biphasic system, the enantioselectivity of the reduction shifted toward the (*R*)-side, and the enantiomeric excess (*R*) increased from 6.6% to 82.5% with the addition of ethanol (1%, v/v). In addition, the effects of the use of ILs as additives in relatively small amounts on the reduction were also studied. The authors found that there is a decline in the enantioselectivity and yield of the reduction with increasing IL concentrations in either organic solvent–water biphasic systems or benzene.

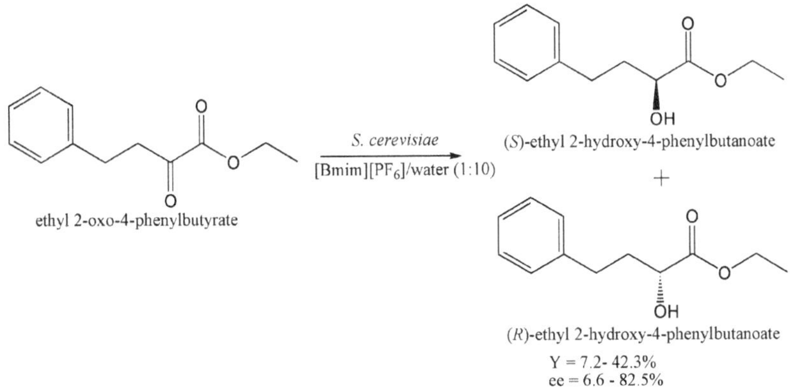

Scheme 7 Asymmetric reduction of EOPB into EHPB by *Saccharomyces cerevisiae*

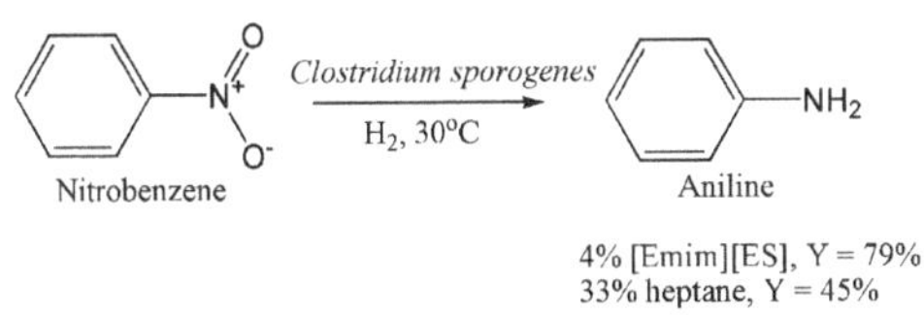

Scheme 8 Reduction of nitrobenzene

Dipeolu et al. investigated nitro reduction by *Clostridium sporogenes* in water-miscible ILs [70]. Water-miscible ILs were investigated as co-solvents. The effects of the ILs between 0.1 and 2% (w/v) on cell growth and corresponding product yields for the reduction of water-insoluble nitrobenzene (Scheme 8) were compared with the results from using organic solvents, such as *n*-heptane and ethanol. The results showed that 2-hydroxy ethyl trimethyl-ammonium dimethyl phosphate ([EtOHNMe$_3$][DMP]) and *N*,*N*-dimethylethanolammonium acetate (DMEAA) increased the growth rate of *C. sporogenes* by as much as 28%, suggesting that they either metabolized or increased the availability of nutrients. By contrast, [Bmim][BF$_4$] and AMMOENG™ 100 inhibited growth. Although 2% w/v [Emim][ES] inhibited growth by 58%, it was sufficiently non-toxic to allow efficient reduction of nitrobenzene using harvested cells, providing aniline yields up to 79%. The high product yield with reactions in [Emim][ES] represented a significant improvement over conventional solvents because the yield was only 8% in aqueous ethanol (4% v/v) and 45% in a biphasic heptane/aqueous system (phase ratio: 0.33). These findings showed that a growth-inhibitory IL can provide benefits for nitro reduction; furthermore, it is not necessary to use completely non-toxic ILs for biocatalysis. The authors recommended that both non-inhibitory and partially inhibitory ILs should be screened for use in biocatalysis.

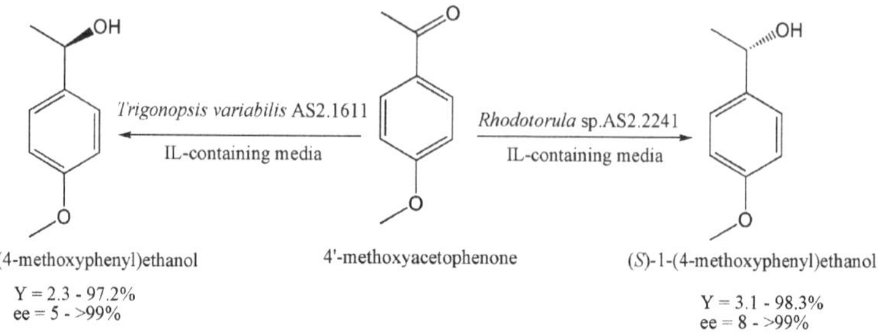

Scheme 9 Reduction of 4′-methoxyacetophenone

Lou et al. investigated the enantioselective reduction of 4′-methoxyacetophenone (MOAP) to (*R*)-1-(4-methoxyphenyl)ethanol (*R*-MOPE) by immobilized *Trigonopsis variabilis* AS2.1611 cells [71] and to (*S*)-MOPE by immobilized cells of *Rhodotorula* sp. AS2.2241 cells (Scheme 9) [47]. The reduction of MOAP to (*R*)-MOPE was successfully achieved in an IL-containing medium. The hydrophilic IL, 1-(2′-hydroxyl)ethyl-3-methylimidazolium nitrate ([C_2OHmim][NO_3]), used as co-solvent, was biocompatible with the cells and led to only a moderate increase in permeability of the cell membrane. Under optimized conditions, the initial reaction rate, the maximum yield, and the product enantiomeric excess were 7.1 μmol h^{-1}, 97.2%, and >99%, respectively, which are much higher than those in aqueous buffers. The presence of IL in an aqueous buffer allowed the cells to tolerate relatively high temperatures and substrate concentrations, and the immobilized cells manifested excellent operational and storage stability [71]. Conversely, the biocatalytic enantioselective reduction of MOAP to (*S*)-MOPE was successfully conducted in a hydrophilic IL-containing system using immobilized *Rhodotorula* sp. AS2.2241 cells. Of all the tested ILs, the best results were observed with [C_2OHmim][NO_3], which showed good biocompatibility with the cells and increased the cell membrane permeability moderately, thus improving the efficiency of the bioreduction. Under the optimized conditions, the initial reaction rate, the maximum yield, and the product enantiomeric excess were 9.8 μmol/h g_{cell}, 98.3%, and >99%, respectively. The established biocatalytic system has proven to be highly effective for the reduction of other aryl ketones. In addition, the cells exhibited excellent operational stability in the presence of [C_2OHmim][NO_3]. Moreover, the ILs can accumulate within the cells, suggesting that ILs are likely to interact with the related enzymes within the cells [47].

3.2 *Oxidation Reactions*

Cornel et al. first reported the microbial oxidation of aromatic compounds in IL-containing media [72]. In their study, toluene was oxidized to toluene *cis*-glycol in a biphasic system containing ILs using recombinant *E. coli* expressing toluene

Scheme 10 Oxidation of toluene

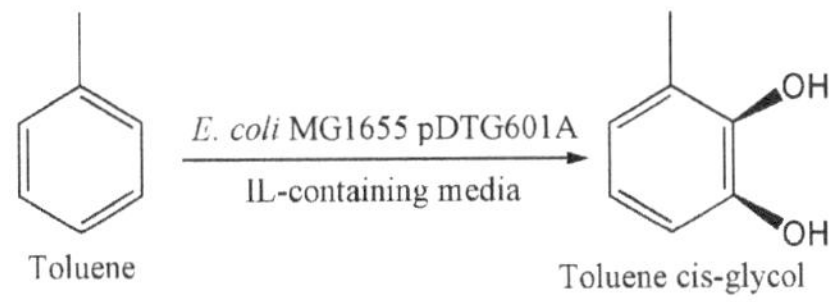

dioxygenase from *Pseudomonas putida* F1 (Scheme 10). Productivity in oxygenase-catalyzed biotransformations is frequently restricted by toxicity of the substrates to the cells. Therefore, the use of ILs in the ILs/aqueous biphasic system can overcome the issue of substrate toxicity along with lower substrate concentrations associated with aromatic compounds by acting as a substrate reservoir. Suitable ILs were first identified by conducting a biocompatibility study with cells. Hydrophilic ILs such as [Omim][Cl], [$NMeOct_3$][Cl], and [$P_{6,6,6,14}$][Cl] caused complete growth inhibition at low phase ratios. Even though [$NMeOct_3$][Tf_2N] and [$P_{6,6,6,14}$][Tf_2N], inhibited the growth of *E. coli* by 39% and 23%, respectively, both ILs could be used to improve toluene dioxygenase-catalyzed conversion of toluene to toluene *cis*-glycol using recombinant *E. coli* cells. In a biphasic system containing [$NMeOct_3$][Tf_2N] or [$P_{6,6,6,14}$][Tf_2N], the maximum toluene concentration tolerated was 75.2 mmol L^{-1}, which was an eightfold increase compared with the control reactions without IL. The increased toluene concentration enabled by using either of these ILs resulted in a similar 2.5-fold increase in toluene *cis*-glycol concentration in a 50-mL scale flask. The product concentrations and specific product yields were improved by 200% and 238%, respectively, in bioreactors with an unrestricted oxygen supply. In addition to the higher product concentrations afforded by use of the ILs, there were additional benefits in terms of the absence of solvent emulsification, potentially improved safety, and lower environmental impact.

Fu et al. investigated the oxidation of terpene botulin into betulinic acid using whole cells of *Armillaria luteo-virens* Sacc in IL-containing media (Scheme 11) [73]. In this study, [Emim][BF_4], [Bmim][BF_4], [Bmim][PF_6], and [Omim][PF_6] were investigated as co-solvents of an aqueous buffer and hexane independently. Several parameters affecting betulinic acid formation in the IL-containing system were investigated. The addition of [Emim][BF_4] in a hexane-containing reaction medium gave rise to better betulinic acid formation in comparison with other ILs used. Under the optimum conditions (5 mL of 50% (v/v) [Emim][BF_4]/hexane co-solvent system containing 0.1 mol L^{-1} butanol, 75 mg L^{-1} betulin and 200 g L^{-1} resting cells), the highest yield of 11.14% was obtained after an 18-h reaction.

Wu et al. investigated the use of two ILs, [Bmim][Tf_2N[and [Bmim][PF_6], in a biphasic system for the 11α-hydroxylation of 16α,17-epoxyprogesterone (EP) catalyzed by *Rhizopus nigricans* [74]. The [Bmim][PF_6]-aqueous biphasic system greatly increased yield to 85% in 24 h at 3 gL^{-1} feeding concentration (compared to less than 50% achieved in a control aqueous system where the substrate was solubilized with ethanol). In contrast, a [Bmim][Tf_2N]-aqueous biphasic system afforded a low yield of product and showed poor biocompatibility. At a substrate concentration of 18 gL^{-1}, a yield of >90% could be obtained in a

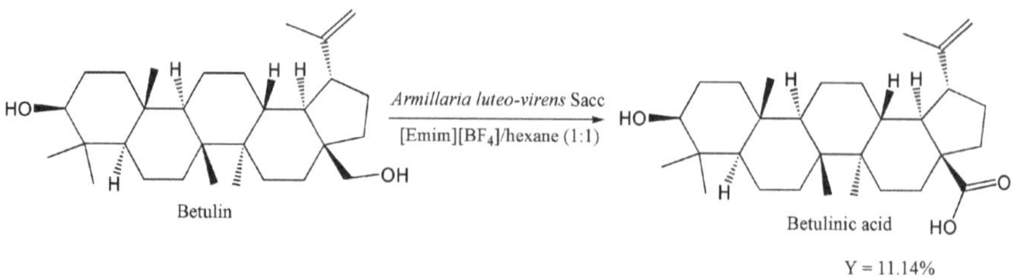

Scheme 11 Oxidation of botulin

[Bmim][PF_6]-aqueous biphasic system. In addition, the conversion of 87% at phase ratio 10 and 75% at phase ratio 5, respectively, could be maintained after three reaction cycles.

Gao et al. reported the asymmetric sulfoxidation of a range of sulfides using a recombinant *E. coli* co-expressing P450pyrI83H monooxygenase and glucose dehydrogenase in an IL/aqueous biphasic system [75]. The IL [$P_{6,6,6,14}$]][Tf_2N] showed good biocompatibility with the *E. coli* and acted as a reservoir for the thioanisole. Therefore, the problems of substrate toxicity and inhibition of sulfoxidation could be efficiently avoided in an IL/aqueous biphasic system. The substrate concentrations had a remarkable improvement, owing to their high distribution in the IL phase, which in turn caused low substrate concentrations in the aqueous phase and further enhanced the yield and enantioselectivity of reactions.

Zhao et al. [76] successfully conducted the biosynthesis of γ-decalactone (GDL) from castor oil solution in an IL-containing cosolvent system using immobilized cells of *Yarrowia lipolytica* G3-3.21 on attapulgite. They found that the immobilized *Y. lipolytica* G3-3.21 cells in *N*-butyl-pyridinium tetrafluoroborate ([BPy]BF_4) solution gave the highest activity of C_{16}-Acyl-CoA oxidase and the maximum yield of GDL. The parameters for catalyst immobilization and oxidation reaction conditions were optimized. Under the optimized conditions, the GDL yield was up to 8.05 gL^{-1} . In addition, after ten reuses, the GDL yield was 7.51 gL^{-1}, corresponding to 93.3% of the initial activity; this suggests good reusability and the potential for industrial applications of whole cell catalysts in IL-containing media.

Melgarejo-Torres et al. evaluated the application of IL in a dispersed phase in a three-phase partitioning (air-water-IL) bioreactor (TPPB) for the production of bicyclic lactone (1*S*,5*R*)-2-oxabicyclo[3.3.0]oct-6-en-3-one from bicyclic ketone bicyclo[3.2.0]hept-2-en-6-one with *E. coli* strain TOP10 pQR239 (Scheme 12) [77]. The partition coefficients of the substrate and product; mass-transfer coefficients of the substrate, product, and oxygen; as well as the deactivation parameters related to the loss of cell viability owing to the presence of the IL, were determined. The use of an IL as a dispersed phase in a TPPB was an adequate strategy for increasing the productivity of lactones compared to the ordinary (water–air) processes, owing to the fact that the IL phase allows a twofold increase in the initial substrate concentration (0.7 gL^{-1}) compared to the water–air system (0.35 gL^{-1}), thus increasing the productivity of the lactones. All evaluated ILs were

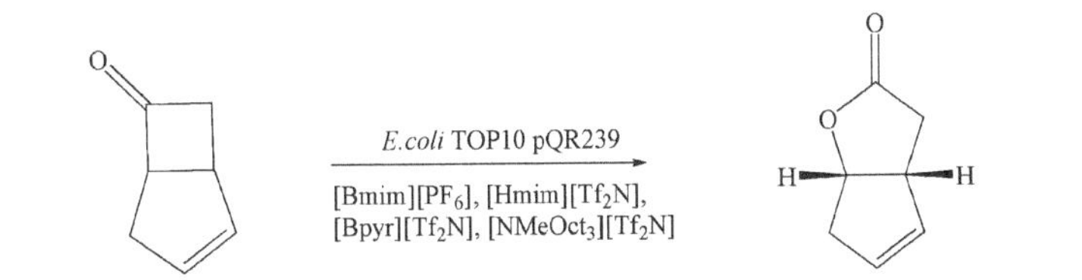

bicyclo[3.2.0]hept-2-en-6-one

(1S,5R)-2-oxabicyclo[3.3.0]oct-6-en-3-one

Scheme 12 Production of bicyclic lactone

thermodynamically suitable for use in the dispersed phase; however, all of them provoked a loss in cell viability. Among the four tested ILs, [NMeOct$_3$][Tf$_2$N] was the most suitable IL for the production of lactones from ketones when using the *E. coli* strain TOP10 pQR239.

3.3 *Hydrolysis Reactions*

Few publications have investigated hydrolysis reactions in IL-containing systems with whole-cell biocatalysts. Chen et al. first reported the production of glycyrrhetic acid 3-*O*-mono-β-*d*-glucuronide (GAMG) by the hydrolysis of glycyrrhizin (GL) by whole-cell catalysts in an IL/buffer biphasic system based on [Bmim][PF$_6$] (Scheme 13) [78]. GAMG is widely used in the pharmaceutical and food industries. However, the low water solubility of both GL and GAMG greatly limits the large-scale production of GAMG.

Three whole-cell biocatalysts were used, including wild-type *Penicillium purpurogenum* Li-3 (w-PGUS) and recombinant strains *E. coli* BL21 and *Pichia pastoris* GS115 expressing the β-D-glucuronidase gene from w-PGUS. The wild-type strain w-PGUS gave highest GAMG yield. After 60 h under optimal reaction conditions, a GAMG yield of 87.63% was achieved in the IL-containing biphasic system, which was a much higher yield than that in the monophasic buffer system (57.3%). In addition, the product GAMG and the byproduct glycyrrhetic acid (GA) partitioned into the aqueous and IL phases, respectively; therefore, the desired product could be isolated from the aqueous phase as long as hydrolysis of the GAMG was complete. Hence, the combination of whole-cell hydrolytic biocatalysis and an IL-containing system provided a promising and economic way for the industrial production of GAMG by regioselective hydrolysis of GL.

Baum et al. studied the application of whole-cell catalysts in ILs for the production of (*S*)-mandelic acid and (*S*)-mandeloamide from benzaldehyde and cyanide [79]. Recombinant *E. coli* that simultaneously expressed an (*S*)-hydroxynitrile lyase (oxynitrilase) from cassava (*Manihot esculenta*) and an arylacetonitrilase from *Pseudomonas fluorescens* EBC191 was used as the catalyst in an IL/buffer biphasic system (Scheme 14).

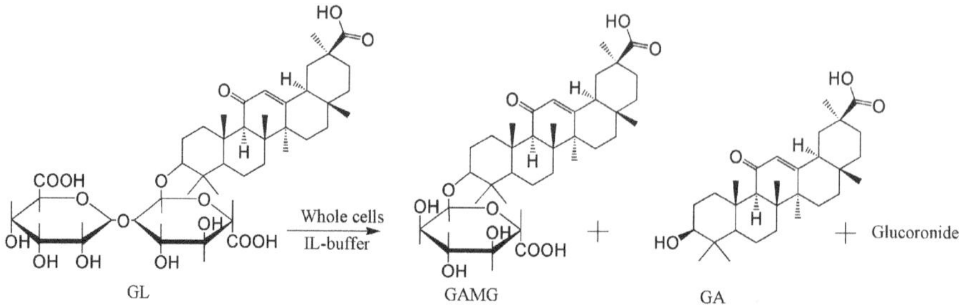

Scheme 13 Hydrolysis of GL to GAMG catalyzed by whole cells in an IL/buffer biphasic system

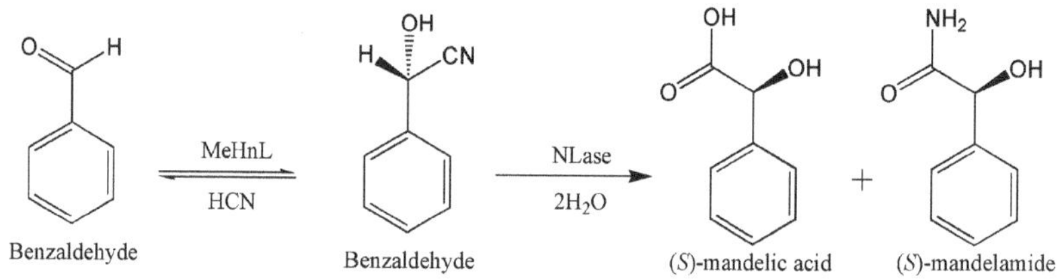

Scheme 14 Bienzymatic synthesis of (*S*)-mandelic acid and (*S*)-mandeloamide from benzaldehyde and HCN by combining the (*S*)-selective oxynitrilase from *Manihot esculenta* (MeHnL) and the non-selective nitrilase from *Pseudomonas fluorescens* EBC191 (NLase)

Benzaldehyde exhibited a pronounced inhibitory effect on the nitrilase activity in concentrations >25 mM. Therefore, an IL/buffer biphasic system could be used for the intended biotransformation. The distribution coefficients of the substrates, intermediates, and products of the reaction were determined. It was found that 1-butyl-1-pyrrolidinium bis(trifluoromethanesulfonyl)imide [BP][Tf_2N] and [Bmim][PF_6] were highly efficient as substrate reservoirs for benzaldehyde. The recombinant *E. coli* strain was active in the presence of [BP][Tf_2N] or [Bmim][PF_6] phases and converted benzaldehyde and cyanide into mandelic acid and mandeloamide. The two-phase systems allowed the conversion of benzaldehyde dissolved in the IL to a concentration of 700 mM with product yields (sum of mandelic acid and mandeloamide) of 87–100%. Between two ILs/buffer biphasic system, the cells were slightly more effective in the presence of [BP]Tf_2N] than in the presence of [Bmim][PF_6]. In both two-phase systems, benzaldehyde and cyanide were converted into (*S*)-mandeloamide and (*S*)-mandelic acid with enantiomeric excesses of >94%. The recombinant *E. coli* cells formed in the two-phase systems with ILs and increased substrate concentrations had higher relative amounts of mandeloamide than in a purely aqueous system with lower substrate concentrations.

The hydrolysis of *rac*-epoxyhexane by yeast *Rhodotorula glutinis* ATCC 201718 containing epoxide hydrolase in its cell membrane in a biphasic system based on an organic solvent or ILs and aqueous buffer (Scheme 15) was investigated by Matsumoto et al. [80]. The effect of organic solvents and ILs on the initial reaction rate of (*R*)-enantiomer (r_{R0}) and the enantiomeric ratio (E) was examined. In the case

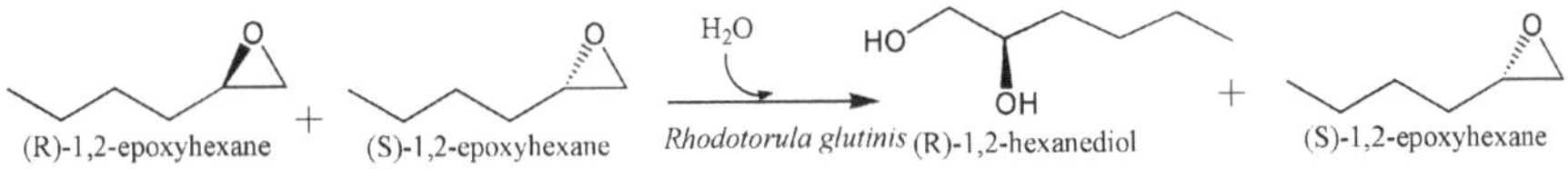

Scheme 15 Hydrolysis reaction of rac-1,2-epoxyhexane

of hydrophobic organic solvents, r_{R0} increased linearly with increasing hydrophobicity of solvent (log*P*) and an enzymatic reaction using alkyl alcohols was not observed. However, the high enantiomeric excess of (*R*)-diol (eep > 99%) without a significant decrease in the reactivity was accomplished by adding 1- heptanol to dodecane. However, there were no correlations between r_{R0} and log*P* of ILs. Moreover, both single solvents showed a small effect on the *E* value. However, a mixture of dodecane and 1-heptanol gave *E* values greater than 100. The kinetic data correlated well with the reaction kinetic model, considering the distribution of epoxides between two liquid phases and the inhibition of diols.

3.4 Transesterification Reactions

Arai et al. [81] investigated the whole-cell catalyzed transesterification of a triglycerides from soybean oil with methanol to produce fatty acyl methyl esters that could be used as biodiesel in IL-containing biphasic systems. Four types of whole-cell biocatalysts – wild-type *Rhizopus oryzae* producing triacylglycerol lipase (w-ROL), recombinant *Aspergillus oryzae* expressing *Fusarium heterosporum* lipase (r-FHL), *Candida antarctica* lipase B (r-CALB), and mono- and diacylglycerol lipase from *A. oryzae* (r-mdlB) – were investigated. The wild-type fungus w-ROL gave the high yield of fatty acid methyl ester in [Emim][BF_4] or [Bmim][BF_4] biphasic systems following a 24-h reaction. Although lipases are known to be severely deactivated by an excess amount of methanol (e.g., 1.5 Mequiv. of methanol against oil) in a conventional system, methanolysis successfully proceeded even with a methanol/oil ratio of 4 in the IL biphasic system, where the ILs would work as a reservoir of methanol to suppress the enzyme deactivation. When only w-ROL was used as a biocatalyst for methanolysis, unreacted monoglyceride remained due to the 1,3-positional specificity of *R. oryzae* lipase. High fatty acid methyl ester conversion was attained by the combined use of two types of whole-cell biocatalysts, w-ROL and r-mdlB. Although native w-ROL became deactivated in [Bmim][BF_4] and lost its reusability, the addition of glutaraldehyde made the biocatalyst more stable and recyclable. Compared to other conventional methods (e.g., alkali-catalyzed methanolysis), this whole-cell-based bioprocess was environmentally friendly and exhibited a high conversion rate. It also demonstrated that ILs are promising candidates for use in solvent systems for producing biodiesel via whole-cell biocatalysts.

Rajapriya et al. [82] reported the synthesis of several caffeic acid esters in IL media by using the freeze-dried mycelia of halotolerant *Aspergillus niger* EXF 4321

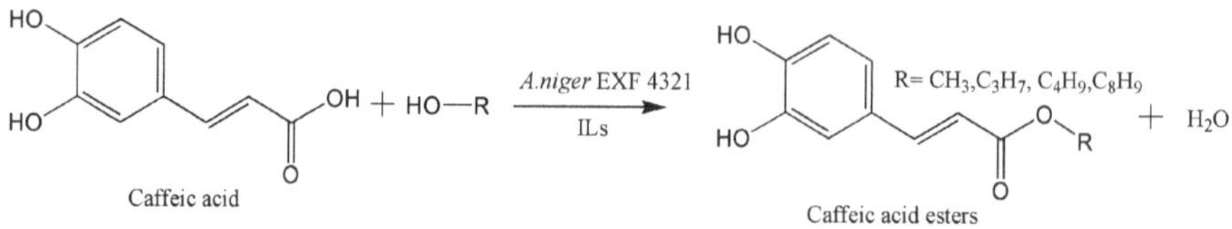

Scheme 16 Halotolerant *A. niger* EXF 4321 catalyzed synthesis of caffeic acid esters

as a biocatalyst (Scheme 16). The high solubility of caffeic acid in [Emim][Tf_2N] is favorable for the synthesis of their corresponding esters by whole-cell catalysis. The whole-cell catalyzed synthesis of caffeic acid phenethyl ester conditions were optimized and bioconversion up to 84% was achieved at a substrate molar ratio of 1:20 (caffeic acid:2-phenyl ethanol), at 30°C for 12 h.

4 Conclusions and Perspective

Using whole cells as a biocatalyst has several advantages, including ease of catalyst preparation and avoidance of the addition of cofactors, which make them highly competitive compared with purified enzymes. The applications of ILs in whole-cell biocatalysis show great potential for industrial applications. Hydrophobic ILs are often used as substrate/product reservoirs to circumvent the issues of low substrate solubility in aqueous media and substrate/product inhibition toward whole-cell catalysts. A number of studies have shown that whole-cell biocatalysis in ILs has a high conversion rate, high stereoselectivity, and favorable IL and biocatalyst reuse.

However, there are some bottlenecks regarding the use of ILs in whole-cell biocatalysis, one of which is the toxicity of ILs toward microbial cells. Although important progress has been made in surveying the toxicity of ILs, additional work is needed to more completely understand the toxicity of the extremely large number of ILs. In addition, research is still needed on whole-cell biocatalysis using ILs in a number of different fields, including microbiology, biochemistry, toxicology, biochemical engineering, phase equilibrium thermodynamics, mass transfer, and process intensification. The design of ILs with better performance characteristics, such as biocompatibility, operational stability, and biodegradability, is necessary to improve the biocatalytic process as well as economic feasibility when applied on an industrial production scale.

Furthermore, the search for or development of new microbial strains that are IL-tolerant can boost the applications of whole-cell biocatalysis in ILs. Although IL-tolerant mechanisms have not been fully explored, a few studies have applied transcriptomic analysis to elucidate the molecular mechanisms of IL tolerance [83] and have successfully demonstrated engineered IL-tolerant microorganisms with potential applications in converting IL-treated biomass into advanced biofuels [84–87]. The results from these studies demonstrate the potential opportunities for discovering genes for IL tolerance. A combined approach that includes ILs designed for reduced toxicity and robust IL tolerant microorganisms will ultimately result in more efficient and economical whole-cell biocatalytic processes [88].

References

1. Welton T (2004) Ionic liquids in catalysis. Coord Chem Rev 248(21):2459–2477
2. Freemantle M (1998) Designer solvents. Chem Eng News 13(76):32–37
3. Liu JF, Jiang GB, Jönsson JÅ (2005) Application of ionic liquids in analytical chemistry. TrAC Trends Anal Chem 24(1):20–27
4. Pandey S (2006) Analytical applications of room-temperature ionic liquids: a review of recent efforts. Anal Chim Acta 556(1):38–45
5. Shamsi SA, Danielson ND (2007) Utility of ionic liquids in analytical separations. J Sep Sci 30(11):1729–1750
6. Sun P, Armstrong DW (2010) Ionic liquids in analytical chemistry. Anal Chim Acta 661(1):1–16
7. Endres F (2004) Ionic liquids: promising solvents for electrochemistry. Z Phys Chem 218(2/2004):255
8. Silvester Debbie S, Compton Richard G (2006) Electrochemistry in room temperature ionic liquids: a review and some possible applications. Z Phys Chem 220(10):1247
9. Fujita K et al (2012) Ionic liquids designed for advanced applications in bioelectrochemistry. RSC Adv 2(10):4018–4030
10. Hasanzadeh M et al (2012) Room-temperature ionic liquid-based electrochemical nano-biosensors. TrAC Trends Anal Chem 41:58–74
11. Zhao H, Xia S, Ma P (2005) Use of ionic liquids as 'green' solvents for extractions. J Chem Technol Biotechnol 80(10):1089–1096
12. Zaijun L et al (2007) Advance of room temperature ionic liquid as solvent for extraction and separation. Rev Anal Chem 26(2):109
13. Keskin S et al (2007) A review of ionic liquids towards supercritical fluid applications. J Supercrit Fluids 43(1):150–180
14. Marciniak A (2010) Influence of cation and anion structure of the ionic liquid on extraction processes based on activity coefficients at infinite dilution. A review. Fluid Phase Equilib 294(1–2):213–233
15. Oppermann S, Stein F, Kragl U (2011) Ionic liquids for two-phase systems and their application for purification, extraction and biocatalysis. Appl Microbiol Biotechnol 89(3):493–499
16. Malik MA, Hashim MA, Nabi F (2011) Ionic liquids in supported liquid membrane technology. Chem Eng J 171(1):242–254
17. Pereiro AB et al (2012) Ionic liquids in separations of azeotropic systems – a review. J Chem Thermodyn 46:2–28
18. Pino V et al (2011) Ionic liquid-based surfactants in separation science. Sep Sci Technol 47(2): 264–276
19. Fontanals N, Borrull F, Marcé RM (2012) Ionic liquids in solid-phase extraction. TrAC Trends Anal Chem 41:15–26
20. Yang Z, Pan W (2005) Ionic liquids: green solvents for nonaqueous biocatalysis. Enzyme Microb Technol 37(1):19–28
21. Durand J, Teuma E, Gómez M (2007) Ionic liquids as a medium for enantioselective catalysis. C R Chim 10(3):152–177
22. van Rantwijk F, Sheldon RA (2007) Biocatalysis in ionic liquids. Chem Rev 107(6):2757–2785
23. Roosen C, Müller P, Greiner L (2008) Ionic liquids in biotechnology: applications and perspectives for biotransformations. Appl Microbiol Biotechnol 81(4):607–614

24. Moniruzzaman M et al (2010) Recent advances of enzymatic reactions in ionic liquids. Biochem Eng J 48(3):295–314
25. Hernández-Fernández FJ et al (2010) Biocatalytic ester synthesis in ionic liquid media. J Chem Technol Biotechnol 85(11):1423–1435
26. Quijano G, Couvert A, Amrane A (2010) Ionic liquids: applications and future trends in bioreactor technology. Bioresour Technol 101(23):8923–8930
27. Itoh T (2017) Ionic liquids as tool to improve enzymatic organic synthesis. Chem Rev 117(15): 10567–10607
28. Egorova KS, Gordeev EG, Ananikov VP (2017) Biological activity of ionic liquids and their application in pharmaceutics and medicine. Chem Rev 117(10):7132–7189
29. Gangu SA, Weatherley LR, Scurto AM (2009) Whole-cell biocatalysis with ionic liquids. Curr Org Chem 13(13):1242–1258
30. Fan L-L, Li H-J, Chen Q-H (2014) Applications and mechanisms of ionic liquids in whole-cell biotransformation. Int J Mol Sci 15(7):12196–12216
31. Xu P et al (2016) Whole-cell biocatalytic processes with ionic liquids. ACS Sustain Chem Eng 4(2):371–386
32. Thuy Pham TP, Cho C-W, Yun Y-S (2010) Environmental fate and toxicity of ionic liquids: a review. Water Res 44(2):352–372
33. Olivier-Bourbigou H, Magna L, Morvan D (2010) Ionic liquids and catalysis: recent progress from knowledge to applications. Appl Catal A Gen 373(1):1–56
34. Park S, Kazlauskas RJ (2003) Biocatalysis in ionic liquids – advantages beyond green technology. Curr Opin Biotechnol 14(4):432–437
35. Kragl U, Eckstein M, Kaftzik N (2002) Enzyme catalysis in ionic liquids. Curr Opin Biotechnol 13(6):565–571
36. Kaar JL et al (2003) Impact of ionic liquid physical properties on lipase activity and stability. J Am Chem Soc 125(14):4125–4131
37. Reichardt C (1994) Solvatochromic dyes as solvent polarity indicators. Chem Rev 94(8): 2319–2358
38. Carmichael AJ, Seddon KR (2000) Polarity study of some 1-alkyl-3-methylimidazolium ambient-temperature ionic liquids with the solvatochromic dye, Nile Red. J Phys Org Chem 13(10):591–595
39. Ropel L et al (2005) Octanol-water partition coefficients of imidazolium-based ionic liquids. Green Chem 7(2):83–90
40. Baker SN, Baker GA, Bright FV (2002) Temperature-dependent microscopic solvent properties of 'dry' and 'wet' 1-butyl-3-methylimidazolium hexafluorophosphate: correlation with (30) and Kamlet-Taft polarity scales. Green Chem 4(2):165–169
41. Wood N et al (2011) Screening ionic liquids for use in biotransformations with whole microbial cells. Green Chem 13(7):1843–1851
42. Kulacki KJ, Lamberti GA (2008) Toxicity of imidazolium ionic liquids to freshwater algae. Green Chem 10(1):104–110
43. Latala A, Nedzi M, Stepnowski P (2009) Toxicity of imidazolium and pyridinium based ionic liquids towards algae. Chlorella vulgaris, Oocystis submarina (green algae) and Cyclotella meneghiniana, Skeletonema marinoi (diatoms). Green Chem 11(4):580–588
44. Santos AG et al (2014) Toxicity of ionic liquids toward microorganisms interesting to the food industry. RSC Adv 4(70):37157–37163
45. Bräutigam S, Bringer-Meyer S, Weuster-Botz D (2007) Asymmetric whole cell biotransformations in biphasic ionic liquid/water-systems by use of recombinant Escherichia coli with intracellular cofactor regeneration. Tetrahedron Asymmetry 18(16):1883–1887
46. Cornmell RJ et al (2008) Accumulation of ionic liquids in Escherichia coli cells. Green Chem 10(8):836–841
47. Lou W-Y et al (2009) Efficient enantioselective reduction of 4′-methoxyacetophenone with immobilized Rhodotorula sp. AS2.2241 cells in a hydrophilic ionic liquid-containing co-solvent system. J Biotechnol 143(3):190–197

48. Ranke J et al (2007) Design of sustainable chemical products the example of ionic liquids. Chem Rev 107(6):2183–2206
49. Evans KO (2006) Room-temperature ionic liquid cations act as short-chain surfactants and disintegrate a phospholipid bilayer. Colloids Surf A Physicochem Eng Asp 274(1):11–17
50. Evans K (2008) Supported phospholipid bilayer interaction with components found in typical room-temperature ionic liquids – a QCM-D and AFM study. Int J Mol Sci 9(4):498
51. Evans KO (2008) Supported phospholipid membrane interactions with 1-butyl-3-methylimidazolium chloride. J Phys Chem B 112(29):8558–8562
52. Xiao Z-J et al (2012) Using water-miscible ionic liquids to improve the biocatalytic anti-Prelog asymmetric reduction of prochiral ketones with whole cells of Acetobacter sp. CCTCC M209061. Chem Eng Sci 84(Supplement C):695–705
53. Petkovic M et al (2012) Unravelling the mechanism of toxicity of alkyltributylphosphonium chlorides in Aspergillus nidulans conidia. New J Chem 36(1):56–63
54. Hartmann DO, Silva Pereira C (2013) A molecular analysis of the toxicity of alkyltributylphosphonium chlorides in Aspergillus nidulans. New J Chem 37(5):1569–1577
55. Hartmann DO et al (2015) Plasma membrane permeabilisation by ionic liquids: a matter of charge. Green Chem 17(9):4587–4598
56. Martins I et al (2013) Proteomic alterations induced by ionic liquids in Aspergillus nidulans and Neurospora crassa. J Proteomics 94(Supplement C):262–278
57. Yu Z, Zhang J, Liu S (2015) Biochemical and gene expression effects of 1-alkyl-3-methylimidazolium tetrafluoroborate on Vibrio qinghaiensis sp.-Q67. J Hazard Mater 300 (Supplement C):483–492
58. Pfruender H et al (2004) Efficient whole-cell biotransformation in a biphasic ionic liquid/water system. Angew Chem Int Ed 43(34):4529–4531
59. Weuster-Botz D (2007) Process intensification of whole-cell biocatalysis with ionic liquids. Chem Rec 7(6):334–340
60. Bräutigam S et al (2009) Whole-cell biocatalysis: evaluation of new hydrophobic ionic liquids for efficient asymmetric reduction of prochiral ketones. Enzyme Microb Technol 45(4): 310–316
61. Zhang B-B et al (2012) Efficient anti-Prelog enantioselective reduction of acetyltrimethylsilane to (R)-1-trimethylsilylethanol by immobilized Candida parapsilosis CCTCC M203011 cells in ionic liquid-based biphasic systems. Microb Cell Fact 11(1):108
62. Cull SG et al (2000) Room-temperature ionic liquids as replacements for organic solvents in multiphase bioprocess operations. Biotechnol Bioeng 69(2):227–233
63. Howarth J, James P, Dai J (2001) Immobilized baker's yeast reduction of ketones in an ionic liquid, [bmim]PF6 and water mix. Tetrahedron Lett 42(42):7517–7519
64. Pfruender H, Jones R, Weuster-Botz D (2006) Water immiscible ionic liquids as solvents for whole cell biocatalysis. J Biotechnol 124(1):182–190
65. Matsuda T et al (2006) An effective method to use ionic liquids as reaction media for asymmetric reduction by Geotrichum candidum. Tetrahedron Lett 47(27):4619–4622
66. Lou W-Y, Zong M-H, Smith TJ (2006) Use of ionic liquids to improve whole-cell biocatalytic asymmetric reduction of acetyltrimethylsilane for efficient synthesis of enantiopure (S)-1-trimethylsilylethanol. Green Chem 8(2):147–155
67. Hussain W, Pollard DJ, Lye GJ (2007) The bioreduction of a β-tetralone to its corresponding alcohol by the yeast Trichosporon capitatum MY1890 and bacterium Rhodococcus erythropolis MA7213 in a range of ionic liquids. Biocatal Biotransformation 25(6):443–452
68. Schroer K, Tacha E, Lütz S (2007) Process intensification for substrate-coupled whole cell ketone reduction by in situ acetone removal. Org Process Res Dev 11(5):836–841
69. Shi Y-G et al (2008) Effect of ionic liquid [BMIM][PF6] on asymmetric reduction of ethyl 2-oxo-4-phenylbutyrate by Saccharomyces cerevisiae. J Ind Microbiol Biotechnol 35(11):1419–1424
70. Dipeolu O, Green E, Stephens G (2009) Effects of water-miscible ionic liquids on cell growth and nitro reduction using Clostridium sporogenes. Green Chem 11(3):397–401

71. Lou W-Y et al (2009) Biocatalytic anti-Prelog stereoselective reduction of 4′-methoxyacetophenone to (R)-1-(4-methoxyphenyl)ethanol with immobilized Trigonopsis variabilis AS2.1611 cells using an ionic liquid-containing medium. Green Chem 11(9): 1377–1384
72. Cornmell RJ et al (2008) Using a biphasic ionic liquid/water reaction system to improve oxygenase-catalysed biotransformation with whole cells. Green Chem 10(6):685–691
73. Ming-liang F et al (2011) Effect of ionic liquid-containing system on betulinic acid production from betulin biotransformation by cultured Armillaria luteo-virens Sacc cells. Eur Food Res Technol 233(3):507
74. Wu D-X et al (2011) 11α-Hydroxylation of 16α,17-epoxyprogesterone by Rhizopus nigricans in a biphasic ionic liquid aqueous system. Bioresour Technol 102(20):9368–9373
75. Gao P et al (2014) Enhancing enantioselectivity and productivity of P450-catalyzed asymmetric sulfoxidation with an aqueous/ionic liquid biphasic system. ACS Catal 4(10):3763–3771
76. Zhao Y, Xu Y, Jiang C (2015) Efficient biosynthesis of γ-decalactone in ionic liquids by immobilized whole cells of Yarrowia lipolytica G3-3.21 on attapulgite. Bioprocess Biosyst Eng 38(10):2045–2052
77. Melgarejo-Torres R et al (2015) Evaluation of ionic liquids as dispersed phase during the production of lactones with E. coli in a three phase partitioning bioreactor. Chem Eng J 279(Supplement C):379–386
78. Chen J-Y et al (2012) Efficient production of glycyrrhetic acid 3-O-mono-β-d-glucuronide by whole-cell biocatalysis in an ionic liquid/buffer biphasic system. Process Biochem 47(6): 908–913
79. Baum S, van Rantwijk F, Stolz A (2012) Application of a recombinant Escherichia coli whole-cell catalyst expressing hydroxynitrile lyase and nitrilase activities in ionic liquids for the production of (S)-mandelic acid and (S)-mandeloamide. Adv Synth Catal 354(1):113–122
80. Matsumoto M et al (2014) Effect of organic solvents and ionic liquids on resolution of 2-epoxyhexane by whole cells of Rhodotorula glutinis in a two-liquid phase system. J Chem Technol Biotechnol 89(4):522–527
81. Arai S et al (2010) Production of biodiesel fuel from soybean oil catalyzed by fungus whole-cell biocatalysts in ionic liquids. Enzyme Microb Technol 46(1):51–55
82. Rajapriya G et al (2018) Aspergillus niger whole-cell catalyzed synthesis of caffeic acid phenethyl ester in ionic liquids. Enzyme Microb Technol 111:67
83. Khudyakov JI et al (2012) Global transcriptome response to ionic liquid by a tropical rain forest soil bacterium, Enterobacter lignolyticus. PNAS 109(32):E2173–E2182
84. Bokinsky G et al (2011) Synthesis of three advanced biofuels from ionic liquid-pretreated switchgrass using engineered Escherichia coli. PNAS 108(50):19949–19954
85. Frederix M et al (2016) Development of an E. coli strain for one-pot biofuel production from ionic liquid pretreated cellulose and switchgrass. Green Chem 18(15):4189–4197
86. Ruegg TL et al (2014) An auto-inducible mechanism for ionic liquid resistance in microbial biofuel production. Nat Commun 5:3490
87. Dickinson Q et al (2016) Mechanism of imidazolium ionic liquids toxicity in Saccharomyces cerevisiae and rational engineering of a tolerant, xylose-fermenting strain. Microb Cell Fact 15(1):17
88. Yu C et al (2016) Ionic liquid-tolerant microorganisms and microbial communities for lignocellulose conversion to bioproducts. Appl Microbiol Biotechnol 100(24):10237–10249

Adv Biochem Eng Biotechnol (2019) 168: 133–176
DOI: 10.1007/10_2018_78

Published online: 22 September 2018

Biopolymer-Based Composite Materials Prepared Using Ionic Liquids

Saerom Park, Kyeong Keun Oh, and Sang Hyun Lee

Contents

S. Park and S. H. Lee (✉)
Department of Biological Engineering, Konkuk University, Seoul, South Korea
e-mail: sanghlee@konkuk.ac.kr

K. K. Oh
Department of Chemical Engineering, Dankook University, Yongin, Gyeonggi, South Korea

Abstract Biopolymer-based composite materials have many potential applications in biomedical, pharmaceutical, environmental, biocatalytic, and bioelectronic fields, owing to their inherent biocompatibility and biodegradability. When used as solvents, ionic liquids can be used to fabricate biopolymers such as polysaccharides and proteins into various forms, including molded shapes, films, fibers, and beads. This article summarizes the processes for preparing biopolymer-based composite materials using ionic liquids. The processes include biopolymer dissolution using ionic liquids, regeneration of the biopolymer by an anti-solvent, formation of shapes, and drying of the regenerated biopolymer. In particular, the preparation and applications of biopolymer blend-based composite materials containing two or more biopolymers are addressed.

Graphical Abstract

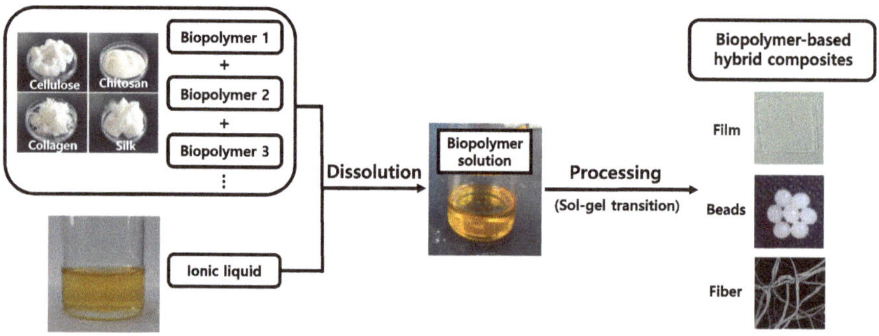

Keywords Biopolymer, Blend, Composite, Gel, Ionic liquid

Abbreviations

[Ac]	Acetate
[Ala]	Alanine
[Amim]	1-allyl-3-methylimidazolium
[Ba]	Benzoate
[Bdmim]	1-butyl-2,3-dimethylimidazolium
[BF_4]	Tetrafluoroborate
[Bmim]	1-butyl-3-methylimidazolium
[BmPy]	*N*-butyl-3-methylpyridinium
[C=C_2mmor]	*N*-allyl-*N*-methylmorpholium
[C=C_2mpip]	*N*-allyl-*N*-methylpiperidium
[C_1mim]	1,3-dimethylimidazolium

[C_1OC_2mim]	1-methoxyethyl-3-methylimidazolium
[C_2C_2im]	1,3-diethylimidazolium
[C_2mmor]	*N*-ethyl-*N*-methylmorpholium
[C_2OHmim]	1-hydroxyethyl-3-methylimidazolium
[C_4mpip]	*N*-butyl-*N*-methylpiperidium
[C_6mim]	1-hexyl-3-methylimidazolium
[CF_3SO_3]	Trifluoromethanesulfonate
[$CH_3CHOHCOO$]	Lactate
[dca]	Dicyanamide
[DEP]	Diethylphosphate
[Emim]	1-ethyl-3-methylimidazolium
[Gly]	Glycine
[H_2NCH_2COO]	Aminoethanoate
[HCOO]	Formate
[Hmim]	1-hydrogen-3-methylimidazolium
[$HOCH_2COO$]	Glycolate
[$HSCH_2COO$]	Thioglycolate
[HSO_4]	Hydrogensulfate
[$MeO(H)PO_2$]	Methylphosphonate
[$MeSO_4$]	Methylsulfate
[mim]	1-methylimidazolium
[Omim]	3-methyl-1-octylimidazolium
[OPr]	Propionate
[phC_1mim]	1-benzyl-3-methylimidazolium
[Py]	Pyridinium
[Pyrr]	Pyrrolidinium
[Ser]	Serine
[Tf_2N]	Bis(trifluoromethylsulfonyl)imide

1 Introduction

Biopolymer-based materials have attracted considerable interest, owing to their potential to decrease society's dependency on fossil fuels [1]. Biopolymers include naturally obtainable polysaccharides such as cellulose, chitin/chitosan, starch, and glycosaminoglycans, and proteins such as collagen, silk, and keratin. The inherent biocompatibility and biodegradability of biopolymers make them particularly useful in the biomedical, pharmaceutical, environmental, biocatalytic, and bioelectronic fields. However, the preparation of biopolymer-based materials still remains a challenge because of the low solubility of biopolymers in conventional solvents. Therefore, new solvents for the dissolution of biopolymers are of interest in developing various biopolymer-based composite materials.

Ionic liquids (ILs) are organic salts that usually melt below 100°C. The interest in ILs stems from their potential application as "green solvents" [2]. Specifically, their nonvolatile character and thermal stability make them attractive alternatives to volatile organic solvents. Moreover, because of their inherent synthetic flexibility, ILs are referred to as "designer solvents" [3]. Since the pioneering results regarding the dissolution and regeneration of cellulose using ILs were reported by Rogers's group [4], various ILs have been developed to dissolve biopolymers. Various biopolymers such as polysaccharides, proteins, and lignocelluloses can be dissolved and regenerated using ILs [5, 6]. The principal merits of the procedures to make biopolymer-based materials using ILs are the high solubility of biopolymers, greener procedures using non-volatile solvents, and easy production of composites containing various biopolymers, synthetic polymers, and functional materials.

The blending of two or more biopolymers is an extremely attractive, inexpensive, and advantageous method to obtain new structural materials with novel physicochemical properties [7, 8]. To prepare biopolymer blends, the development of solvents that can co-dissolve two or more biopolymers is the most important aspect. Various biopolymer blend-based materials can be prepared using ILs because of the high solvating power of ILs for biopolymers.

This review focuses on the role of ILs as solvents to make biopolymer-based materials. The process of using ILs to fabricate biopolymer-based materials in various forms such as molded shapes, films, fibers, and beads is summarized. In particular, the preparation and applications of biopolymer blend-based composite materials containing two or more biopolymers are addressed. The general procedures to make biopolymer-based materials include biopolymer dissolution using ILs, regeneration of the biopolymer by an anti-solvent, formation of shapes, and drying of the regenerated biopolymer.

2 Dissolution of Biopolymers Using ILs

To prepare biopolymer-based composite materials using ILs, the first consideration is the choice of the most efficient IL for the dissolution of various biopolymers. Many ILs that can dissolve biopolymers, such as polysaccharides, proteins, lignocelluloses, and nucleic acids, have been developed. Binary mixtures of ILs and ILs containing hydrophilic co-solvents, such as dimethylsulfoxide (DMSO), dimethylformamide (DMF), and dimethylacetamide (DMAc), were also investigated to enhance the biopolymer solubility and decrease the solution viscosity. Tables 1 and 2 show some examples of ILs to dissolve biopolymers.

Table 1 Solubility of microcrystalline cellulose in ILs

Ionic liquid	Solubility (wt%)	Dissolution temperature
[Bmim][Ac]	15.5 [9]	70°C
[Bmim][$HSCH_2COO$]	13.5 [9]	70°C
[Bmim][HCOO]	12.5 [9]	70°C
[Bmim][Ba]	12.0 [9]	70°C
[Bmim][H_2NCH_2COO]	12.0 [9]	70°C
[Bmim][$HOCH_2COO$]	10.5 [9]	70°C
[Bmim][$CH_3CHOHCOO$]	9.5 [9]	70°C
[Bmim][Ac]	29.3 [10]	120°C
[C_1OC_2mim][Ac]	28.0 [10]	120°C
[C_2OHmim][Ac]	18.3 [10]	120°C
[Bdmim][Ac]	17.4 [10]	120°C
[C_2mmor][Ac]	16.5 [10]	120°C
[phC_1mim][Ac]	14.6 [10]	120°C
[C=C_2mmor][Ac]	13.9 [10]	120°C
[C=C_2mpip][Ac]	9.5 [10]	120°C
[C_4mpip][Ac]	3.3 [10]	120°C

Table 2 Solubility of biopolymers in ILs

Biopolymer	Ionic liquids	Solubility (wt%)
Polysaccharide		
Chitin (shrimp shell)	[Emim][Ac] [11]	>10 (100°C)
α-Chitin (crab)	[Bmim][Ac] [12]	6 (110°C)
β-Chitin (squid pen)	[Bmim][Ac] [12]	6 (110°C)
Chitosan	[Emim][Ac] [13]	>10 (70°C)
	[Bmim][Ac] [12]	12 (110°C)
	[Bmim][Cl] [12]	10 (110°C)
	[Amim][Cl] [12]	8 (110°C)
	[C_1mim][Cl]/[Hmim][Cl] (9:1) [14]	17.5 (110°C)
Chondroitin sulfate	[Hmim][HSO_4] [15]	>4 (100°C)
Chondroitin sulfate (imidazolium salt)	[Emim][Ba] [16]	9.9 (35°C)
	[Bmim][Ba] [16]	5.7 (35°C)
Heparin	[Emim][Ac] [17]	>0.7 (80°C)
Heparin (imidazolium salt)	[Emim][Ba] [16]	7 (35°C)
	[Bmim][Ba] [16]	7 (35°C)
Heparan sulfate (imidazolium salt)	[Emim][Ba] [16]	3 (35°C)
	[Bmim][Ba] [16]	2.8 (35°C)
Hyaluronic acid (imidazolium salt)	[Emim][Ba] [16]	10 (35°C)
	[Bmim][Ba] [16]	10 (35°C)
Starch	[Amim][Cl] [18]	15 (80°C)
	[Bmim][Cl] [19]	15 (80°C)

(continued)

Table 2 (continued)

Biopolymer	Ionic liquids	Solubility (wt%)
Agarose	[Bmim][Cl] [20]	16 (70°C)
	[Bmim][$MeSO_4$] [20]	5 (70°C)
	[Omim][Cl] [20]	4.5 (70°C)
	[BmPy][Cl] [20]	13 (70°C)
Carrageenan	[Bmim][Cl] [21]	>15 (100°C)
Guar gum	[Bmim][Cl] [22]	>25 (50°C)
Tamarind gum	[Bmim][Cl] [23]	>10 (80°C)
Xanthan gum	[Bmim][Cl] [24]	>50 (100°C)
Protein		
Collagen	[Bmim][Cl] [25]	>1 (100°C)
Collagen (acid soluble)	[Emim][Ac] [26]	7.4 (45°C)
	[Emim][Ac]/Na_2HPO_4 (1 wt%) [26]	10.5 (45°C)
Gelatin	[Amim][Cl] [27]	>8 (100°C)
Silk (*B. mori*)	[Emim][Cl] [28]	23.3 (100°C)
	[Bmim][Cl] [28]	13.2 (100°C)
	[Bdmim][Cl] [28]	8.3 (100°C)
	[Emim][Gly] [17]	26.3 (100°C)
	[Emim][Ala] [17]	>20 (100°C)
	[Emim][Ser] [17]	>20 (100°C)
	[Emim][Ac] [29]	>10 (95°C)
Wool keratin	[Amim][dca] [30]	47.5 (130°C)
	[Bmim][Cl] [30]	25 (130°C)
	[Amim][Cl] [30]	20 (130°C)
	[choline][$HSCH_2COO$] [30]	22.5 (130°C)
Lignocellulose		
Xylan (beechwood)	[Emim][Ac] [31]	5 (70°C)
Xyloglucan	[Emim][Ac] [32]	>7 (100°C)
Kraft lignin	[Py][Ac] [33]	>50 (90°C)
	[C_1mim][Ac] [33]	>50 (90°C)
	[Pyrr][Ac] [33]	>50 (90°C)
Kraft lignin (softwood)	[Bmim][$MeSO_4$] [34]	26 (50°C)
	[C_2C_2im][$MeSO_4$] [34]	26 (50°C)
	[C_6mim][CF_3SO_3] [34]	22 (70°C)
	[Bmim][Cl] [34]	13 (75°C)
	[Bdmim][BF_4] [34]	14 (75°C)
Lignin (Indulin AT)	[C_1mim][$MeSO_4$] [33]	>50 (90°C)
	[Bmim][CF_3SO_3] [33]	>50 (90°C)
	[Emim][Ac] [33]	>30 (90°C)
	[Amim][Cl] [33]	>30 (90°C)
	[Bmim][Cl] [33]	>10 (90°C)
	[Bmim][BF_4] [33]	4 (90°C)

2.1 *Polysaccharides*

2.1.1 Cellulose

Cellulose, a linear polysaccharide of glucose residues linked by β-(1 → 4)-glycosidic bonds, is the most abundant renewable biopolymer. It has excellent thermal and mechanical properties. Cellulose offers excellent biocompatibility and is considered a promising material for biotechnological applications. Cellulose is highly crystalline as a result of an extensive hydrogen bonding network, making it insoluble in most conventional organic and aqueous solvents. The first publication on cellulose dissolution in ILs was reported by the Rogers group in 2002 [4]. Cellulose was dissolved at 25% in [Bmim][Cl] by microwave irradiation and the dissolved cellulose was regenerated by the addition of an anti-solvent, such as water, ethanol, or acetone. These results have opened up new paths for the preparation of various cellulose-based composites. To date, many reports related to cellulose dissolution have been published [9, 10]. Xu et al. studied the effect of an anion of an IL on the dissolution of microcrystalline cellulose (Table 1). They concluded that the hydrogen-bond accepting strength played an important role; a linear correlation was proposed between the solubility of cellulose in wt% and the Kamlet-Taft β parameter representing the hydrogen bond basicity [35]. Lu et al. studied the effect of a cation of an IL on the dissolution of microcrystalline cellulose (Table 1). It was found that acidic protons on the heterocyclic rings of the cations are essential for the dissolution of cellulose in ILs, but the van der Waals interaction of the cation with cellulose is not important [36]. The dissolution rate and solubility of cellulose in ILs may be enhanced by using microwave irradiation [37–39] and ultrasonication [40, 41]. A more detailed discussion of cellulose dissolution in ILs can be found in earlier reviews [9, 10].

2.1.2 Chitin/Chitosan

Chitin, a co-polymer with more than 50% *N*-acetyl-glucosamine and *N*-glucosamine units, is one of the most abundant polysaccharides, with an annual production second only to cellulose [1]. Chitin has been used for biomedical applications such as tissue engineering scaffold and wound healing, owing to its biodegradability and biocompatibility. However, the use of chitin has been limited because chitin forms strong inter- and intramolecular hydrogen bonds that are not easily broken by common molecular solvents. Although chloride-containing ILs such as [Bmim][Cl], [Emim][Cl], and [Amim][Cl] have been reported to be good solvents to dissolve cellulose, they have not dissolved chitins with satisfactory results. [Bmim][Ac] and [Emim][Ac] were reported to be good solvents for chitins [11, 12]. The solubilities of α-chitin from crab and β-chitin from squid pen in [Bmim][Ac] were both 6%, whereas the chitins were partially soluble or insoluble in [Bmim][Cl]. [Emim][Ac] was able to dissolve more than 10% of chitin from shrimp shells. However, conflicting results were also reported. [Emim][Cl] dissolved

10% chitin from crab shell at 100°C, although the dissolution process took days [17]. These different results may be caused by different molecular weights (MW) and degrees of acetylation (DAC) of the chitin samples.

Chitosan is a fully or partially deacetylated product of chitin. It has gained tremendous interest in many fields, such as pharmaceuticals, foods, textiles, daily chemicals, and papermaking, owing to its biocompatibility, biodegradability, hygroscopicity, antimicrobial activity, and fiber/film-forming properties [6]. Chitosan was well dissolved in various cellulose-dissolving ILs. The solubilities of chitosan (MW = 97 kDa, DAC = 5%) from crab in [Amim][Cl], [Bmim][Cl], and [Bmim][Ac] were 8%, 10%, and 12%, respectively [12]. [Hmim][HSO_4] could also dissolve over 4% chitosan (MW = 22 kDa, DAC = 15%) [15]. Recently, a binary mixture of [C_1mim][Cl] and [Hmim][Cl] (9:1) was developed for chitosan dissolution, and the solubility of chitosan (DAC = 10%) in this mixture was 17.5% [14]. Many solubility results for chitosan dissolution in ILs have been reported. However, the solubility of chitosan in ILs was highly dependent on the molecular weight, DAC, and the source of chitosan.

2.1.3 Glycosaminoglycans

Glycosaminoglycans such as chondroitin sulfate, heparin, heparan sulfate, and hyaluronic acid are long unbranched polysaccharides consisting of a repeating disaccharide unit. They are widely used in the biomedical sector, such as for implant coating materials and as components of tissue engineering scaffolds. However, they are intractable solutes that can only be dissolved in water and a few highly polar organic solvents with undesirable properties, such as formamide and pyridine. The Linhardt group developed benzoate-based ILs to dissolve various glycosaminoglycans. [Emim][Ba] and [Bmim][Ba] proved to be good solvents for the dissolution of glycosaminoglycans. The solubilities of chondroitin sulfate, heparin, heparan sulfate, and hyaluronic acid as imidazolium salts in [Emim][Ba] were 9.9%, 7%, 3%, and 10%, respectively [16]. Recently, [Hmim][HSO_4] was used to dissolve chondroitin sulfate and the solubility was higher than 4% [15]. [Emim][Ac] was also shown to be able to dissolve more than 0.7% of heparin [42].

2.1.4 Other Polysaccharides

Starch is an omnipresent polysaccharide found in vegetable sources [10]. Among industrial materials, starch is one of the cheapest and most cost-effective. It is currently used industrially for coating and sizing in paper, textiles, and carpets, as well as in binders, adhesives, absorbents, and encapsulants [19]. Several studies reported notable solubility of starch in ILs. [Bmim][Cl] dissolved 15% of starch at 80°C. [Bmim][dca] dissolved more than 10% of starch at 80°C [19]. [Amim][Cl] was a very efficient solvent for starch. A solution of 50% starch in [Amim][Cl] was obtained at 100°C [43].

Agarose and carrageenan are algal polysaccharides, generally extracted from seaweed. Agarose is composed of a repeating unit of agarobiose, which is a disaccharide made up of D-galactose and 3,6-anhydro-L-galactose. Agarose has many applications in various fields, such as the food industry, pharmaceutical formulations, electrophoresis, and tissue engineering scaffolds. However, the large number of hydroxyl groups makes agarose insoluble in cold water and common organic solvents [20]. The solubilities of agarose in [Bmim][Cl], [Omim][Cl], [BmPy][Cl], and [Bmim][$MeSO_4$] were 16%, 4.5%, 13%, and 5% at 70°C. The solubility of agarose in [Bmim][Ac] was over 6% at 70°C [44]. Carrageenan consists of alternating 1,3-linked α-D-galactopyranose and 1,4-linked β-(3,6-anhydro-)D-galactopyranose. Carrageenan is used mostly as a stabilizer and structure provider in the food and ice cream industries. Three types of carrageenan (κ-, ι-, λ-) were dissolved in [Bmim][Cl], with solubilities of more than 15% at 100°C [21].

Guar and tamarind gum are polysaccharides obtained from seeds of plants, and xanthan gum is produced by *Xanthamonas campestris*. They are widely used as thickening, stabilizing, emulsifying, and gelling agents in the food and pharmaceutical industries. Guar gum classified as galactomannan consists of a β-1,4-mannose backbone with randomly distributed α-1,6-galactose moieties at the C-6 position. Guar gum was dissolved in [Bmim][Cl] up to 25% at 80°C. [Emim][Ac], [Amim][Cl], and [Bmim][MeO(H)PO_2] could dissolve 5% guar gum at 50°C [22]. Tamarind gum is composed of β-(1,4)-D-glucan backbone substituted with side chains of α-(1,4)-D-xylopyranose and (1,6)-linked [β-D-galactopyranosyl-(1,2)-α-D-xylopyranosyl] to glucose residues. [Bmim][Cl], [Bmim][Br], [choline][acrylate], [choline][caproate], and [choline][caprylate] could dissolve 10% tamarind gum at 80–100°C [23]. Xanthan gum has a cellulose-type main-chain ((1→4)-β-glucan) with trisaccharide side-chains attached to alternate glucose units in the main-chain. [Bmim][Cl] dissolved xanthan gum up to 50% at 100°C [24].

2.2 Proteins

Collagen, a fibrous structural protein with excellent mechanical properties, is one of the most abundant proteins in nature. It comprises a right-handed bundle of three parallel left-handed helices. Collagen has many applications in the commercial fields of food, cosmetics, and medicine, owing to its good biocompatibility and biodegradability. [Bmim][Cl] was shown to dissolve 1% native collagen fibers at 100°C [25]. The solubility of acid-soluble collagen was 7.4% in [Emim][Ac] at 45°C, and the addition of 1% Na_2HPO_4 to [Emim][Ac] enhanced the solubility up to 10.5% [26]. [Hmim][Ac] dissolved 4% collagen at 40°C [45]. Gelatin is derived from collagen through acid or alkaline hydrolysis. Owing to its abundance, biocompatibility, low toxicity, biodegradability, and adhesiveness, gelatin is widely used in the food, pharmaceutical, cosmetic, and photographic industries [46]. [Amim][Cl] could dissolve 8% gelatin at 100°C [27].

Natural silk fibers have outstanding mechanical properties that rival the most advanced synthetic polymers. A single strand of natural cocoon silk fiber contains two silk fibroin cores surrounded by a protective, glue-like sericin coating. The fibroin cores consist of heavy and light chains. The hydrophobic nature and hydrogen bonding of the heavy chain make dissolution of silk a difficult task [17, 28]. Silk fibers have been widely used in biomedical applications as a suture material for repairing wounds. Silk is an attractive tissue engineering scaffold because of its slow degradation, excellent mechanical properties, and biocompatibility [47]. *Bombyx mori* silk fibroin was dissolved in [Emim][Cl], [Bmim][Cl], and [Bdmim][Cl]. The solubility of silk fibroin was 23.3% in [Emim][Cl] at 100°C. [Bmim][Ac] dissolved 10% silk at 95°C [29]. Amino acid-based ILs also proved to be good solvents for silk. [Emim][Gly], [Emim][Ala], and [Emim][Ser] could dissolve more than 20% of silk at 100°C [17].

Wool is a fibrous protein (approximately 95 wt% pure keratin) that consists of approximately 11–17% cysteine. This protein is insoluble in water and common organic solvents, owing to the tight packing of the α-helices and β-sheets present in the polypeptide structure of wool keratin [30]. [Bmim][Cl] and [Amim][Cl] easily dissolved keratin fiber at 130°C [48, 49]. [Amim][dca] and [Choline] [thioglycolate] were also used to dissolve wool keratin. [Amim][dca] dissolved 47.5% keratin at 130°C [30]. During the dissolving process of keratin, the α-helix structure was disrupted but the β-sheet structure remained.

2.3 Lignocellulose

Lignocelluloses, such as agricultural residues, waste paper, forestry waste, and energy crops, have long been recognized as sustainable sources of biopolymer-based composites. Lignocelluloses consist of three major biopolymers with distinct chemical, physical, and structural properties: cellulose, hemicellulose, and lignin. Cellulose is the most abundant renewable biopolymer. The cellulose content in lignocelluloses ranges from 41% to 50%. Hemicelluloses are heterogeneous, branched polymers of pentoses, hexoses, and acetylated sugars, with xylans being the most predominant hemicellulose. Hemicelluloses are as common as cellulose, but with a lower molecular weight and wood content of 25–35%. Lignin is an aromatic network polymer composed of phenylpropanoid units, which aids in binding cellulose and hemicellulose together. Lignin is more hydrophobic than cellulose and hemicelluloses, with a wood content of 18–35% [1, 50]. The applications of lignocellulosic biomass in the field of biopolymer-based composites have been only focused on the use of cellulose. However, hemicelluloses and lignins also have many potential applications.

The xylan from beechwood was dissolved in [Emim][Ac] with a solubility of 5% at 70°C [31]. Xyloglucan, which is known to bind tightly to cellulose surfaces in the lignocellulosic biomass, was also dissolved in [Emim][Ac]. Commercially available lignins include alkaline lignin, Kraft lignin, and Organosolv lignin. In fact, these lignins are significantly different from native lignin. However, they can be used to make various composite materials. Currently, lignins are used as binding and dispersing agents. Various ILs can dissolve lignins. ILs containing $[Ac]^-$ or $[Cl]^-$ can easily dissolve lignin. [Emim][Ac] and [Amim][Cl] were able to dissolve more than 30% of lignin (Indulin AT) at 90°C. [C_1mim][$MeSO_4$] and [Bmim][CF_3SO_3] were more efficient solvents to dissolve lignins than ILs containing $[Ac]^-$ or $[Cl]^-$ [2].

The dissolution of the entire lignocellulosic biomass in ILs has also been studied. [Amim][Cl], [Emim][Cl], [Bmim][Cl], [bzmim][Cl], and [Emim][Ac] were successfully used to completely dissolve lignocellulosic biomasses [10]. The ability of ILs to dissolve lignocellulosic biomass was highly dependent on the biomass type, particle size, and heating method, among other factors. Most research on the full dissolution of lignocellulosic biomass has been focused on the pretreatment of biomass to remove lignins.

2.4 *Effects of Co-solvents for Biopolymer Dissolution*

The addition of co-solvents to ILs can increase the solubility of biopolymers and decrease the viscosity of the dissolved biopolymer solution. Xu et al. dissolved 15% cellulose in [Bmim][Ac]/DMSO (2.54:1 molar ratio) at 25°C, whereas cellulose was insoluble in [Bmim][Ac] at 25°C. At the same molar ratio, [Bmim][Ac]/DMF and [Bmim][Ac]/DMAc also dissolved cellulose up to 12.5% and 5.5%, respectively [51]. Zhao et al. investigated the effects of DMSO, DMF, methanol, and water as cosolvents on cellulose dissolution in [Bmim][Ac] by molecular dynamics simulations and quantum chemistry calculations. The results showed that the dissolution of cellulose in [Bmim][Ac]/cosolvent systems is mainly determined by the hydrogen bond interactions between the $[Ac]^-$ anions and the hydroxyl protons of cellulose. The strong preferential solvation of $[Ac]^-$ by the protic solvents such as methanol and ethanol can compete with the cellulose–$[Ac]^-$ interaction, resulting in decreased cellulose solubility. However, no preferential solvation was observed in aprotic solvents such as DMSO and DMF. The dissociated $[Ac]^-$ would readily interact with cellulose to improve the dissolution of cellulose [52].

3 Regeneration of Biopolymers

The biopolymers dissolved in an IL can be precipitated when an anti-solvent such as water, ethanol, methanol, or acetone is added into the biopolymer solution, owing to the lowered solubility of the biopolymer in the IL/anti-solvent mixture. The precipitated biopolymer is called a regenerated biopolymer, whose morphology generally becomes rougher than that of the native biopolymer. For example, the thermal stability and crystallinity of the cellulose regenerated by using IL as a solvent and water as an anti-solvent were decreased, whereas the chemical structures were retained [6]. Before the precipitation process using an anti-solvent proceeds, a pretreatment process of the biopolymer solution can sometimes be helpful to regenerate the biopolymer.

3.1 Anti-solvents

The type of anti-solvent used to coagulate a biopolymer dissolved in IL can influence the characteristics of the regenerated biopolymer. Therefore, the choice of anti-solvent for biopolymer regeneration is an important issue to obtain the desired biopolymer composites. The most commonly used anti-solvents are water, methanol, and ethanol. Biopolymer hydrogel and organogel can be prepared via sol–gel transition of the biopolymer solution using water and an organic solvent, respectively, as anti-solvents. Hydrogel and organogel are interchangeable by a simple solvent exchange process. For example, cellulose organogel regenerated from a cellulose solution by using ethanol as an anti-solvent can be changed to cellulose hydrogel by repeated washing with water [53, 54].

3.1.1 Anti-solvents for Polysaccharides Dissolved in ILs

Cellulose dissolved in IL can be regenerated by using various anti-solvents. The anti-solvent must be miscible with the IL and a precipitant for cellulose [55]. Water, methanol, ethanol, acetone, and acetonitrile can be used to regenerate cellulose from cellulose solution. Water was found to be the most efficient anti-solvent among water, ethanol, and acetone when the cellulose was regenerated from cellulose solutions in various ILs, such as [Amim][Cl], [Bmim][Cl], [Emim][Cl], [Bmim][Ac], [Emim][Ac], and [Emim][DEP] [55]. Water was effective for breaking the bonds between the IL and cellulose in solution, subsequently helping to form the bonds between cellulose molecules [4].

Various properties of cellulose regenerated from a cellulose solution in ILs are dependent on the type of anti-solvent. The transmittance of cellulose film regenerated from [Emim][Ac] solution increased in the following sequence of anti-solvents: acetonitrile < ethanol < water. When DMSO or DMF were used as anti-solvents, the cellulose regenerated from [Bmim][Ac] solution could not form films [56]. The regeneration of cellulose in water from [Emim][Ac]/DMSO (3:7 wt ratio) solution resulted in a higher degree of crystallinity than regeneration in methanol, whereas the cellulose hydrogel regenerated in methanol exhibited larger pores than that regenerated in water [57].

The chitin dissolved in [Emim][Ac] could be regenerated with water, methanol, and ethanol [12, 17]. Methanol used as an anti-solvent resulted in a regenerated chitin film with a plane and featureless surface, with water-engendered furrows on the top surface. The chitin regenerated in methanol showed relatively low crystallinity compared to that in water [12].

3.1.2 Anti-solvents for Proteins Dissolved in ILs

Collagen dissolved in [Bmim][Cl] was regenerated with water, ethanol, acetone, and methanol [25, 58]. The triple helical structure of collagen was partly destroyed during the dissolution and regeneration. The film-forming ability and thermostability of the regenerated collagen was highly dependent on the type of anti-solvent. The collagen films regenerated with water and ethanol showed smooth and compact surfaces. The films regenerated with acetone and methanol presented rough and porous surfaces. Better film-forming ability can be obtained with water and ethanol [25]. Collagen dissolved in [Emim][Ac] or [Emim][Ac] containing 1% Na_2HPO_4 could be also regenerated with water [26, 59].

Silk dissolved in [Bmim][Cl] was regenerated by methanol and acetonitrile, whereas water could not regenerate silk owing to the high solubility of silk in the mixture of water and [Bmim][Cl]. The structure of the regenerated silk films was highly dependent on the anti-solvent. Acetonitrile yielded a convoluted surface structure with low crystallinity, whereas methanol yielded a transparent film with a high degree of crystallinity [28]. Only methanol could be used as an anti-solvent to form silk fiber by wet-spinning of the silk solution in [Emim][Cl] [17]. Silk dissolved in [Bmim][Ac] was regenerated with ethanol, and then sequential treatment with 80% methanol induced the formation of β-sheets [29].

Wool keratin was regenerated from wool keratin solution in [Bmim][Cl] and [Amim][Cl] by using water, methanol, and ethanol as anti-solvents. A portion of the disulfide bonds in wool keratin were broken during the dissolution. The regenerated wool keratin films exhibited a β-sheet structure and the disappearance of the α-helix structure. The thermal stability of regenerated wool keratin films decreased slightly compared to natural wool fibers. The regenerated wool keratin film precipitated from methanol exhibited a high degree of crystallinity compared to other solvents [48].

3.1.3 Anti-solvents for Lignocellulose Dissolved in ILs

Cellulose, xylan, and lignin co-dissolved in [Emim][Ac] or [Bmim][Cl] were regenerated by water and an aqueous ethanol solution [32, 48, 56, 57]. The forms of the coagulated cellulose/xylan/lignin composites were highly dependent on the ethanol concentration. For example, more than 90% ethanol could not precipitate lignin when the concentration of lignin was higher than cellulose [57]. Spruce wood powder dissolved in [Bmim][Cl] could also be regenerated by aqueous ethanol solution. However, spruce wood could not be regenerated into gel when the dissolution time was too long.

4 Processing to Prepare Biopolymer-Based Composite Materials

Biopolymers can be fabricated into various forms such as molded shapes, films, fibers, and beads by using ILs. The processes to fabricate biopolymer-based composite materials include biopolymer dissolution using IL, co-dissolution or

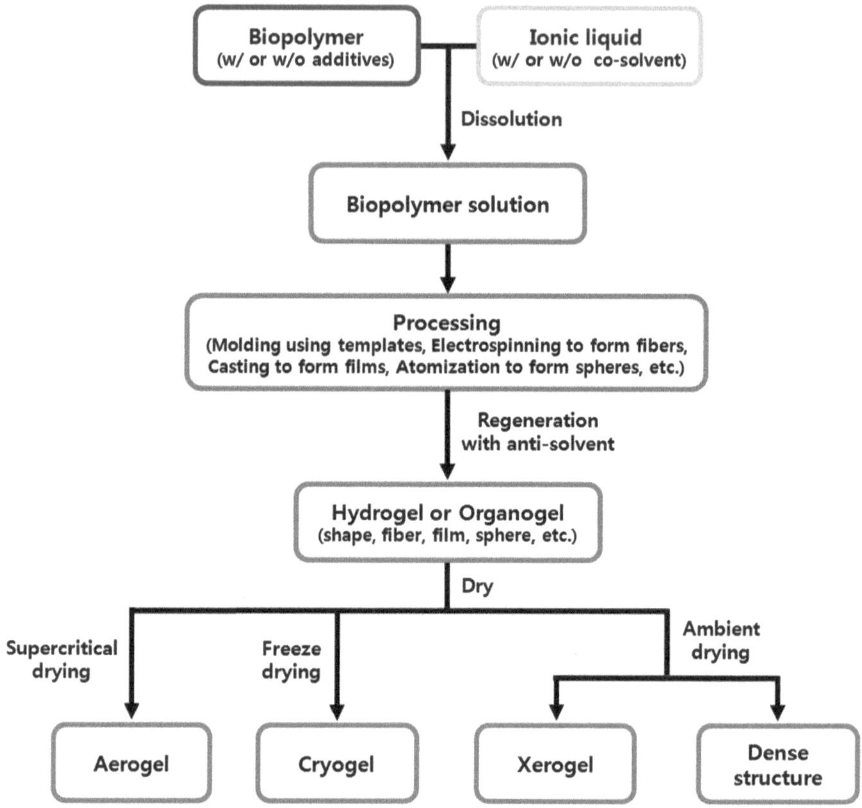

Fig. 1 Process to prepare biopolymer-based composite materials using ILs

dispersion of additives in biopolymer solution, pretreatment of biopolymer solution, regeneration of a biopolymer into hydrogel or organogel using anti-solvents, and drying of hydrogel or organogel into aerogel, cryogel, xerogel, or a dense structure (Fig. 1).

4.1 Pretreatment of Biopolymer Solutions Before Regeneration

Solidification of biopolymer solutions by cooling can be carried out before the regeneration of biopolymers. This pretreatment process may be useful to prepare biopolymer composites of molded shapes. A simple standing of biopolymer solution at room temperature as a pretreatment process can sometimes form gels without regeneration process. In these gels, ILs act as one component to fabricate gel materials together with the biopolymer. The gels are termed "organogels" when the dispersed phase is an organic solvent and "hydrogels" when the dispersed phase is aqueous. Therefore, biopolymer gel materials containing ILs are classified as "ion gels." A detailed discussion of ion gels prepared using ILs can be found in earlier review [60]. Ion gels can be changed to biopolymer hydrogels or organogels by a solvent exchange process. For example, a mixture of cellulose and xanthan gum in [Bmim][Cl] could form an ion gel after incubation for 1 day at room temperature [61]. The cellulose and xanthan gum could be regenerated from the ion gel by using ethanol or water as an anti-solvent. Cooling the biopolymer solution at 4°C or sub-zero temperatures was also used as a pretreatment process. A mixture of cellulose and agar in [Bmim][Cl]/DMSO (1:1 wt ratio) could form an ion gel with incubation at 4°C [62]. A cyclic freezing–thawing process of biopolymer solution was also used as a pretreatment process prior to biopolymer regeneration. A mixture of cellulose and polyvinyl pyrrolidone in [Amim][Cl] was changed to an ion gel by repeated freezing–thawing cycles. This pretreatment process could increase the hardness, gumminess, and resilience of the resultant hydrogel [63, 64]. When a wood solution in [Amim][Cl] was pretreated with a cyclic freezing–thawing process, the intensity, specific surface, crystallinity, and thermostability of the resultant wood aerogel could be changed [65]. To maintain the molded shape of a biopolymer solution, spraying water or storing the solution in a moisture-saturated environment were also investigated [32, 66].

4.2 *Processing to Prepare Various Shapes*

Generally, regeneration of biopolymers can be performed by a sol–gel transition method. During the regeneration step, the form of the prepared biopolymer composites can be determined. Biopolymer-based composite hydrogels or organogels regenerated with water or organic solvents, respectively, can be changed to various dried gels by different drying methods.

4.2.1 Molded Shapes

Regenerated biopolymer hydrogels with molded shapes can be simply prepared by pouring the biopolymer solution into molds, followed by gelation of the solution through immersion in an anti-solvent. However, biopolymer hydrogels prepared using ILs generally showed heterogeneous segregated structures in gels, resulting in significant shrinkage of the gel during the regeneration process. Therefore, the preparation of shape-persistent hydrogels is very important to fabricate biopolymer hydrogels of various shapes for use in biomedical fields, such as tissue engineering scaffolds. Kimura et al. prepared a tough hydrogel by a step-wise solvent exchange process from a homogeneous [Emim][MeO(H)PO_2] solution of cellulose exposed to methanol vapor. The original shape and volume did not change during the solvent exchange process [67]. This result suggests that the preparation of an organogel using a slow regeneration process followed by solvent exchange with water to prepare the hydrogel is useful to fabricate biopolymer-based composites with various molded shapes.

4.2.2 Films and Membranes

Biopolymer-based films and membranes can be used as basic materials in environmental, biomedical, and bioelectronic fields. Biopolymer-based hydrogel films are typically prepared by simply pouring the biopolymer solution into a mold followed by regeneration [32, 59]. To obtain a thin hydrogel film, casting of a biopolymer solution onto a flat surface using rod-coating or spin-coating is needed [37, 50, 68]. The prepared hydrogel films can be dried under high temperatures or vacuum to obtain thin films.

4.2.3 Fibers

Fibers and fibrous membranes have a number of potential biomedical applications because of their flexibility, permeability, high liquid retention, and high surface area. Biopolymer-based composite fibers have been prepared by dry-jet wet-spinning and electrospinning using ILs as solvents [1]. In a typical dry-jet wet-spinning process using ILs, the spin dope is prepared from a cellulose solution. The spinning is performed by extruding the spin dope across an air gap into a coagulation bath. As the fiber is formed, the anti-solvent removes ILs from the fiber. In the extraction process, it is possible to draw the fiber to enhance its properties. Magnetite-embedded cellulose fibers, cellulose/keratin fibers, and cellulose/chitosan fibers were prepared with this method [14, 49, 69]. The thickness of the fibers prepared with dry-jet wet-spinning was in the range of hundreds of micrometers. Electrospinning has been recognized as a simple and versatile method for producing ultrathin fibers with an extremely high surface area. The electrospinning process using ILs can be defined as a dry-jet wet-electrospinning process, which forms biopolymer-based ultrathin fibers fabricated by collecting the jet in a grounded coagulation bath. This process is stable at atmospheric pressure, with no need for gas recovery or a fire-safety system [1]. Although the high viscosity of the ILs is a presumed disadvantage for electrospinning, nanoscale (about 500 nm) fibers could be observed. Our group solved the high viscosity problem by using cosolvents such as DMF and DMAc. As the weight ratio of the cosolvent to [Emim][Ac] increased, the spinnability improved, with stable whipping and splaying motions. Regardless of the cosolvent type, the higher cosolvent concentration resulted in finer fiber diameter, better web uniformity, higher crystallinity, and better thermal stability [70].

4.2.4 Beads

Biopolymer-based beads can be used as adsorbents to remove toxic chemicals, chromatographic resin, and enzyme support. Milliscale and microscale biopolymer-based composite beads were prepared by using ILs. Millimeter-sized beads were usually prepared by drop-wise addition of the biopolymer solution into an anti-solvent bath under vigorous stirring. Syringe pumps are useful to obtain regular beads. The needle size of the syringe, rate of dropping, viscosity of the biopolymer solution, types of anti-solvents and ILs, and concentration of the biopolymer can influence the size, shape regularity, and surface morphology of the prepared beads [8, 71].

Controlling the viscosity of the biopolymer solution is very important to obtain a uniform bead size. The viscosity of the biopolymer solution can be decreased without changing the biopolymer solubility by adding aprotic solvents, such as DMSO and DMF. The choice of anti-solvent is also important. The strength and hardness of the hydrogel beads are dependent on the anti-solvent. Approximately 2- to 3-mm biopolymer-based hydrogel beads were obtained by this simple dropping method [8, 58, 71–73]. The size of the prepared biopolymer-based hydrogel beads could be decreased to less than 1 mm after the drying process, owing to significant shrinking of hydrogels during drying. Microscale beads are usually prepared by a sol–gel transition using an emulsion of oil/biopolymer solution with the help of a surfactant. The biopolymer solution is dispersed in oil with surfactants such as Span 80 and Tween 80 under vigorous stirring. Ultrasonication of this mixture can help to enhance the dispersity. Then, an anti-solvent is added to the emulsion of oil/biopolymer solution to regenerate the biopolymer. The size and regularity of the biopolymer microbeads are influenced by the ratio of oil to biopolymer solution, concentration of biopolymer, type of surfactant, and type of anti-solvent. The size of the obtained microbeads has ranged from 10 to 70 μm [54, 74, 75].

4.3 Drying Methods

Biopolymers can be shaped into forms such as molded shapes, films, fibers, and beads using ILs. Subsequently, the fabricated biopolymer-based materials can be changed to dried forms by various drying methods. The morphology and properties of these dried objects can be different, ranging from homogeneous materials with very high densities to porous foams with very low densities. To control the morphology of the dried materials, a well-controlled drying process is very important. A biopolymer-based hydrogel or organogel can be changed to an aerogel, cryogel, and xerogel by supercritical drying, freeze-drying, and ambient drying, respectively. It can also be changed to a highly dense structure with high volume shrinkage by heating or vacuum drying (Fig. 1). Buchtová and Budtova investigated the effects of the drying method on the density, porosity, specific surface area, and morphology of cellulose regenerated from [Emim][Ac]/DMSO [76]. It was possible to tune the properties of the porous cellulose materials by varying the cellulose concentration and drying method. Depending on the properties and morphology, each material may have a wide range of applications. For example, cellulose aerogels and cryogels can be used in the biomedical field as matrices for controlled drug release or in tissue engineering.

5 Biopolymer-Based Composite Materials

5.1 Biopolymer Blends

The blending of different biopolymers is a simple and attractive method to obtain new materials with new physicochemical properties. The properties of biopolymer blends can be controlled by changing the blending ratio of two or more biopolymers. In the preparation of biopolymer blends, it is important to consider whether all biopolymers can be co-dissolved in the used ILs, because the dissolution condition of each biopolymer is usually different [8]. The anti-solvent should also be carefully selected to regenerate all the biopolymers, owing to the different solubilities of biopolymers in the mixture of ILs and anti-solvent. Table 3 shows various biopolymer blends prepared using ILs.

5.1.1 Cellulose/Polysaccharide Blend Materials

Cellulose blended with polysaccharides such as agar [8], agarose [8], carrageenan [8, 21, 64, 78, 79], chitin [51], chitosan [8, 14, 79, 81, 82], guar gum [79], heparin [39], starch [64, 78, 79], and xanthan gum [61, 78] created novel materials to satisfy special applications. The properties of cellulose-based materials can be simply changed by blending various polysaccharides. For example, the surface charge of cellulose can be changed to positive or negative by adding chitosan or carrageenans, respectively. Mannans and starch can change the hydrophilicity of cellulose-based materials.

Cellulose/chitosan blend materials have been prepared in various forms, such as hydrogel beads, films, and fibers. Sun et al. prepared cellulose/chitosan beads by using [Bmim][Cl] as a solvent for biopolymers and water as an anti-solvent. Cellulose/chitosan beads showed good adsorption capacity for heavy metals such as copper and zinc [81]. Our group prepared electrospun fibers by co-dissolution of both cellulose and chitosan in [Emim][Ac] and regeneration with ethanol. A cellulose/chitosan blend fiber mat showed antibacterial effects for *Escherichia coli* [82]. Xiao et al. used [C_1mim][Cl] and [C_1mim][Cl]/[Hmim][Cl] (9:1) to dissolve cellulose and chitosan. The prepared cellulose/chitosan membranes and fibers retained the native structure of chitosan during the dissolution and regeneration processes [14]. Mundsinger et al. prepared cellulose/chitin blend fibers by

Table 3 Blended biopolymer materials

Major biopolymer	Minor biopolymer	Dissolution solvent	Dissolution condition	Anti-solvent	Product form (treatment condition)	Ref.
Agarose	Chitosan	[Bmim][Cl]	100°C	MeOH	Molded shape (dry, 70°C)	[77]
Agar	Cellulose	[Bmim][Cl] + DMSO	95°C	Water	Film (freeze-dry)	[62]
Agar	Starch	[Bmim][Cl] + DMSO	95°C	Water	Film (freeze-dry)	[62]
Agar	Zein protein	[Bmim][Cl] + DMSO	95°C	Water	Film (freeze-dry)	[62]
Cellulose	Carboxymethyl cellulose	[Emim][Ac]			Fiber	[78]
Cellulose	Carrageenan	[Emim][Ac]			Fiber	[78]
Cellulose	Carrageenan	[Emim][Ac] + DMF	80°C, 2 h	20% EtOH	Hydrogel bead	This work
Cellulose	Carrageenan	[Bmim][Cl]	100°C, 10 h		Molded shape (vacuum dry, 60°C)	[21]
Cellulose	Carrageenan	[Amim][Cl]	100°C, 8 h	Water	Hydrogel	[79]
Cellulose	Carrageenan	[Amim][Cl]	100°C, 12 h	Water	Hydrogel (dry/freeze-dry)	[64]
Cellulose	Chitin	[Emim][OPr]	110°C, 1 h	Water	Film (dry, RT)	[80]
Cellulose	Chitin	[Emim][OPr]	110°C, 1 h	60% sucrose	Fiber (dry, 120°C)	[80]
Cellulose	Chitin	[Bmim][Cl]	Microwave	Water	Bead (dry/freeze-dry)	[81]
Cellulose	Chitosan	[Emim][Ac]		EtOH	Electrospun fiber, Film	[82]
Cellulose	Chitosan	[Bmim][Cl]	Microwave	Water	Bead (dry/freeze-dry)	[81]
Cellulose	Chitosan	[Amim][Cl]	100°C, 8 h	Water	Hydrogel	[79]
Cellulose	Chitosan	[C_1mim][Cl] + [Hmim][Cl]		MeOH	Film (dry, 60°C), Fiber (dry, RT)	[14]
Cellulose	Collagen	[Emim][Ac]	60°C, 4 h	Water	Film (dry, RT)	[59]
Cellulose	Guar gum	[Amim][Cl]	100°C, 8 h	Water	Hydrogel	[79]

(continued)

Table 3 (continued)

Major biopolymer	Minor biopolymer	Dissolution solvent	Dissolution condition	Anti-solvent	Product form (treatment condition)	Ref.
Cellulose	Heparin	[Bmim][Cl] + [Emim][Ba]	Microwave	EtOH	Electrospun fiber (vacuum dry)	[39]
Cellulose	Lignin	[Emim][Ac]	90°C, 3 h	Water	Film (dry, RT, 3 h)	[68]
Cellulose	Lignin	[Emim][Ac]	80°C, 3 h	Water	Film	[50]
Cellulose	Lignin	[Emim][Ac] + DMF	70°C	EtOH	Electrospun fiber	[53]
Cellulose	Lignin	[Bmim][Cl]	130°C, 4 h	Aqueous EtOH	Film ($scCO_2$ dry)	[83]
Cellulose	Locust bean gum	[Emim][Ac]			Fiber	[78]
Cellulose	Silk	[Bmim][Cl]	90°C	MeOH	Film (dry, RT)	[84]
Cellulose	Silk	[Amim][Cl]	90°C	MeOH	Film	[85]
Cellulose	Silk	[Amim][Cl]	90°C	Water	Film	[85]
Cellulose	Soy protein isolate	[Amim][Cl]	80°C	Water	Film (vacuum dry, 50°C)	[86]
Cellulose	Starch	[Emim][Ac]			Fiber	[78]
Cellulose	Starch	[Amim][Cl]	100°C, 8 h	Water	Hydrogel	[79]
Cellulose	Starch	[Amim][Cl]	100°C, 12 h	Water	Hydrogel (dry/freeze-dry)	[64]
Cellulose	Starch, lignin	[Amim][Cl]	80°C	Water	Film (dry, RT)	[87]
Cellulose	Tragacanth gum	[Emim][Ac]			Fiber	[78]
Cellulose	Wool keratin	[Bmim][Cl]	100–130°C, 10 h	MeOH	Film, Fiber (dry, RT)	[49]
Cellulose	Xanthan gum	[Emim][Ac]			Fiber	[78]
Cellulose	Xanthan gum	[Bmim][Cl]	100°C, 9 h	Water	Hydrogel	[61]
Cellulose	Xanthan gum	[Bmim][Cl]	100°C, 9 h	EtOH	Film (dry, RT, 4 days)	[61]
Cellulose	Xylan	[Emim][Ac]			Fiber	[78]
Cellulose	Xylan	[Emim][Ac]	80°C, 3 h	Water	Film	[50]

Cellulose	Xyloglucan	[Emim][Ac]	100°C, 1 h	Water	Film (dry, 40°C)	[32]
Cellulose	Xylan, Lignin	[Emim][Ac]	80°C, 3 h	Water	Film	[50]
Cellulose	Xylan, Lignin	[Emim][Ac]	90°C, 3 h	Water	Film (dry, RT, 3 h)	[68]
Cellulose	Xylan, Lignin	[Bmim][Cl]	130°C, 4 h	Aqueous EtOH	Film ($scCO_2$ dry)	[83]
Cellulose	Xylan, Lignin, Chitosan	[Emim][Ac]	90°C, 3 h	Water	Film (dry, RT, 3 h)	[68]
Chitosan	Chondroitin sulfate	[Hmim][HSO_4]	100°C, 10 min	Water	Hydrogel	[15]
Chitosan	Silk	[Bmim][Ac]	95°C	EtOH, MeOH	Hydrogel	[88]
Collagen	Alginate	[mim][Ac]	40°C	3M $CaCl_2$	Film (dry, RT)	[45]
Collagen	Cellulose	[Bmim][Cl]	100°C	Water	Hydrogel bead	[58]

wet-spinning using [Emim][OPr] as a solvent and 60% sucrose solution as an anti-solvent [80]. The cellulose/chitin blend fiber can be used for medical or hygienic applications because it has enhanced water retention capacity compared to pure cellulose fiber.

Wendler et al. used cellulose derivatives, mannans, starch, carrageenan, and xanthan gum as blend polysaccharides for cellulose fiber as a matrix [78]. [Emim][Ac] is a powerful solvent for wet-spinning of cellulose/polysaccharide blend fibers. The fiber morphology, hydrophobicity/hydrophilicity, and surface properties are highly dependent on various polysaccharides blended with cellulose, compared with those of cellulose fiber. Liu and Huang prepared tea residue cellulose-based hydrogels blended with κ-carrageenan, chitosan, guar gum, and soluble starch by using [Amim][Cl] as a dissolving solvent [79]. Chitosan and guar gum improved the thermostability and mechanical characteristics of the biopolymer blend hydrogels, whereas κ-carrageenan and soluble starch improved the equilibrium swelling ratio and sodium salicylate loading. The cell compatibility and non-cytotoxicity of the biopolymer blend hydrogels were also demonstrated. The same group also prepared pineapple peel cellulose-based hydrogels blended with κ-carrageenan and soluble starch using the same procedure [64]. The addition of κ-carrageenan increased the hardness of the hydrogels, whereas soluble starch played an opposite role. Soluble starch increased the springiness and cohesiveness of the hydrogels.

The Linhardt group prepared cellulose/heparin blend materials to enhance blood compatibility, making it unnecessary to chemically couple heparin to cellulose. To fabricate cellulose/heparin blend materials, a binary mixture of [Bmim][Cl] and [Emim][Ba] as a solvent and ethanol as an anti-solvent were used. A mixture of heparin containing [Emim][Ba] and cellulose containing [Bmim][Cl] was prepared; the resulting solution could be fabricated into various shapes and forms, such as films, membranes, micro- or nanofibers, or any other molded shape [1, 39]. Binary mixtures of ILs are useful to prepare biopolymer blends because the mutual solubilities of ILs are usually very high.

5.1.2 Cellulose/Protein Blend Materials

Biomaterials made from natural polysaccharides and proteins have become increasingly popular in the biomedical field, owing to their good biocompatibility and tunable biodegradability [85]. In addition, combinations of polysaccharides and proteins may mimic the naturally occurring environment of certain tissues. However, the low miscibility of polysaccharides with proteins in the same solvent presents challenges. Recently, cellulose-based materials blended with proteins such as collagen [58, 59], silk [84, 85], soy protein isolate [86], and wool keratin [49] were prepared using ILs.

Zhang et al. prepared a cellulose/collagen blend film by using [Emim][Ac] as a solvent and water as an anti-solvent [59]. Collagen was successfully composited with cellulose and showed a higher denaturation temperature than that of native collagen. This was explained by the hydrogen bond interactions between collagen and cellulose. Wang et al. also prepared cellulose/collagen blend hydrogel beads using [Bmim][Cl] as a solvent and water as an anti-solvent [58].

DeFrates et al. prepared cellulose/*Bombyx mori* silk blend films using [Amim][Cl] as a solvent and methanol as an anti-solvent. [Amim][Cl] could effectively dissolve both cellulose and silk, as well as preserve the structure and integrity of silk. The thermal and physical properties of the cellulose/silk blend films could be tuned by varying the ratio of silk to cellulose. The blended films tended to be more thermally stable, which could be due to the presence of hydrophobic-hydrophobic or electrostatic interactions between the silk and cellulose [85]. Tian et al. also prepared a cellulose/silk blend film using [Bmim][Cl] as a solvent and methanol as an anti-solvent. The cellulose/silk blend film exhibited better strength and toughness than that of a regenerated cellulose film [84].

Wu et al. prepared a cellulose/soy protein isolate blend film using [Amim][Cl] as a solvent and water as an anti-solvent [86]. It was found that cellulose/soy protein isolate blends have excellent miscibility in all weight ratios of cellulose to soy protein isolate. With an increase in the cellulose content of blend films, the tensile strength, elongation at break, water resistance, and thermal stability of the blend films all increased. Xie et al. prepared a cellulose/wool keratin blend film by using [Bmim][Cl] as a solvent and methanol as an anti-solvent [49]. The thermal stability of the regenerated wool keratin was slightly superior to that of natural wool keratin fibers.

5.1.3 Wood Component-Based Blend Materials

Wood component-based biopolymer blends were prepared using [Emim][Ac], [Bmim][Cl], or [Amim][Cl] [32, 50, 53, 65, 68, 78, 83, 89]. Bendaoud et al. prepared cellulose/xyloglucan blends by dissolution in [Emim][Ac] and regeneration in water, taking inspiration from the outstanding mechanical properties of plant cell walls [32]. The cellulose/xyloglucan blend films were optically transparent and homogeneous at the micrometer scale. Enhanced mechanical properties were obtained by adding xyloglucan to cellulose, which was explained by the formation of a co-continuous nanostructure of a hard cellulose domain and soft xyloglucan domain. Wu et al. prepared cellulose/starch/lignin films using [Amim][Cl] as a solvent and water as an anti-solvent [87]. The contents of cellulose, lignin, and starch had a significant influence on the mechanical properties of blend films.

Our group prepared "synthetic wood" blend films containing three major components of natural wood by the co-dissolution of cellulose, xylan, and lignin in [Emim][Ac] and regeneration in water [50, 68]. The composition and characteristics of the synthetic wood films were highly controllable and predictable through variations of the concentration of each component in the wood solution (Fig. 2). The water vapor solubility of the wood films was increased when the xylan content was increased and the content of lignin was decreased. The biodegradability of cellulose in the synthetic wood films was enhanced when the xylan content was increased and the content of lignin was decreased. The synthetic wood films showed smoother

Fig. 2 Synthetic wood hydrogels (**a–d**) and thin films with various compositions (**e–h**): (**a**, **e**) cellulose only, (**b**, **f**) cellulose/xylan (5/5 wt% in [Emim][Ac]), (**c**, **g**) cellulose/lignin (5/5), (**d**, **h**) cellulose/xylan/lignin (5/3/2) [50]

surface textures, higher water resistance, and higher tensile strengths than the cellulose films formed by the same methods [50, 68]. The wood component solution containing cellulose, xylan, and lignin could also be electrospun to microsized fiber mats [89].

5.1.4 Other Biopolymer-Based Blend Materials

Most biopolymer blend materials prepared by using ILs were limited to the cellulose-based materials, owing to the excellent thermal and mechanical properties of cellulose. However, other biopolymer-based blend materials were also recently reported.

Trivedi et al. prepared agarose/chitosan blends using [Bmim][Cl] as a solvent and methanol as an anti-solvent [77]. The agarose/chitosan blends had good thermal and conformational stability, compatibility, and strong hydrogen bonding interactions between the agarose and chitosan. The blend materials could be decorated with Ag nanoparticles, and agarose/chitosan/Ag nanoparticles have the potential to be used in biotechnology and biomedical applications owing to the good antimicrobial activity of Ag nanoparticles. Shiamsuri et al. prepared agar/cellulose, agar/starch, and agar/zein protein blend aerogels [62]. Cellulose, starch, or zein protein was dissolved in [Bmim][Cl]/DMSO (1:1 weight ratio) and then blended with the agar solution in water, followed by gelation at low temperature and regeneration in water. This study showed the successful use of a [Bmim][Cl]/DMSO mixture as a solvent for the fabrication of agar-based blend materials.

Nunes et al. prepared chitosan/chondroitin sulfate hydrogels using [Hmim][HSO_4] as a solvent and water as an anti-solvent [15]. The preparation of chitosan/

chondroitin sulfate hydrogels was carried out in homogeneous [Hmim][HSO_4] solution for short dissolution times. The blend hydrogels showed excellent stabilities in a wide pH range, large swelling capacities, as well as a lack of cytotoxicity toward the normal VERO and diseased HT29 cells. The blend hydrogel could be applied in the medical, pharmaceutical, and environmental fields. Silva et al. prepared chitosan/silk hydrogels by using [Bmim][Ac] as a solvent and methanol as an anti-solvent [88]. Chitosan/silk blend hydrogels could be successfully prepared with short dissolution times of both biopolymers. The characteristics of hydrogel and the porous structure could be altered by changing the silk content.

Iqbal et al. prepared collagen/alginate hydrogels by using [Hmim][Ac] as a solvent and 3M $CaCl_2$ as an anti-solvent [45]. The thermal stability of the blend film increased with collagen loading. The fluid uptake ability of hydrogels improved to 40% with the addition of collagen. However, the tensile strength decreased with the addition of collagen. The blend hydrogels were evaluated as highly hemocompatible. The collagen/alginate blend hydrogels were suggested to be used for skin dressing.

5.2 *Cellulose-Based Composites*

Various biopolymer-based composite materials have been prepared using ILs. To prepare biopolymer-based composite materials, additives are simply dispersed in biopolymer solutions and then the biopolymer solutions containing additives are regenerated with an anti-solvent. Cellulose was well fabricated with inorganic materials, including montmorillonite [90, 91] and hydroxyapatite [92, 93], synthetic polymers such as polyvinyl alcohol (PVA) [64, 94], polyethylene glycol (PEG) [64], polypyrrole [95], and polyvinyl pyrrolidone (PVPP) [63], conductive carbon materials such as carbon nanotubes [40, 96, 97] and graphene oxide [98], and enzymes such as laccase [99], lipase [8, 71], and papain [100]. Table 4 shows some examples of cellulose-based composite materials prepared using ILs.

Mahmoudian et al. prepared cellulose/montmorillonite composite film using [Bmim][Cl] [90]. They also prepared cellulose/montmorillonite composite fiber via wet-spinning using [Emim][Ac] [91]. Considerable improvements in thermal stability and char yield were observed in composite fibers compared to pure regenerated cellulose fibers. The tensile strength and Young's modulus were also enhanced. X-ray diffraction and scanning electron microscopy (SEM) results showed good dispersion of montmorillonite as well as enhanced interactions between montmorillonite and the cellulose matrix. Tsioptsias et al. prepared porous cellulose/nanohydroxyapatite composite using [Bmim][Cl] as a solvent and poly (methyl methacrylate) as a porogen [92]. [Bmim][Cl] solution showed better dispersion of nanohydroxyapatite than *N,N*-dimethylacetamide/LiCl solution.

Hu et al. prepared cellulose/PVA, cellulose/PEG, and cellulose/PVA/PEG composite hydrogels using [Amim][Cl] as a solvent [64]. The addition of PVA or PEG decreased the hardness of the composite hydrogels, whereas the composite

Table 4 Cellulose-based composite materials

Additive material	Dissolution solvent	Dissolution condition	Product form (treatment condition)	Ref.
Montmorillonite	[Emim][Ac]	80°C, 4 h	Fiber (vacuum, 50°C, 3 h)	[91]
Montmorillonite	[Bmim][Cl]	85°C, 4 h	Film (30°C, 3 h)	[90]
Nanohydroxyapatite	[Bmim][Cl]	95°C, 5 h	Molded shape (40°C)	[92]
Nanohydroxyapatite	[Bmim][Cl]	95°C, 5 h	Film (vacuum, 60°C)	[93]
Polyvinyl alcohol	[Bmim][Ac]	70°C	Film (freeze-dry)	[94]
Polyvinyl alcohol	[Amim][Cl]	100°C, 12 h	Hydrogel (dry/freeze-dry)	[64]
Polyethylene glycol	[Amim][Cl]	100°C, 12 h	Hydrogel (dry/freeze-dry)	[64]
Polyethylene glycol, Polyvinyl alcohol	[Amim][Cl]	100°C, 12 h	Hydrogel (dry/freeze-dry)	[64]
Polypyrrole	[Bmim][Cl]	100°C, 12 h	Hydrogel	[95]
Polyvinyl pyrrolidone	[Amim][Cl]	100°C, 12 h	Hydrogel (50°C/ freeze-dry)	[63]
MWCNT	[Amim] [Cl] +DMSO (1:2)	70°C	Electrospun fiber (vacuum)	[96]
MWCNT	[Amim][Cl]	100°C, 45 min	Fiber	[97]
SWCNT	[Bmim][Br]	Sonication	Film (60°C, 3 h)	[40]
Graphene oxide	[Amim][Cl]	100°C, 5 h	Film	[98]
Laccase	[Bmim][Cl]	Microwave	Film	[99]
Lipase	[Emim][Ac]	80°C, 2 h	Hydrogel bead	[76]
Lipase	[Emim][Ac]	80°C, 1 h	Hydrogel bead	[71]
Papain, Fe_3O_4	[Amim][Cl]	100°C, 5 h	Hydrogel (freeze-dry)	[100]
Fe_3O_4	[Amim][Cl]	100°C, 5 h	Bead (freeze-dry)	[100]
Fe_3O_4	[Emim][Cl]	Microwave	Fiber (air dry)	[69]

hydrogels showed enhanced loading ratio for sodium salicylate. The addition of the PVA and PEG combination could lower the equilibrium swelling ratio of the hydrogels. Peng et al. also prepared cellulose/PVA composite hydrogel membranes using [Bmim][Ac] as a solvent [94]. The swelling and water-absorbing properties of the cellulose/PVA hydrogel were significantly improved. Liang et al. prepared a cellulose/polypyrrole composite hydrogel using [Bmim][Cl] [95]. The cellulose/polypyrrole composite hydrogel doped with *p*-toluenesulfonate showed high electrical conductivity. The equilibrium swelling ratio decreased with increasing cellulose content in the hydrogels. The cellulose/polypyrrole composite hydrogels exhibited significantly enhanced mechanical properties compared with cellulose hydrogel. Hu et al. prepared cellulose/PVPP composite hydrogels using [Amim][Cl] [63]. The addition of PVPP increased the mechanical properties and thermal stability of the composite hydrogels, as well as decreased the equilibrium swelling ratio.

Zhang et al. prepared cellulose/multiwalled carbon nanotube (MWCNT) composite fibers via dry-jet wet-spinning using [Amim][Cl] [97]. To enhance the dispersity of MWCNTs in the cellulose solution, the mixture of cellulose solution and MWCNT was ground in an agate mortar for 15 min. Well-dispersed and aligned MWCNTs were observed in the cellulose fibers. The composite fibers exhibited improved mechanical properties and thermal stability, owing to the good interaction between cellulose and MWCNTs. Chen et al. prepared bacterial cellulose (BC)/MWCNT nanofibers via electrospinning using [Amim][Cl]/DMSO (1:2, weight ratio) [96]. SEM and transmission electron microscopy results showed that the MWCNTs were embedded and well aligned along the fiber axis. The tensile strength and modulus of the BC/MWCNT nanofibers were approximately three times higher than those of the BC nanofibers. Li et al. prepared a cellulose/single-walled carbon nanotube (SWCNT) composite using [Bmim][Br] [40]. The obtained cellulose/SWCNT composite can be dispersed in water, forming a stable solution with excellent biocompatibility. It was also found that cellulose/SWCNT scaffolds could promote the growth of HeLa cells. Therefore, the composites have potential applications in biomaterial scaffolds and intracellular drug delivery systems. Liu et al. prepared tea residue cellulose/graphene oxide hydrogel using [Amim][Cl] [98]. The cellulose/graphene oxide hydrogels showed higher thermal stability and enhanced textural properties. As an adsorbent, the cellulose/graphene oxide hydrogels had enhanced adsorption capacity for methylene blue (MB).

The direct encapsulation of enzymes into cellulose-based materials is considered to be a very difficult task because cellulose-dissolving solvents fully inactivate enzymes. Turner et al. successfully encapsulated laccase as an active form in cellulose by dispersion of laccase in a cellulose solution using [Bmim][Cl] and regeneration with water [99]. The inactivation of laccase during the process was reduced by precoating the laccase with hydrophobic [Bmim][Tf_2N]. Our group entrapped lipase from *Candida rugosa* into cellulose hydrogel by using biocompatible [Emim][Ac], which can retain the activity of lipase [8]. Liu et al. entrapped papain into tea cellulose using [Amim][Cl] as a solvent [100]. The papain-embedded cellulose hydrogel was additionally coated with magnetic Fe_3O_4 and showed paramagnetic behavior.

Sun et al. prepared magnetically active cellulose/Fe_3O_4 composite fiber via a dry-jet wet-spinning process using [Emim][Cl] [69]. The fiber texture was found to be related to the overall magnetite concentration, cellulose concentration, and molecular weight of cellulose. It was found that increasing the degree of polymerization (DP) of cellulose and/or cellulose concentration resulted in more robust fibers; conversely, the addition of magnetite particles weakened the overall mechanical properties.

5.3 Blended Biopolymer-Based Composites

Composite materials based on biopolymer blends consisting of two or more biopolymers have many potential applications, owing to their biocompatibility, biodegradability, and tunable properties. Various biopolymer blends have been fabricated with lipase [8, 71], PEG [64, 68], magnetic particles [42, 54, 75], MWCNTs [68], porphyrin [66], poly(3-octylthiophene) [89], and charcoal [101]. Table 5 shows various biopolymer blends containing additives.

Our group successfully entrapped lipase into various cellulose/biopolymer blend hydrogels [8]. Lipase-entrapped cellulose/biopolymer blend hydrogel beads were prepared by the co-dissolution of biopolymers in [Emim][Ac] and dispersion of lipase in biopolymer solution followed by regeneration using water. Cellulose/biopolymer blend hydrogels proved to be good supports for the entrapment of enzymes and have many potential applications, including drug delivery, biosensors, biofuel cells, and tissue engineering.

Hu et al. prepared cellulose/carrageenan/PEG and cellulose/starch/PEG composite hydrogels [64]. The addition of the carrageenan and PEG combination or starch and PEG combination could markedly increase the equilibrium swelling ratio of the composite hydrogels, compared with cellulose hydrogel. Sodium salicylate-loaded cellulose/carrageenan/PEG composite hydrogel was useful for extending the fast release phase of sodium salicylate.

Peng et al. prepared cellulose/chitosan/Fe_3O_4 magnetic composite microspheres via a sol–gel transition of [Bmim][Cl] solution emulsified with vacuum pump oil using Tween 80 [75]. The composite microspheres exhibited an efficient adsorption capacity of Cu(II) from an aqueous solution, owing to the favorable chelating groups in the structure. These composite microspheres are expected to be a promising candidate for practical use in heavy metal ion removal. Liu et al. similarly prepared cellulose/chitosan/Fe_3O_4 magnetic composite microspheres via a sol–gel transition of [Bmim][Cl] solution emulsified with vacuum pump oil using Span 80 [54]. The composite microspheres were regular spheres with a mean diameter of approximately 10 μm. Glucose oxidase could be successfully immobilized on the composite microspheres via cross-linking using glutaraldehyde. The immobilized enzyme presented higher thermostability, wider range of optimal pH, and improved storage stability compared with the free enzyme. The immobilized enzyme could be successfully recovered by a magnetic field. These composite microspheres showed potential applications in the field of biocatalysis.

Recently, our group prepared various cellulose/biopolymer/Fe_3O_4 composite hydrogel microspheres such as cellulose/carrageenan, cellulose/chitosan, cellulose/lignin, and cellulose/soluble starch containing magnetic particles via sol–gel transition of an emulsified [Emim][Ac] solution (Fig. 3). The prepared composite microspheres showed regular spherical shapes and retained magnetic properties. The mean diameters of the composite microspheres ranged from 30 μm to 50 μm. Crystal violet (CV), methyl orange, bovine serum albumin (BSA), and pepsin were adsorbed on various cellulose/biopolymer/Fe_3O_4 composite microspheres to investigate the

Table 5 Blended biopolymer-based composite materials

Major biopolymer	Minor biopolymer	Additive material (dispersion method)	Dissolution solvent	Dissolution condition	Anti-solvent	Product form (treatment)	Ref.
Cellulose	Heparin	Charcoal	[Bmim] [Cl] + [Emim][Ba]		EtOH	Bead (dry, RT)	[101]
Cellulose	Xylan, Lignin	Porphyrin	[Emim][Ac]	70°C, 3 h	Water	Film	[66]
Cellulose	Carrageenan	Polyethylene glycol	[Amim][Cl]	100°C, 12 h	Water	Hydrogel (dry/freeze-dry)	[64]
Cellulose	Starch	Polyethylene glycol	[Amim][Cl]	100°C, 12 h	Water	Hydrogel (dry/freeze-dry)	[64]
Cellulose	Xylan, Lignin	Polyethylene glycol	[Emim][Ac]	90°C, 3 h	Water	Film (dry, RT, 3 h)	[68]
Cellulose	Carrageenan	Polyethylene glycol, polyvinyl alcohol	[Amim][Cl]	100°C, 12 h	Water	Hydrogel (dry/freeze-dry)	[64]
Cellulose	Carrageenan	Fe_3O_4 (sonication in IL)	[Emim][Ac]	100°C, 1 h	EtOH	Microsphere	This work
Cellulose	Chitosan	Fe_3O_4	[Bmim][Cl]	100°C, 0.5 h	EtOH	Microsphere	[75]
Cellulose	Chitosan	Fe_3O_4 (sonication in IL)	[Emim][Ac]	100°C, 1 h	EtOH	Microsphere	This work
Cellulose	Chitosan	Fe_3O_4	[Bmim][Cl]	100°C, 0.5 h	EtOH	Microsphere	[54]
Cellulose	Heparin	Fe_3O_4	[Emim][Ac]	80°C	Water	Electrospun fiber	[42]
Cellulose	Lignin	Fe_3O_4 (sonication in IL)	[Emim][Ac]	100°C, 1 h	EtOH	Microsphere	This work
Cellulose	Starch	Fe_3O_4 (sonication in IL)	[Emim][Ac]	100°C, 1 h	EtOH	Microsphere	This work
Cellulose	Xylan, Lignin	MWCNT (grind in IL)	[Emim][Ac]	90°C, 3 h	Water	Film (RT, 3 h)	[68]
Cellulose	Xylan, Lignin	Poly(3-octylthiophene)	[Emim][Ac]			Electrospun film	[89]

(continued)

Table 5 (continued)

Major biopolymer	Minor biopolymer	Additive material (dispersion method)	Dissolution solvent	Dissolution condition	Anti-solvent	Product form (treatment)	Ref.
Cellulose	Agar	Lipase	[Emim][Ac]	80°C	Water	Hydrogel bead	[8]
Cellulose	Agarose	Lipase	[Emim][Ac]	80°C, 12 h	Water	Hydrogel bead	[8]
Cellulose	Carrageenan	Lipase	[Emim][Ac]	80°C, 12 h	Water	Hydrogel bead	[8]
Cellulose	Chitosan	Lipase	[Emim][Ac]	80°C, 12 h	Water	Hydrogel bead	[8]
Cellulose	Lignin	Lipase	[Emim][Ac]	80°C, 1 h	Water	Hydrogel bead	[71]
Cellulose	Xylan	Lipase	[Emim][Ac]	80°C, 1 h	Water	Hydrogel bead	[71]
Cellulose	Xylan, Lignin	Lipase	[Emim][Ac]	80°C, 1 h	Water	Hydrogel bead	[71]

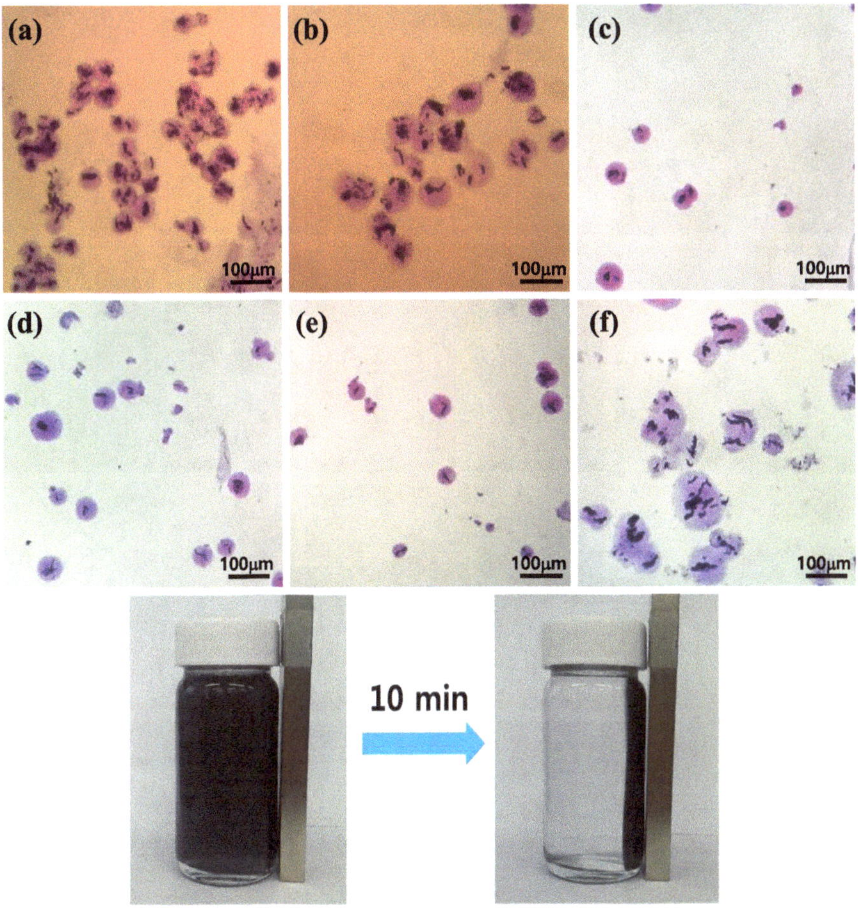

Fig. 3 Optical microscopy images and magnetic properties of the biopolymer-based magnetic microspheres: (**a**) cellulose, (**b**) cellulose/chitosan (3/1 wt% in [Emim][Ac]), (**c**) cellulose/carrageenan (3/1), (**d**) cellulose/alkali lignin (3/1), (**e**) cellulose/Organosolv lignin (3/1), (**f**) cellulose/starch (3/1) [unpublished work]

surface properties of the composites. The cellulose/chitosan, cellulose/carrageenan, cellulose/lignin, and cellulose/starch microspheres were more positively charged, negatively charged, hydrophobic, and hydrophilic, respectively, compared with cellulose microspheres. Furthermore, the lipase from *Candida rugosa* was successfully immobilized on various blended biopolymer-based microspheres. The immobilized lipase showed higher specific activity and thermal stability than free lipase. In particular, the thermal stabilities of the lipase immobilized on cellulose/starch/Fe_3O_4 and cellulose/alkali lignin/Fe_3O_4 were much higher than that of free lipase. Owing to the adjustable properties of blended biopolymer-based magnetic composite microspheres, they can be used as supports to adsorb various materials such as enzymes, biomedicines, drugs, and heavy metals. Moreover, their

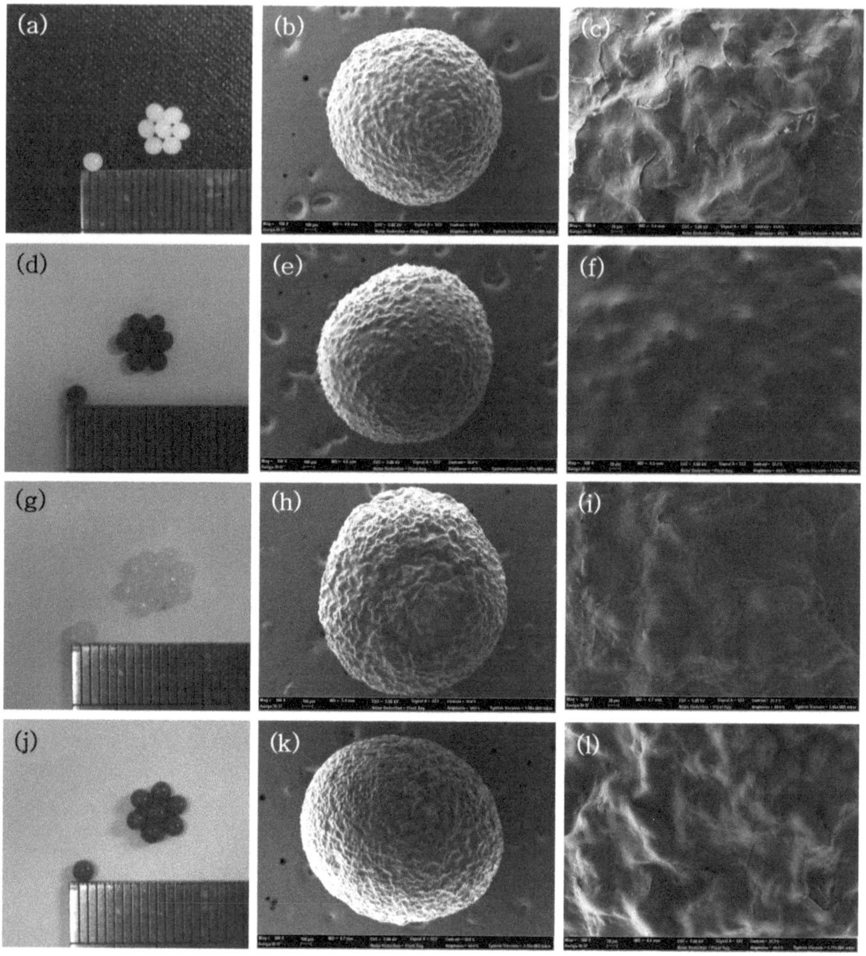

Fig. 4 Photographic images and scanning electron microscopy images of various synthetic wood hydrogel beads containing lipase: (**a–c**) cellulose bead, (**d–f**) cellulose/alkali lignin (4/2 wt% in [Emim][Ac]) bead, (**g–i**) cellulose/xylan (4/2) bead, (**j–l**) cellulose/xylan/alkali lignin (4/1/1) bead [71]

biocompatibility, biodegradability, and recoverability facilitate various applications in the chemical, biological, and environmental engineering fields.

Our group prepared various composite materials based on synthetic wood blends containing cellulose, xylan, and lignin by using [Emim][Ac] as a solvent to dissolve the three major wood components. Synthetic wood-based hydrogels have potential applications in biomedical fields owing to their low cost, biodegradability, and biocompatibility [72]. In particular, the controllable properties of synthetic wood hydrogels are useful for lipase immobilization. Lipase from *Candida rugosa* was entrapped in synthetic wood hydrogel beads by dissolving wood components with lipase in [Emim][Ac], followed by reconstitution with water (Fig. 4). The lipase entrapped in cellulose/xylan/lignin beads in a 5:3:2 ratio showed the highest activity; interestingly, this ratio is very similar to that in natural wood. The lipase entrapped in various synthetic wood hydrogel beads showed increased thermal and pH stability. The synthetic wood hydrogel beads can be used to immobilize various enzymes for applications in the biomedical, bioelectronic, and biocatalytic fields. Synthetic wood composite films were also fabricated with PEG and MWCNTs [68]. The surface of the synthetic wood/PEG composite film was stickier and more hydrophilic than the synthetic wood film, with an average contact angle of 32°, owing to the humectant properties of PEG. The synthetic wood/MWCNT maintained a very high resistance (megaohms), even at high loading, suggesting that it would make a suitable candidate for a flexible high-dielectric insulator composite. An integrated artificial photosynthetic system was also developed with a synthetic wood/porphyrin composite film [66]. Similar to the natural light harvesting by chloroplasts, porphyrins encapsulated in a synthetic wood matrix enabled the utilization of light energy toward the photochemical regeneration of NADH cofactors and enzymatic chemical synthesis. The porous synthetic wood composite provided a microenvironment for porphyrin encapsulation and also allowed effective photosynthesis because of the inclusion of the redox-active lignin component. Photoluminescent synthetic wood fibers were prepared by electrospinning [Emim][Ac] solution containing wood components and poly(3-octylthiophene) [89] (Fig. 5). These synthetic wood/poly(3-octylthiophene)

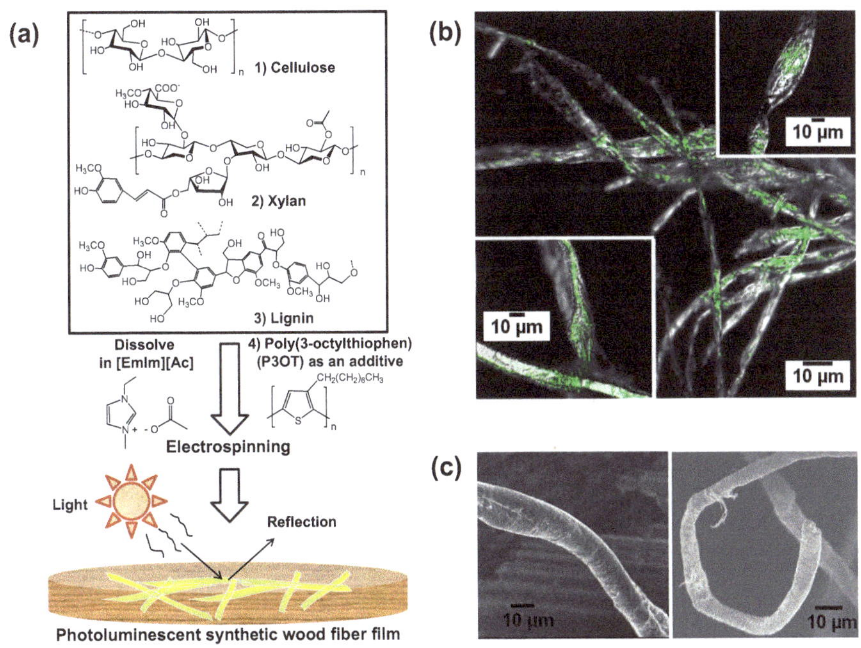

Fig. 5 Schematic representation of electrospinning from a synthetic wood/P3OT/IL solution (**a**), confocal laser scanning microscopy images of electrospun synthetic wood/P3OT fibers with an excitation wavelength of 488 nm (**b**), and scanning electron microscopy images of synthetic wood/P3OT composite fibers (**c**) [89]

composite fiber films showed orange-color emissions under ultraviolet irradiation and high intensity at 540 nm. The promise of utilizing these composite fiber films in a wide variety of applications can be realized in applications such as biopolymer-based solar cells, luminescent devices, chemical and optical sensors, and light-emitting diodes.

The Linhardt group prepared cellulose/heparin blend composites that showed anticoagulant activity for biomedical applications. The cellulose/heparin/charcoal composites were prepared using a mixture of [Bmim][Cl] and [Emim][Ba] to enhance the biocompatibility and blood compatibility of activated charcoal beads [102]. An field emission SEM image of cellulose/heparin/charcoal composites showed a smooth, uniformly coated surface with a large number of small and nanosized pores (Fig. 6). The surface morphology indicated that the composite is potentially capable of inhibiting the adsorption of proteins while permitting small drug molecules to adsorb to the underlying charcoal bead. The activated partial thromboplastin time results and adsorption efficiency of phenytoin compared to BSA showed that the coating of activated charcoal with cellulose/heparin could be useful for direct hemoperfusion to remove small-sized free-diluted and protein-

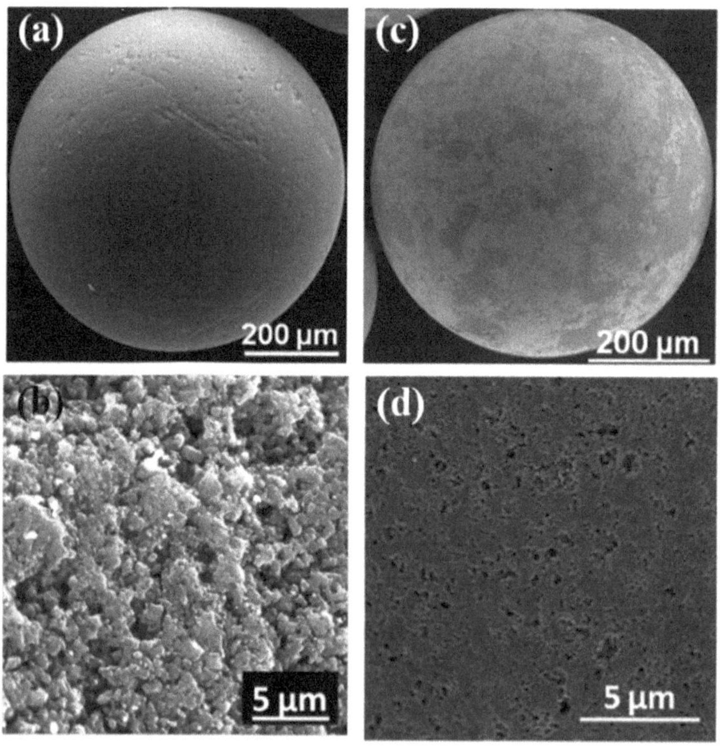

Fig. 6 SEM images of charcoal beads: (**a**, **b**) uncoated charcoal bead, (**c**, **d**) cellulose/heparin/charcoal composite bead [102]

bound toxins. The same group also prepared magnetically responsive heparin-immobilized cellulose nanofiber composites by wet-wet electrospinning using [Emim][Ac] [42]. Three types of nanofibers—cellulose/Fe_3O_4/heparin monofilament fiber, cellulose/Fe_3O_4 core–shell fiber with heparin covalently immobilized on the fiber surface, and cellulose/Fe_3O_4 core–shell fiber with heparin physically immobilized on the fiber surface—were prepared. The anticoagulant activity of immobilized heparin on the composite fiber surfaces was evaluated and confirmed by antifactor Xa and antifactor IIa assays.

6 Applications of Biopolymer-Based Composite Materials

Biopolymer-based materials prepared using ILs have many potential applications. Although several studies regarding possible applications have been reported, the untapped potential of biopolymer-based materials is still tremendous. Briefly, biopolymer-based materials can be used as adsorbents for heavy metals or dye removal, enzyme supports, carriers for drug delivery, scaffolds for tissue engineering, food packaging films, and basic materials for biomedical and bioelectrical applications.

6.1 Applications as Adsorbents

Heavy metals in wastewater may cause long-term risks to humans and the ecosystem [81]. As a method to remove toxic metal ions, the use of biopolymer-based adsorbents gained interest owing to their nontoxic and biodegradable properties. Cellulose/collagen hydrogel beads [58], tannin-immobilized cellulose/collagen beads [103], cellulose/chitosan beads [81], and magnetic cellulose/chitosan microspheres [75] were prepared with ILs and used for heavy metal ion adsorption. Blending cellulose with collagen enhanced the adsorption capacity for Cu(II) and cellulose/collagen beads showed good reusability after repeated adsorption and desorption. Magnetic cellulose/chitosan beads could be easily regenerated with HCl and reused repeatedly for Cu(II) adsorption up to five cycles. Cellulose/chitosan beads could also be used for the adsorption of Zn(II), Cr(VI), Ni(II), and Pb(II) ions.

Synthetic dyes such as MB, CV, and methyl orange are water pollutants from the textile dyeing industry and are not easily degraded. Synthetic dyes can contaminate water resources and thus harm aquatic life; through the food chain, they can then enter human bodies and cause adverse effects. Biopolymer-based adsorbents can be used to remove synthetic dyes. Cellulose/graphene oxide hydrogels [98] and cellulose/sepia ink hydrogels [38] have been prepared with ILs and used to adsorb MB. The MB adsorption capacity was enhanced by adding graphene oxide or sepia ink to cellulose. Our group prepared cellulose/carrageenan hydrogel beads. The CV adsorption was increased by increasing the carrageenan content in the

Fig. 7 Photographic images for the cellulose/carrageenan hydrogel beads of before adsorption (**a**) and after adsorption of CV (**b**) [unpublished work]

cellulose/carrageenan hydrogel beads. The adsorption capacity for CV of cellulose/carrageenan hydrogel beads was 200.8 mg/g, whereas that of cellulose hydrogel beads was only 0.5 mg/g (Fig. 7). This result shows that biopolymer blending can be effectively used to modify the properties of biopolymers.

Biopolymer-based materials have been used as enzyme supports, owing to their biocompatibility with enzymes. In particular, biopolymer-based composite hydrogels prepared using ILs can be effectively used for enzyme immobilization because the blending of biopolymers and additives can make new functional groups on the surface of the hydrogel without chemical modification. Our group prepared various cellulose/lignin hydrogel beads using [Emim][Ac] and the biopolymer blend hydrogels were used to immobilize lipase [73]. The lignin in the cellulose/lignin hydrogel beads enhanced the activity and stability of the adsorbed lipase as the lignin content increased. In particular, the interaction between lipase and the thiol group in alkali lignin improved the thermal stability of the immobilized lipase. BC/chitosan hydrogels and microcrystalline cellulose/chitosan hydrogels were also prepared for lipase immobilization [72]. The loaded content and activity of lipase were higher in the BC/chitosan hydrogel beads than the microcrystalline cellulose/chitosan hydrogel beads. The half-life of the lipase cross-linked on BC/chitosan beads was 22.7 times higher than that of free lipase. The Rogers group prepared cellulose/polyamine composite films and beads, which provided high loading of primary amines on the surface, allowing the direct conjugation of enzymes. These composites were used to immobilize laccase and lipase [104]. They also prepared cellulose/polyamidoamine films to immobilize laccase. The immobilization yield was increased by adding 1,3-phenylene diisocyanate as a linker [37].

6.2 Biomedical Applications

Biopolymer-based composites can be used as carriers for controlled drug release. The loaded drug content and drug release rate can be changed by blending biopolymers and preparing composites with additives. Hu et al. prepared a salicylate-loaded cellulose-based composite hydrogel [63]. It was found that PVPP addition could lower the salicylate load ratios of freeze-dried hydrogels but increase the salicylate load ratios of oven-dried hydrogels. Oven-drying processing and PVPP addition were promising approaches for slowing the salicylate release. The addition of PEG, PVA, carrageenan, and starch could increase the salicylate load ratio of the hydrogels [64]. The drug loading and releasing ratios were controllable by changing the additives [79].

Some biopolymers, such as chitosan and lignin, have been shown to have antibacterial properties. Blending these biopolymers can simply make biocompatible and antibacterial materials. For example, our group prepared cellulose/chitosan fiber mats via electrospinning [82]. The cellulose/chitosan fibers showed excellent antibacterial activity on *E. coli*. The loading of antibiotics in biopolymer-based composites is also a useful method for antibacterial materials. Tsioptsias and Panayiotou prepared a cellulose/nanohydroxyapatite composite and encapsulated amoxicillin [92]. Iqbal et al. prepared a collagen/alginate/gentamicin hydrogel, which showed antibacterial properties. The biocompatibility was assessed through hemolysis and the MTT assay on rat mesenchymal stem cells, which showed satisfactory results [45]. These biopolymer-based antibacterial materials can be used for wound dressing.

Biopolymers have been widely used for the fabrication of biomedical devices and tissue engineering scaffolds. Biopolymer-based composite materials are attracting great attention, owing to their unique properties such as biodegradability, biocompatibility, and mechanical properties. In particular, biopolymer blends have much potential because various physicochemical properties can be easily controlled with the help of ILs. Singh et al. prepared a chitin/MWCNT composite scaffold, which showed very good neuron adhesion as well as support for the synaptic function of neurons [105]. The addition of MWCNT into chitin improved the electrical conductivity and the assisted oxygen plasma treatment introduced more oxygen species onto the composite surface. The biocompatible and electrically conducting chitin/MWCNT composite scaffold can be used for tissue engineering of neurons. Silva et al. prepared various chitosan/silk hydrogels using [Bmim][Ac] [88]. The chitosan/silk hydrogel showed biocompatibility and no cytotoxicity during the culture of human dermal fibroblasts. In vitro assays demonstrated that these hydrogels supported the adhesion and growth of primary human dermal fibroblasts. Hence, chitosan/silk hydrogels may be promising biomaterials for skin tissue engineering. Zadegan et al. fabricated a cellulose–nanohydroxyapatite composite for bone tissue engineering applications [93]. Cellulose was reinforced with bioactive nanohydroxyapatite for replacement or healing of bones. The composite was biocompatible with no toxicity, and human osteoblast cells could attach to the surface of the composite. This composite can be used for bone tissue engineering.

6.3 Other Applications

Biopolymer-based films have tremendous potential to substitute for nonbiodegradable plastics in the food and packaging industry. Wu et al. prepared cellulose/soy protein isolate blend films [86]. The blend films have excellent miscibility in all weight ratios of cellulose to soy protein isolate. The addition of cellulose results in a decrease in the water vapor permeability. All films have high gas barrier capacities. Cellulose/starch/lignin films were also prepared [87]. The composite films showed relatively excellent mechanical properties in both the dry and wet states, owing to the mutual property supplementation of the different components. The composite films had good thermal stability and high gas barrier capacity. Mahmoudian et al. prepared cellulose/montmorillonite composite films [90]. The presence of montmorillonite enhanced the thermal stability and tensile strength. The composite films exhibited improved gas barrier properties and water absorption resistance compared to regenerated cellulose film.

The direct electrochemistry of redox proteins recently has gained considerable interest for use in constructing biofuel cells and biosensors. Various systems have been developed to enhance direct electron transfer and protein stability because the electroactive centers of redox proteins are embedded in the molecules. For this purpose, host materials should be able to immobilize proteins well and promote the direct electrochemistry. Biopolymers can provide a favorable microenvironment for redox proteins and enzymes to fabricate excellent biosensors. In addition, ILs allow the efficient direct electron transfer of various proteins. Therefore, biopolymer/ILs composite systems may represent unique materials that can open up new opportunities for direct electrochemistry. A more detailed discussion of the application of ILs in bioelectrochemistry can be found in an earlier review [106].

7 Conclusions and Prospects

Biopolymer-based materials have gained considerable interest for their inherent biocompatibility and biodegradability. However, the processes to prepare biopolymer-based materials have been limited because of the low solubility of biopolymers in general organic solvents. Recently, the high dissolving power of ILs for various biopolymers has opened up new paths to fabricate biopolymer-based composite materials. The dissolution of biopolymers using ILs and regeneration of biopolymers via sol–gel transition are very simple processes to perform, which can be used to fabricate biopolymer-based composites of various shapes. As new ILs that can dissolve various biopolymers are developed, innovative materials with novel structures, physical properties, and chemical functions will be developed in the future. In particular, ILs are very useful solvents to make biopolymer blends because ILs can co-dissolve two or more biopolymers. A tremendous number of combinations of biopolymer blends can be used to prepare materials with tightly controlled properties. Blended biopolymer-based composites have many potential applications, including as adsorbents, enzyme supports, scaffolds for tissue engineering, carriers for drug delivery, food packaging materials, and basic materials for biomedical devices and bioelectronics.

Acknowledgements This work was supported by the Basic Science Research Program through the National Research Foundation of Korea (NRF) funded by the Ministry of Education (2015R1D1A1A01060206 and 2018R1D1A1B07050163). This research was also supported by the Technology Development Program to Solve Climate Changes of the NRF funded by the Ministry of Science and ICT (2017M1A2A2087627 and 2017M1A2A2087647).

References

1. Lee SH, Miyauchi M, Dordick JS, Linhardt RJ (2010) Preparation of biopolymer-based materials using ionic liquids for the biomedical applications. In: Malhotra SV (ed) Ionic liquids application: pharmaceutical, therapeutics, and biotechnology. ACS symposium series, vol 1038. American Chemical Society, Washington, DC, pp 115–134
2. Lee SH, Doherty TV, Linhardt RJ, Dordick JS (2009) Ionic liquid-mediated selective extraction of lignin from wood leading to enhanced enzymatic cellulose hydrolysis. Biotechnol Bioeng 102:1368–1376
3. Freemantle M (1998) Designer solvents - ionic liquids may boost clean technology development. Chem Eng News 76:32–37
4. Swatloski RP, Spear SK, Holbrey JD, Rogers RD (2002) Dissolution of cellose with ionic liquids. J Am Chem Soc 124:4974–7975
5. Yang X, Wang Q, Yu H (2014) Dissolution and regeneration of biopolymers in ionic liquids. Russ Chem Bull 63:555–559
6. Yang X, Qiao C, Li Y, Li T (2016) Dissolution and resourcfulization of biopolymers in ionic liquids. React Funct Polym 100:181–190
7. Chang C, Zhang L (2011) Cellulose-based hydrogels: present status and application prospects. Carbohydr Polym 84:40–53
8. Kim MH, An S, Won K, Kim HJ, Lee SH (2012) Entrapment of enzymes into cellulose–biopolymer composite hydrogel beads using biocompatible ionic liquid. J Mol Catal B-Enzym 75:68–72
9. Wang H, Gurau G, Rogers RD (2012) Ionic liquid processing of cellulose. Chem Soc Rev 41:1519–1537
10. van Osch DJGP, Kollau LJBM, van den Bruinhorst A, Asikainen S, Rocha MAA, Kroon MC (2017) Ionic liquids and deep eutectic solvents for lignocellulosic biomass fractionation. Phys Chem Chem Phys 19:2636–2665
11. Oh DX, Shin S, Lim C, Hwang DS (2013) Dopamine-mediated sclerotization of regenerated chitin in ionic liquid. Materials 6:3826–3939
12. Wu Y, Sasaki T, Irie S, Sakurai K (2008) A novel biomass-ionic liquid platform for the utilization of native chitin. Polymer 49:2321–2327
13. Heckel T, Konieczna DD, Wilhelm R (2013) An ionic liquid solution of chitosan as organocatalyst. Catalysts 3:914–921
14. Xiao W, Chen Q, Wu Y, Wu T, Dai L (2011) Dissolution and blending of chitosan using 1,3-dimethylimidazolium chloride and 1-H-3-methylimidazolium chloride binary ionic liquid solvent. Carbohydr Polym 83:233–238

15. Nunes CS, Rufato KB, Souza PR, de Almeida EAMS, da Silva MJV, Scariot DB, Nakamura CV, Rosa FA, Martins AF, Muniz EC (2017) Chitosan/chondroitin sulfate hydrogels prepared in [Hmim][HSO_4] ionic liquid. Carbohydr Polym 170:99–106
16. Murugesan S, Wiencek JM, Ren RX, Linhardt RJ (2006) Benzoate-based room temperature ionic liquids—thermal properties and glycosaminoglycan dissolution. Carbohydr Polym 63:268–271
17. Mantz RA, Fox DM, Green III JM, Fylstra PA, De Long HC, Trulove PC (2007) Dissolution of biopolymers using ionic liquids. Zeitschrift für Naturforschung A 62:275–280
18. Xu Q, Wang Q, Liu L (2008) Ring-opening graft polymerization of L-lactide onto starch granules in an ionic liquid. J Appl Polym Sci 107:2704–2713
19. Biswas A, Shogren RL, Stevenson DG, Willett JL, Bhowmik PK (2006) Ionic liquids as solvents for biopolymers: acylation of starch and zein protein. Carbohydr Polym 66:546–550
20. Singh T, Trivedi TJ, Kumar A (2010) Dissolution, regeneration and ion-gel formation of agarose in room-temperature ionic liquids. Green Chem 12:1029–1035
21. Prasad K, Kaneko Y, Kadokawa J (2009) Novel gelling systems of κ-, ι- and λ-carrageenans and their composite gels with cellulose using ionic liquid. Macromol Biosci 9:376–382
22. Lacroix C, Sultan E, Fleury E, Charlot A (2012) Functional galactomannan platform from convenient esterification in imidazolium-based ionic liquids. Polym Chem 3:538–546
23. Sharma M, Mondal D, Mukesh C, Prasad K (2014) Preparation of tamarind gum based soft ion gels having thixotropic properties. Carbohydr Polym 102:467–471
24. Izawa H, Kadokawa J (2010) Preparation and characterizations of functional ionic liquid-gel and hydrogel materials of xanthan gum. J Mater Chem 20:5235–5241
25. Meng Z, Zheng X, Tang K, Liu J, Ma Z, Zhao Q (2012) Dissolution and regeneration of collagen fibers using ionic liquid. Int J Biol Macromol 51:440–448
26. Hu Y, Liu L, Dan W, Dan N, Gu Z (2013) Evaluation of 1-ethyl-3-methylimidazolium acetate based ionic liquid systems as a suitable solvent for collagen. J Appl Polym Sci 15:2245–2256
27. Qiao C, Li T, Zhang L, Yang X, Xu J (2014) Rheology and viscosity scaling of gelatin/1-allyl-3-methylimidazolium chloride solution. Korea-Aust Rheol J 26:169–175
28. Phillips DM, Drummy LF, Conrady DG, Fox DM, Naik RR, Stone MO, Trulove PC, De Long HC, Mantz RA (2004) Dissolution and regeneration of *Bombyx mori* silk fibroin using ionic liquids. J Am Chem Soc 126:14350–14351
29. Silva SS, Popa EG, Gomes ME, Oliveira MB, Nayak S, Subia B, Mano JF, Kundu SC, Reis RL (2013) Silk hydrogels from non-mulberry and mulberry silkworm cocoons processed with ionic liquids. Acta Biomater 9:8972–8982
30. Idris A, Vijayaraghavan R, Rana UA, Pattia AF, MacFarlane DR (2014) Dissolution and regeneration of wool keratin in ionic liquids. Green Chem 16:2857–2864
31. Castro MC, Arce A, Soto A, Rodríguez H (2015) Influence of methanol on the dissolution of lignocellulose biopolymers with the ionic liquid 1-ethyl-3-methylimidazolium acetate. Ind Eng Chem Res 54:9605–9614
32. Bendaoud A, Kehrbusch R, Baranov A, Duchemin B, Maigret JE, Falourd X, Staiger MP, Cathala B, Lourdin D, Leroy E (2017) Nanostructured cellulose-xyloglucan blends via ionic liquid/water processing. Carbohydr Polym 168:163–172
33. Achinivu EC, Howard RM, Li G, Gracz H, Henderson WA (2014) Lignin extraction from biomass with protic ionic liquids. Green Chem 16:1114–1119
34. Pu Y, Jiang N, Ragauskas AJ (2007) Ionic liquid as a green solvent for lignin. J Wood Chem Technol 27:23–33
35. Xu A, Wang J, Wang H (2010) Effects of anionic structure and lithium salts addition on the dissolution of cellulose in 1-butyl-3-methylimidazolium-based ionic liquid solvent systems. Green Chem 12:268–275
36. Lu B, Xu A, Wang J (2014) Cation does matter: how cationic structure affects the dissolution of cellulose in ionic liquids. Green Chem 16:1326–1335

37. Bagheri M, Rodríguez H, Swatloski RP, Spear SK, Daly DT, Rogers RD (2008) Ionic liquid-based preparation of cellulose-dendrimer films as solid supports for enzyme immobilization. Biomacromolecules 9:381–387
38. Dai H, Huang H (2016) Modified pineapple peel cellulose hydrogels embedded with sepia ink for effective removal of methylene blue. Carbohydr Polym 148:1–10
39. Viswanathan G, Murugesan S, Pushparaj V, Nalamasu O, Ajayan PM, Linhardt RJ (2006) Preparation of biopolymer fibers by electrospinning from room temperature ionic liquids. Biomacromolecules 7:415–418
40. Li L, Meng L, Zhang X, Fu C, Lu Q (2009) The ionic liquid-associated synthesis of a cellulose/SWCNT complex and its remarkable biocompatibility. J Mater Chem 19:3612–3617
41. Mikkola J, Kirilin A, Tuuf J, Pranovich A, Holmbom B, Kustov LM, Murzin DY, Salmi T (2007) Ultrasound enhancement of cellulose processing in ionic liquids: from dissolution towards functionalization. Greem Chem 9:1229–1237
42. Hou L, Udangawa WMRN, Pochiraju A, Dong W, Zheng Y, Linhardt RJ, Simmons TJ (2016) Synthesis of heparin-immobilized, magnetically addressable cellulose nanofibers for biomedical applications. ACS Biomater Sci Eng 2:1905–1913
43. Xu Q, Kennedy JF, Liu L (2008) An ionic liquid as reaction media in the ring opening graft polymerization of e-caprolactone onto starch granules. Carbohydr Polym 72:113–121
44. Trivedi TJ, Bhattacharjya D, Yu J, Kumar A (2015) Functionalized agarose self-healing ionogels suitable for supercapacitors. ChemSusChem 8:3294–3303
45. Iqbal B, Muhammad N, Jamal A, Ahmad P, Khan ZUH, Rahim A, Khan AS, Gonfa G, Iqbal J, Rehman IU (2017) An application of ionic liquid for preparation of homogeneous collagen and alginate hydrogels for skin dressing. J Mol Liq 243:720–725
46. Zhang L, Qiao C, Ding Y, Cheng J, Li T (2012) Rheological behavior of gelatin/1-allyl-3-methylimidazolium chloride solutions. J Macromol Sci Part B-Phys 51:747–755
47. Gupta MK, Khokhar SK, Phillips DM, Sowards LA, Drummy LF, Kadakia MP, Naik RR (2007) Patterned silk films cast from ionic liquid solubilized fibroin as scaffolds for cell growth. Langmuir 23:1315–1319
48. Li R, Wang D (2013) Preparation of regenerated wool keratin films from wool keratin–ionic liquid solutions. J Appl Polym Sci 127:2648–2653
49. Xie H, Li S, Zhang S (2005) Ionic liquids as novel solvents for the dissolution and blending of wool keratin fibers. Green Chem 7:606–608
50. Kim SH, Kim MH, Kim JH, Park S, Kim H, Won K, Lee SH (2015) Preparation of artificial wood films with controlled biodegradability. J Appl Polym Sci 132:42109
51. Xu A, Zhang Y, Zhao Y, Wang J (2013) Cellulose dissolution at ambient temperature: role of preferential solvation of cations of ionic liquids by a cosolvent. Carbohydr Polym 92:540–544
52. Zhao Y, Liu X, Wang J, Zhang S (2013) Insight into the cosolvent effect of cellulose dissolution in imidazolium-based ionic liquid systems. J Phys Chem B 117:9042–9049
53. Ahn Y, Kang Y, Park B, Ku MK, Lee SH, Kim H (2013) Influence of lignin on rheological behaviors and electrospinning of polysaccharide solution. J Appl Polym Sci 131:40031
54. Liu Z, Wang H, Li B, Liu C, Jiang Y, Yu G, Mu X (2012) Biocompatible magnetic cellulose–chitosan hybrid gel microspheres reconstituted from ionic liquids for enzyme immobilization. J Mater Chem 22:15085–15091
55. Elhi F, Aid T, Koel M (2016) Ionic liquids as solvents for making composite materials from cellulose. Proc Est Acad Sci 65:255–266
56. Muginova SV, Myasnikova DA, Kazarian SG, Shekhovtsova TN (2016) Evaluation of novel applications of cellulose hydrogel filmsreconstituted from acetate and chloride of1-butyl-3-methylimidazolium by comparing their optical,mechanical, and adsorption properties. Mater Today Commun 8:108–117
57. Johns MA, Bernardes A, De Azevêdo ER, Guimarães FEG, Lowe JP, Gale EM, Polikarpov I, Scott JL, Sharma RI (2017) On the subtle tuneability of cellulose hydrogels: implications for binding of biomolecules demonstrated for CBM 1. J Mater Chem B 5:3879–3887

58. Wang J, Wei L, Ma Y, Li K, Li M, Yu Y, Wang L, Qiu H (2013) Collagen/cellulose hydrogel beads reconstituted from ionic liquid solution for Cu(II) adsorption. Carbohydr Polym 98:736–743
59. Zhang M, Ding C, Chen L, Huang L (2014) The preparation of cellulose/collagen composite films using 1-ethyl-3-methylimidazolium acetate as a solvent. Bioresources 9:756–771
60. Marr PC, Marr AC (2016) Ionic liquid gel materials: applications in green and sustainable chemistry. Green Chem 18:105–128
61. Setoyama M, Yamamoto K, Kadokawa J (2014) Preparation of cellulose/xanthan gum composite films and hydrogels using ionic liquid. J Polym Environ 22:298–303
62. Shamsuri AA, Abdullah DK, Dail R (2012) Fabrication of agar/biopolymer blend aerogels in ionic liquid and co-solvent mixture. Cell Chem Technol 46:45–52
63. Hu X, Hu K, Zeng L, Zhao M, Huang H (2010) Hydrogels prepared from pineapple peel cellulose using ionic liquid and their characterization and primary sodium salicylate release study. Carbohydr Polym 82:62–68
64. Hu X, Wang J, Huang H (2013) Impacts of some macromolecules on the characteristics of hydrogels prepared from pineapple peel cellulose using ionic liquid. Cellulose 20:2923–2933
65. Li J, Lu Y, Yang D, Sun Q, Liu Y, Zhao H (2011) Lignocellulose aerogel from wood-ionic liquid solution (1-allyl-3-methylimidazolium chloride) under freezing and thawing conditions. Biomacromolecules 12:1860–1867
66. Lee M, Kim JH, Lee SH, Lee SH, Park CB (2011) Biomimetic artificial photosynthesis by light-harvesting synthetic wood. ChemSusChem 4:581–586
67. Kimura M, Shinohara Y, Takizawa J, Ren S, Sagisaka K, Lin Y, Hattori Y, Hinestroza JP (2015) Versatile molding process for tough cellulose hydrogel materials. Sci Rep 5:16266
68. Simmons TJ, Lee SH, Miao J, Miyauchi M, Park T, Bale SS, Pangule R, Bult J, Martin JG, Dordick JS, Linhardt RJ (2011) Preparation of synthetic wood composites using ionic liquids. Wood Sci Technol 45:719–733
69. Sun N, Swatloski RP, Maxim ML, Rahman M, Harland AG, Haque A, Spear SK, Daly DT, Rogers RD (2008) Magnetite-embedded cellulose fibers prepared from ionic liquid. J Mater Chem 18:283–290
70. Ahn Y, Hu D, Hong JH, Lee SH, Kim HJ, Kim H (2012) Effect of co-solvent on the spinnability and properties of electrospun cellulose nanofiber. Carbohydr Polym 89:340–345
71. Park S, Kim SH, Won K, Choi JW, Kim YH, Kim HJ, Yang Y, Lee SH (2015) Wood mimetic hydrogel beads for enzyme immobilization. Carbohydr Polym 115:223–229
72. Kim HJ, Jin JN, Kan E, Kim KJ, Lee SH (2017) Bacterial cellulose-chitosan composite hydrogel beads for enzyme immobilization. Biotechnol Bioprocess Eng 22:89–94
73. Park S, Kim SH, Kim JH, Yu H, Kim HJ, Yang Y, Kim H, Kim YH, Ha SH, Lee SH (2015) Application of cellulose/lignin hydrogel beads as novel supports for immobilizing lipase. J Mol Catal B-Enzym 119:33–39
74. Du K, Yan M, Wang Q, Song H (2010) Preparation and characterization of novel macroporous cellulose beads regenerated from ionic liquid for fast chromatography. J Chromatogr A 1217:1298–1304
75. Peng S, Meng H, Quyang Y, Chang J (2014) Nanoporous magnetic cellulose−chitosan composite microspheres: preparation, characterization, and application for Cu(II) adsorption. Ind Eng Chem Res 53:2106–2113
76. Wu R, Wang X, Li F, Li H, Wang Y (2009) Green composite films prepared from cellulose, starch and lignin in room-temperature ionic liquid. Bioresour Technol 100:2569–2574
77. Liu Z, Huang H (2016) Preparation and characterization of cellulose composite hydrogels from tea residue and carbohydrate additives. Carbohydr Polym 147:226–233
78. Buchtová N, Budtova T (2016) Cellulose aero-, cryo- and xerogels: towards understanding of morphology control. Cellulose 23:2585–2595
79. Sun X, Peng B, Ji Y, Chen J, Li D (2009) Chitosan(chitin)/cellulose composite biosorbents prepared using ionic liquid for heavy metal ions adsorption. AICHE J 55:2062–2069

80. Park T, Jung YJ, Choi S, Park H, Kim H, Kim E, Lee SH, Kim JH (2011) Native chitosan/cellulose composite fibers from an ionic liquid via electrospinning. Macromol Res 19:213–215
81. Wendler F, Persin Z, Stana-Kleinschek K, Reischl M, Ribitsch V, Bohn A, Fink H, Meister F (2011) Morphology of polysaccharide blend fibers shaped from NaOH, N-methylmorpholine-N-oxide and 1-ethyl-3-methylimidazolium acetate. Cellulose 18:1165–1178
82. Wu R, Wang X, Wang Y, Blan X, Li F (2009) Cellulose/soy protein isolate blend films prepared via room-temperature ionic liquid. Ind Eng Chem Res 48:7132–7136
83. DeFrates K, Markiewicz T, Callaway K, Xue Y, Stanton J, la Cruz DS, Hu X (2017) Structure–property relationships of Thai silk–microcrystalline cellulose biocomposite materials fabricated from ionic liquid. Int J Biol Macromol 104:919–928
84. Mundsinger K, Müller A, Beyer R, Hermanutz F, Buchmeiser MR (2015) Multifilament cellulose/chitin blend yarn spun from ionic liquids. Carbohydr Polym 131:34–40
85. Tian D, Li T, Zhang R, Wu Q, Chen T, Sun P, Ramamoorthy A (2017) Conformations and intermolecular interactions in cellulose/silk fibroin blend films: a solid-state NMR perspective. J Phys Chem B 121:6108–6116
86. Park T, Jung YJ, Park H, Choi S, Kim E, Lee SH, Kim JH (2011) Photoluminescent synthetic wood fibers from an ionic liquid via electrospinning. Macromol Res 19:317–320
87. Trivedi TJ, Rao KS, Kumar A (2014) Facile preparation of agarose–chitosan hybrid materials and nanocomposite ionogels using an ionic liquid via dissolution, regeneration and sol–gel transition. Green Chem 16:320–330
88. Aaltonen O, Jauhiainen O (2009) The preparation of lignocellulosic aerogels from ionic liquid solutions. Carbohydr Polym 75:125–129
89. Silva SS, Santos TC, Cerqueira MT, Marques AP, Reys LL, Silva TH, Caridade SG, Mano JF, Reis RL (2012) The use of ionic liquids in the processing of chitosan/silk hydrogels for biomedical applications. Green Chem 14:1463–1470
90. Mahmoudian S, Wahit MU, Ismail AF, Yussuf AA (2012) Preparation of regenerated cellulose/montmorillonite nanocomposite films via ionic liquids. Carbohydr Polym 88:1251–1257
91. Mahmoudian S, Wahit MU, Ismail AF, Balakrishnan H, Imran M (2015) Bionanocomposite fibers based on cellulose and montmorillonite using ionic liquid 1-ethyl-3-methylimidazolium acetate. J Mater Sci 50:1228–1236
92. Tsioptsias C, Panayiotou C (2008) Preparation of cellulose-nanohydroxyapatite composite scaffolds from ionic liquid solutions. Carbohydr Polym 74:99–105
93. Zadegan S, Hossainalipour M, Ghassai H, Rezaie HR, Naimi-Jamal MR (2010) Synthesis of cellulose–nanohydroxyapatite composite in 1-n-butyl-3-methylimidazolium chloride. Ceram Int 36:2375–2381
94. Peng H, Wang S, Xu H, Hao X (2017) Preparation, properties and formation mechanism of cellulose/polyvinyl alcohol bio-composite hydrogel membranes. New J Chem 41:6564–6573
95. Liang X, Qu B, Li J, Xiao H, He B, Qian L (2015) Preparation of cellulose-based conductive hydrogels with ionic liquid. React Funct Polym 86:1–6
96. Chen P, Yun YS, Bak H, Cho SY, Jin H (2010) Multiwalled carbon nanotubes-embedded electrospun bacterial cellulose nanofibers. Mol Cryst Liq Cryst 519:169–178
97. Zhang H, Wang Z, Zhang Z, Wu J, Zhang J, He J (2007) Regenerated-cellulose/multiwalled-carbon-nanotube composite fibers with enhanced mechanical properties prepared with the ionic liquid 1-allyl-3-methylimidazolium chloride. Adv Mater 19:698–704
98. Liu Z, Li D, Dai H, Huang H (2017) Enhanced properties of tea residue cellulose hydrogels by addition of graphene oxide. J Mol Liq 244:110–116
99. Turner MB, Spear SK, Holbrey JD, Rogers RD (2004) Production of bioactive cellulose films reconstituted from ionic liquids. Biomol Ther 5:1379–1384
100. Liu Z, Li D, Dai H, Huang H (2017) Preparation and characterization of papain embedded in magnetic cellulose hydrogels prepared from tea residue. J Mol Liq 232:449–456
101. Park T, Govindaiah P, Hwang T, Kim E, Choi S, Kim JH (2011) Biocompatible charcoal composites prepared by ionic liquids for drug detoxification. Macromol Res 19:734–738

102. Park T, Lee S, Simmons TJ, Martin JG, Mousa SA, Snezhkova EA, Sarnatskaya VV, Nikolaev VG, Linhardt RJ (2008) Heparin–cellulose–charcoal composites for drug detoxification prepared using room temperature ionic liquids. Chem Commun (40):5022–5024
103. Zhang M, Ding C, Chen L, Huang L, Yang H (2014) Tannin-immobilized collagen/cellulose bead as an effective adsorbent for Cu(II) in aqueous solutions. J Biobaased Mater Bioenergy 8:610–617
104. Turner MB, Spear SK, Holbrey JD, Daly DT, Rogers RD (2005) Ionic liquid-reconstituted cellulose composites as solid support matrices for biocatalyst immobilization. Biomacromolecules 6:2497–2502
105. Singh N, Chen J, Koziol KK, Hallam KR, Janas D, Patil AJ, Strachan A, Hanley J, Rahatekar S (2016) Chitin and carbon nanotube composites as biocompatible scaffolds for neuron growth. Nanoscale 8:8288–8299
106. Fujita K, Murata K, Masuda M, Nakamura N, Ohno H (2012) Ionic liquids designed for advanced applications in bioelectrochemistry. RSC Adv 2:4018–4030

Adv Biochem Eng Biotechnol (2019) 168: 177–198
DOI: 10.1007/10_2018_63

Published online: 10 May 2018

Advances in Processing Chitin as a Promising Biomaterial from Ionic Liquids

Julia L. Shamshina, Oleksandra Zavgorodnya, and Robin D. Rogers

Contents

Abstract Chitin isolated through microwave-assisted dissolution using ionic liquids is a high molecular weight (MW) polymer that can be manufactured into materials of different architectures (e.g., fibers, films, microspheres, nanostructured materials) to be used as wound care dressings, drug delivery devices, scaffolds, etc. However,

J. L. Shamshina (✉) and O. Zavgorodnya
Mari Signum, Mid-Atlantic, Rockville, MD, USA
e-mail: Julia.Shamshina@MariSignum.com

R. D. Rogers (✉)
525 Solutions, Inc., Tuscaloosa, AL, USA
e-mail: Robin.Rogers@525Solutions.com

because of differences from traditional isolation methods and, thus, differences in polymer length and degree of deacetylation, it could exhibit bio-related properties that differ from those of traditionally 'pulped' chitin. Here we present the initial assessments of bio-related chitin properties in order to provide a useful scientific basis for clinical applications: biocompatibility, cytotoxicity (intracutaneous reactivity), wound healing efficacy, histological evaluation of the wounds treated with chitin dressing, and antibacterial activity. We also provide the studies that outline potential applications of chitin as a raw polymer for preparation of biomaterials.

Graphical Abstract

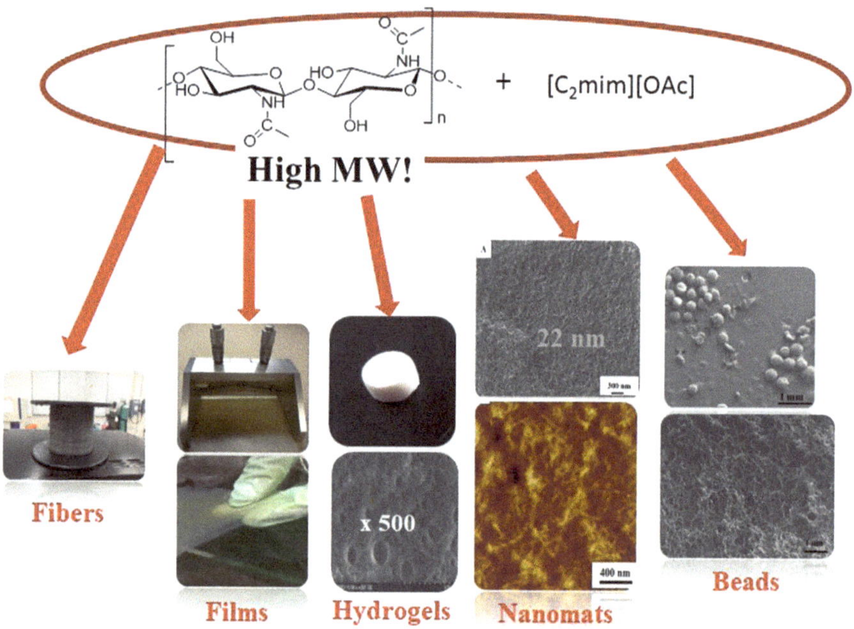

Keywords Antibacterial properties, Biocompatibility, Chitin, Chitin films for drug delivery, Cytotoxicity, Histological evaluation, Ionic liquids, Microbeads for drug delivery, Nanomats for scaffolding, Wound healing rat model

1 Introduction

1.1 'Plastic' Problem

Our reliance on plastics is the source of many societal problems. Of all the threats posed by our dependence on petroleum, pollution from the use of synthetic plastics is one of the most imminent. Plastics use has resulted in multiple 'garbage patches' in the world's oceans, with total floating plastic weight exceeding 250,000 tons [1]. Given that primary manufacture of plastics maintains its current level, recycling and environmental decontamination is a never-ending task.

Renewable replacements for plastics have so far been inadequate. Applications for biodegradable, synthetic plastics such as poly(L-lactic acid) (PLA) are limited by the fact that there are so few examples of such polymers compared to the nearly unlimited range of petrochemically-derived ones. Besides, although advertised as degradable, PLA is impervious to UV-light irradiation: its complete degradation to carbon dioxide and water requires high humidity and elevated temperature (at or above 60°C) [2]. In landfills, under regular environmental conditions, PLA is known to be extremely stable. In addition, PLA is hydrophobic, often requiring structural modification prior to making materials [3, 4].

Recently, people have begun using solid waste materials (e.g., coffee grounds) as fillers for disposable products [5]. This approach is promising for replacing plastics in some applications but, from an engineering material standpoint, use is restricted to non-structural fill applications.

The real answer to the successful replacement of plastics must enable the preservation of the low cost, versatility and ease of processing. Perhaps, even more important, the same consistent supply, economies of scale, and market acceptance enjoyed by synthetic plastics, must be available for its replacement.

1.2 Going Back in Time?

At the same time, biomass is a permanent source of renewable feedstock (e.g., wood from trees, animal residues) [6], and could compete with oil as an abundant resource. The suitability of biopolymers (such as starches, sugars, cellulose, lignin, chitin, and polypeptides) as replacements for plastics is evident from the fact that the high volume uses of plastics are mostly replacements for materials that were traditionally made from wood, paper, or cloth.

Indeed, using Nature's biopolymers for *plastics* manufacture can be traced back to 1856, when English chemist Alexander Parkes was granted the original patent on Parkesine [7]. Parkesine, the very first plastic, was made of ca. 70% of nitrated cellulose and ca. 30% camphor, both being derived from natural components. The Hyatt brothers added shellac, the refined form of a secretion of insect lac, to this composition in 1870, and obtained first patent for Celluloid [8]. This discovery is officially recognized as the birth of the plastics industry in the U.S. With time, polymers derived from Nature have been replaced by those from cheap petroleum and fossil fuel, a change driven by the cost-effectiveness of synthetic plastics; at the present time the plastics industry is a USD 654.38 billion market [9].

Plastics rose to dominance through the availability of cheap oil as a feedstock and open markets for its products, and they enjoy technological maturity and an entrenched economy of scale. For any replacement to be successful, these two factors must first be met. In this regard, biopolymers hold promise, although substantial investment in research and development is required to create a market of a competitive size. To achieve some progress in this direction, we can use modern biotechnology tools and provide breakthrough technologies to set up safer, cleaner, and more efficient industrial manufacturing processes. One such solution involves the use of ionic liquids.

1.3 Ionic Liquids to Unlock Biopolymers as a Source for New Biomaterials

Although biomass is a permanent source of renewable feedstock [6], it is, regrettably an underutilized resource, mainly because biopolymers are highly crystalline and insoluble in conventional solvents. Manufacturing approaches for biomaterials include dissolving biopolymers in inorganic bases [10–13], strong acids [14, 15], or highly polar fluorinated solvents [16, 17], which are often classified as unsuitable for use in medical applications or even change biopolymer structurally during product manufacturing. Thus, it is not surprising that 'purely biopolymeric' materials have neither been comprehensively studied nor extensively produced and marketed.

In exploitation of full biopolymers' potential, ionic liquids (ILs, salts that melt at or below 100°C [18]) present a transformational change. ILs allow the dissolution/deconstruction and biopolymer recovery from any biomass feedstock. The initial discovery was made by Rogers in 2002 on the example of dissolving cellulose [19]. The pioneering work has been expanded toward the dissolution and regeneration of wood in various ILs (among which 1-ethyl-3-methylimidazolium acetate $[C_2mim][OAc]$ exhibits the best dissolving ability because of the high basicity of the acetate anion) [20–25]. Since then it has been shown that ILs are suitable for dissolution of a wide range of biomasses comprising a great variety of chemical structures: carbohydrate biopolymers (e.g., cellulose [21], hemicellulose [25, 26], chitin [27]), phenolic biopolymers (e.g., lignin [25]), as well as peptides [28], proteins [29], and polynucleotides. More effective dissolutions have been achieved through the use of additives, including molecular solvents, other ILs, and solid salts, as well as catalysts (e.g., polyoxometalates, POMs [30]).

A vast range of biomass sources [21] and biopolymers [31] have proved to be amenable to processing using IL-based technology, and, indeed, actual dissolution that preserves biopolymers' high molecular weight (MW) allows for the manipulation of otherwise insoluble biopolymers and their easy processing with straightforward modification of the materials architecture. Solutions of a single polymer or a mixture of polymers, obtained from a suitable biomass, can be used to prepare fibers [27, 32–34], films [34, 35], beads [34, 36], and nanomats [37–39], using wet or dry jet-wet spinning [32–34], electrospinning [37–39], molding [40], and casting [34, 35] processes. The process allows for flexibility in producing the composite materials through incorporation of additives and/or blending of polymers, and the possibility of surface modification though either covalent or ionic functionalization.

By using certain loading of biopolymers into IL, formulations with standard viscosities and properties similar to conventional technologies can be easily achieved, overcoming the challenge for manufacturers to deliver a product using conventional equipment.

1.4 Chitin as a Raw Polymer for Biomaterials

Chitin, a biopolymer isolated from crustacean biomass, and the second most abundant natural polymer on Earth after cellulose [41] (Fig. 1), has a remarkable compatibility with living tissue and provides a non-protein matrix for 3D tissue growth [42]. The reason for this is that the *N*-acetylglucosamine moiety in chitin structurally resembles glycosaminoglycan macromolecules (GAGs, composed of *N*-acetylglucosamine or *N*-acetylgalactosamine units alternating with uronic sugar units such as galactose, glucuronic acid, or iduronic acid; see Fig. 1). GAGs are present in the extracellular matrix and control the regulation and binding of different proteins (e.g., enzymes, growth factors, cytokines, adhesion proteins, etc.) [43]. Structural similarities between GAGs and chitin polymers support related bioactivities [44]: chitin is a wound-healing accelerator [45], regulator of the inflammatory mediator secretion (interleukin-8, prostaglandin E, interleukin-1) [46], and a stimulator of cell proliferation [47] and tissue organization. It is also biodegradable (the degradation of chitin in the human body on the example of implanted chitin fabric was reported to take 12 weeks post-surgery [48]), non-toxic, and non-allergenic.

Because of these properties, chitin can be used for producing different biomaterials, including sutures, wound healing gauges, drug delivery vehicles, and scaffolds [49, 50]. There are several successful examples of commercial chitin-based biomaterials including Beschitin W by Unitika, Ltd. Japan [51, 52], Sieve-Patch® by Marine Polymer Technologies [53], Excel Arrest® by LLC Haemostasis Co. [54] to name but a few. Academic studies show chitin to have a tremendous potential as a suture material [55], where its antibacterial properties are an asset [56].

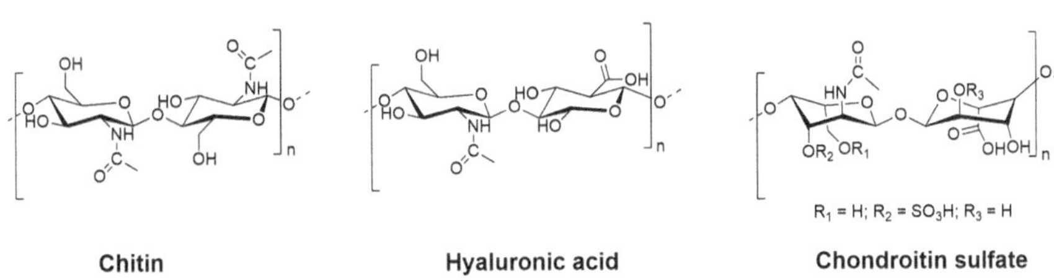

Fig. 1 Structural similarities between chitin (left) and two glycosaminoglycans (GAGs), hyaluronic acid (center) and chondroitin sulfate (right)

2 Chemical and Biomedical Properties of IL-Extracted Chitin

Traditional isolation of chitin polymer from crustacean biomass is conducted through a pulping process, which includes two major steps: demineralization (conducted using highly concentrated acids (e.g., HCl)) and deproteinization (conducted using highly concentrated hydroxides (e.g., NaOH)) at high temperatures (80–100°C) for prolonged periods of time [57, 58]. Pulping processes generate large amounts of waste, which has become a public and governmental concern, resulting in no chitin-producing plants that use acid/base treatment in North America [59]. Moreover, such harsh conditions degrade the quality of the isolated chitin through deacetylation and depolymerization, resulting in the lack of a reproducibly high-quality polymer product.

For the most demanding applications, chitin purity, that is, the isolated polymer, should be solely chitin, 100% poly-(*N*-acetylglucosamine), and this is of utmost importance. To determine the purity of commercial chitin and IL-extracted chitin, we have developed a ^{13}C solid state (SS) technique using Cross-Polarization Magic Angle Spinning Nuclear Magnetic Resonance (CP/MAS SS NMR). Although the primary contaminants after isolation could include residual calcium carbonate, proteins, silica, and heavy metals, there was no significant variation in purity independently with the origin of the chitin (PG-chitin: Sigma Aldrich catalog # C7170 vs chitin extracted with ionic liquid) [60, 61].

In addition to purity, the quality of the polymer is related to its two main characteristics: MW and degree of acetylation (DA). The MW of chitin determines most (if not all) of the material properties (stiffness, strength, viscoelasticity, toughness, and viscosity) of chitin-based products. If the MW is too low, the mechanical properties are generally insufficient for the polymer material to have any useful commercial applications. Harsh acid/base treatment for isolation of commercial chitin greatly decreases its MW. The extraction of chitin from biomass using the [C_2mim][OAc] is a physical rather than chemical dissolution process and therefore preserves the superior MW of the native biopolymer [27].

Because the high MW of IL-extracted chitin renders it insoluble, even in dimethylacetamide/lithium chloride complex (DMAc/LiCl), gel permeation chromatography (GPC) measurements have been impossible to obtain, and we have instead studied the relative MWs of commercial vs extracted chitin by comparison of their relative viscosities in IL solutions. We observed that the viscosity of the IL solution of the chitin extracted with ionic liquid is significantly higher than that of commercially available chitin and was the main parameter directing chitin processing into materials with different shapes and properties, for example, fibers [32–34], films [35], beads, and nanomats [37–39]). These architectures made of IL-extracted chitin can be used as wound care dressings (e.g., films, gel dressings, woven or non-woven textile patches), drug delivery devices (patches, beads), scaffolds (nanomaterials) to name but a few.

We have also shown that IL extraction does not result in any deacetylation. Namely, the DA of various chitin samples has been evaluated by ^{13}C CP–MAS solid-state NMR [60, 61], and no deacetylation was found: integration of the CH3 peak to the C3, C5 peaks gave a 1:2 ratio, indicating that the IL-extracted chitin has not been deacetylated during the IL processing, whereas commercial products were deacetylated to some extent.

We believe that the IL extraction method resulting in higher MWs with lower DA gives a reproducible polymer with distinctive and useful bio-related properties. The basic tests for an initial evaluation of biomaterials, conducted in order to provide useful scientific basis for clinical application, refer to cytotoxicity and irritation or intracutaneous reactivity. Preliminary efficacy tests and histological evaluations are more advanced, usually conducted for chitin wound care dressings and directed to assessment of an actual healing response of a wound. Finally, antibacterial properties of chitin and allergy tests are required for many biomaterials.

Yet, compared to chitin materials produced from chitin isolated using acid/base pulping, biomaterials made of IL-extracted chitin are still not mass-manufactured and there is little information available on their bio-related properties. Below we summarize what is known about the bio-related properties of extracted chitin, based on work conducted in our group and the available literature.

2.1 *Biocompatibility*

Biocompatibility is a mandatory requirement for biomaterials to be accepted by regulatory agencies [62]. By medical definition, biocompatible materials are materials that are not harmful to living organisms, including cells, organs, and tissues, and thus are not toxic, injurious, or physiologically reactive and do not cause immunological rejection. From the testing perspective for applying materials as wound care dressings, the biocompatibility is determined for the Medical Devices category [63] through an irritation/intracutaneous reactivity test (in accordance with ISO 10993-10: 2010 Standard, Biological Evaluation of Medical Devices, Part 10: Tests for Irritation and Skin Sensitization, Pages 11–14) [64]. The test allows determines the effect of any chemicals leaching from the test articles and capable of causing local irritation in dermal tissues.

We have conducted irritation studies using chitin fibers [32], with the aim of valuating local irritant effects to rabbit skin following a single application of test samples (chitin patches prepared from chitin fibers). First, chitin fibers were pulled and dried prior the test, as described in [32], and then were placed into both polar (normal saline, NS) and nonpolar (cottonseed oil, CSO) solvent to prepare both polar and nonpolar extracts. According to the test protocol, 2 g of chitin fibers were placed into 20 mL of extraction solvent and incubated at 50°C for 72 h, with agitation. After 72 h, the liquid (NS and CSO) was aseptically decanted into a sterile glass vessel and then maintained at room temperature and used within 24 h of preparation. As

Table 1 ISO intracutaneous reactivity test – dermal observation[a]

	Test	Control
Normal saline		
Rabbit 1	0.3	0
Rabbit 2	0	0
Rabbit 3	0.1	0
Avg.	0.1	0
Results	0.1	
Cottonseed oil (CSO)		
Rabbit 1	1	1
Rabbit 2	1	1
Rabbit 3	1	1
Avg.	1	1
Results	1	

[a]Comparative results with test-average control. Negative values reported as 0. Table adapted from [32]

negative controls, NS and CSO, which had no contact with the chitin materials, were used and incubated under identical conditions.

Assessments of the extracts from each material were conducted on three rabbits of the New Zealand breed by injecting them intracutaneously with the leachables from chitin samples as well as control solutions. None of the animals showed abnormal clinical signs during the 72-h test period and there were no significant dermal reactions observed at the injected sites on the skin of the rabbits at 24, 48, and 72 h. The intracutaneous irritation index was calculated (Table 1), and yielded 0.0–0.1, resulting in the conclusion that the extracts of the chitin material could be classified as 'non-irritants' to the skin of rabbits after comparison with the control.

2.2 Cytotoxicity

Cytotoxicity refers to the ability of material to destroy living cells. Considering that materials prepared with the help of IL could be cytotoxic, and might interfere with wound healing, study of their cytotoxicity is important. Although we were unable to find many studies for chitin materials prepared from [C_2mim][OAc] IL, cytotoxicity assessment was conducted for porous chitin-based gels prepared from chitin solubilized in 1-butyl-3-imidazolium acetate ([C_4mim][OAc]) [65]. The materials were produced via initial dissolution of commercially available chitin in [C_4mim][OAc], followed by molding, gelation, and coagulation in ethanol. Gels were Soxlet-washed with ethanol and then super-critically dried.

Cytotoxicity test was performed against the mouse fibroblast-like (L929) cell line [60]. The chitin materials showed good cell viability, indicating that IL processing did not introduce toxicity into the final materials. Specifically, materials were tested for cytotoxicity using *Dulbecco's modified Minimum Essential Medium (DMEM)*

Elution Test using L-929 mouse fibroblast cells by a method compliant with the requirements specified in ISO 10993-5 [66]. The purpose of this study was to evaluate the ability of a chitin extract to elicit a cytotoxic response in cultured cells.

The test articles were prepared using extracts from gels in DMEM, 10% fetal bovine serum (FBS), and 1% antibiotic/antimycotic (A/B) solution. Samples were extracted at 37°C for 24 h and the extract was inoculated onto the cell line and incubated at 37°C in a humidified atmosphere with 5% CO_2 in the air, until 90% confluence. Positive controls (latex extracts) were run in parallel with the test articles. Cultures were evaluated for cytotoxic effects by microscopic observation after 24, 48, and 72 h.

The cytotoxicity scores for the test articles were evaluated according to ISO, where scores of '0,' '1,' or '2' indicate 'non-toxic' material and scores of '3' or '4' are considered 'toxic.' In this test, the positive control should display a moderate to strong cytotoxic reaction. Criteria for evaluating cytotoxicity included morphologic changes in cells, such as granulation, crenation, or rounding, and loss of viable cells from the monolayer by lysis or detachment. Although results of the study were not expressed in numbers, the authors stated that "the produced materials have extremely low cytotoxicity levels."

2.3 Wound Healing Model

The purpose of wound healing studies is to evaluate wound healing response after a topical post-wounding application of novel wound treatment products. Normally, the course of wound healing is carried out by means of animal models, in vitro, to evaluate efficacy of the treatments applied to wounds. A rat wound healing model is often used to compare wound healing response after topical application of a novel wound treatment product to that of a control article.

Although numerous reports have demonstrated the accelerating effects of chitin on wound healing, materials prepared with the help of IL have not been extensively tested. We have conducted wound healing model studies using chitin fibers [32] prepared using IL [C_2mim][OAc], as described in [32]. For holding the fibers in place and to secure devices to the skin, we used a hypoallergenic in vitro tested 3 M Tegaderm Transparent Waterproof Dressing that protects the wound from external contaminants (e.g., bacteria, viruses) but allows transmission of oxygen and moisture.

Chitin fibers were cut into pieces and adhered to the surface of the 3 M system (Fig. 2) to create 'patches' for wound healing assessment. In each of 30 rats (Sprague Dawley), two wounds were created (on the left and right side, respectively), and chitin fiber-containing patches were placed onto the wounds.

The assessment of wounds has been conducted through the measurement of the length (along the longest axis, L) and width (smaller axis, perpendicular to L, W) of wound sites on days 0, 3, 7, 10, and 14 post-wounding, and comparing them to that

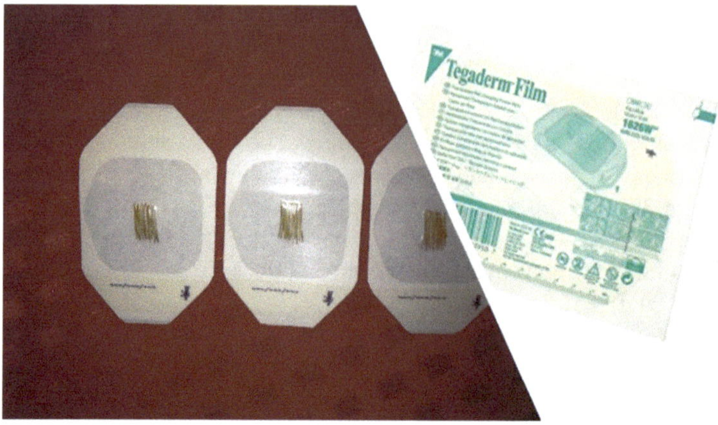

Fig. 2 Patches from chitin (left) that have been prepared using 3 M Tegaderm Transparent Dressing adhesive system, type 1,624 W (right). Adapted from [32]

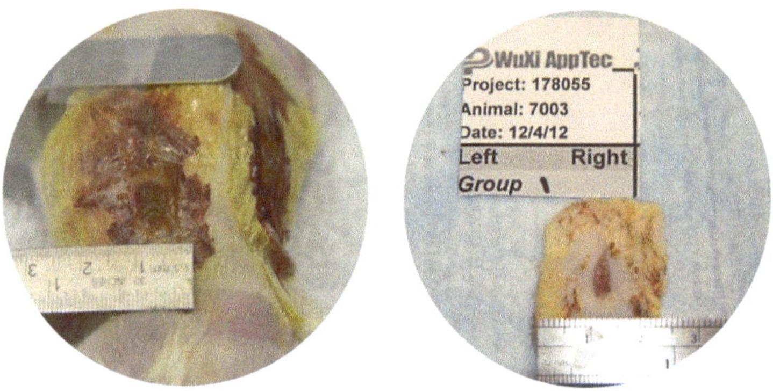

Fig. 3 Representative image of the wounds sites taken on days 3 and 7 post-wounding

seen with the control. Figure 3 shows representative photographs of the wound closure process, taken on days 3 and 7.

Assessment of wound healing has shown that the chitin patch product results in fast healing of the wound (Table 2). Moreover, wounds dressed with our chitin patches demonstrated so-called ‘early healing’ with ca. 30% wound closure occurring in the first 3 days. The initial healing rate was substantially faster than that for wounds treated with commercially available Opsite® dressing, marketed by Smith and Nephew. (Please note that data for Opsite® were extracted from article [67].) Wounds covered with chitin showed 35% healing after 3 days, as opposite to Opsite® dressing that showed 3% healing. After 1-week post-wounding (mid-healing, day 7), results in terms of wound closure were relatively close for both Opsite® and chitin fibers, where chitin-covered wounds were 70% healed and those covered with Opsite® 58% healed. Chitin-covered wounds were completely closed after 10 days, faster than

Table 2 Wound closure results summary[a]

	Measurement			
	Day 3	Day 7	Day 10	Day 14
Treatment	Mean ± Std. Dev.	Mean ± Std. Dev.	Mean ± Std. Dev.	Mean ± Std. Dev.
Chitin	42.1 ± 9.7	19.7 ± 11.7	0.6 ± 1.2	0
Opsite®	Not available			
	Wound closure %			
Treatment	Day 3	Day 7	Day 10	Day 14
Chitin	35	70	99	100
Opsite®	3	58	82	100

[a]Data for Opsite® extracted from plot "Rate of closure of wounds treated with different dressings" published in [62]. Table adapted from [32]

those dressed with commercial wound dressing Opsite®, which required 14 days for complete healing. The results seem even more substantial if one considers that the entire 14-day study was done using a single chitin-alginate patch per wound site per animal.

2.4 Histological Evaluation

Histology is the study of the microanatomy of cells and tissues of animals that allows for the clinical assessment of the wound. Once again, materials prepared with the help of IL have not been tested and we have conducted histological evaluation of chitin biomaterials using chitin fibers [32] prepared from [C_2mim][OAc] IL (Fig. 4) [32].

Overall, the tissue responses found at day 7 and day 14 at the wound sites treated with chitin were consistent with normal healing of a full thickness dermal wound. On day 7, wound sites covered with chitin-alginate patches displayed early healing (fibrosis with neovascularization, inflammation, and epidermal proliferation) and on day 14 the wound sites displayed matured dermal wounds. Although on day 7 the repair process was still ongoing with the epidermal tissue completely covering the dermal wounds dressed with the chitin-alginate, the epidermal tissue was only partially covering the wound dressed with the chitin fiber. On day 14 there was contraction of the wound dressed with the chitin (Fig. 4).

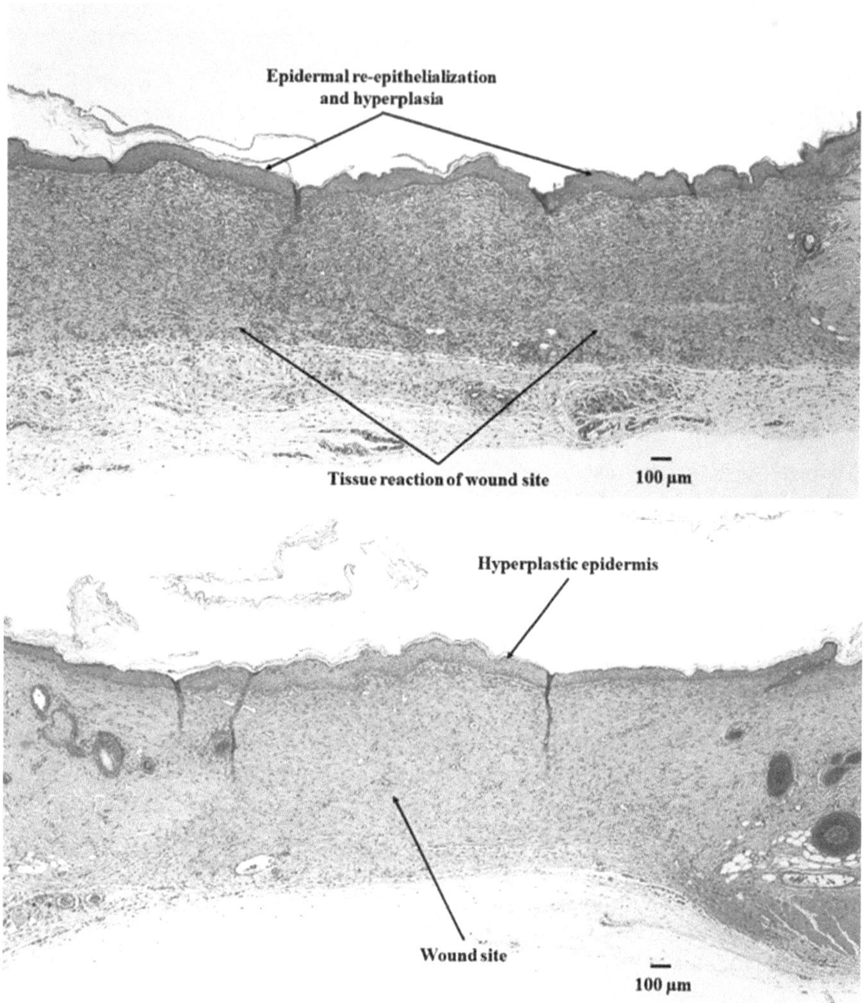

Fig. 4 Representative photos of chitin-covered wound site on day 7 (top) and day 14 (bottom). Tissue sections stained with hematoxylin and eosin (H&E), 40× magnification. Adapted from [32]

2.5 Antibacterial Properties

Although chitin and especially chitosan are known by their antibacterial properties, the extent of their activity against bacterial colonization depends on the chitin isolation method and its MW. The MW of native chitin is usually greater than 1,300 kDa [68] whereas commercial chitin has a MW of 100–1,200 kDa [69] because of the harsh isolation process. High temperature and concentrated acid treatment in the isolation of chitin from shellfish result in substantial MW decrease.

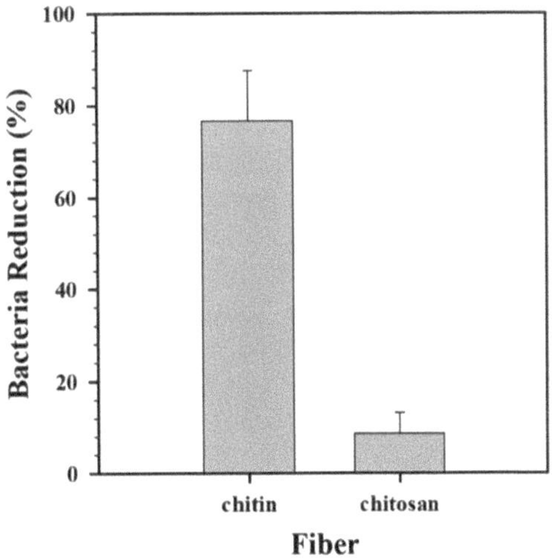

Fig. 5 Results on antibacterial activity of chitin fibers (left) spun from IL-extracted chitin in $[C_2mim][OAc]$ and chitin fibers that underwent surface deacetylation (right). Activity was determined against *E. coli* after 24 h using the shake flask method (pH 7.3 ± 0.2) [71]

To determine antibacterial properties of biomaterials made of IL-extracted chitin, a high-MW polymer has been extracted directly from crustacean shells [32]. Chitin then has been re-dissolved in $[C_2mim][OAc]$ IL and directly dry-wet jet spun into macrofibers [32]. A portion of these fibers was set aside and a second portion surface-deacetylated using aqueous NaOH to obtain the primary amine (the functional group of chitosan) on the surface but preserve the MW of the polymer because of the main fiber core being chitin 'as spun' [33]. Both methods produced high mechanical strength materials whose antibacterial properties depended on surface structure.

The antibacterial activity of the chitin fibers was evaluated by a shake flask test with Gram-negative *Escherichia coli* as the bacteria model. Test results revealed that, in the absence of any other antibacterial agents, 'as spun' chitin fibers had demonstrated (albeit weakly) antibacterial activity against *E. coli* with about 23% bacterial reduction after 24 h of incubation (Fig. 5). The same fibers after surface treatment with sodium hydroxide suppressed bacterial growth by 94% after the same incubation time because of substantial deacetylation and the presence of quaternizable amine groups rather than amide groups in chitin [70].

3 Case Studies

3.1 Chitin Films for Drug Delivery

Drug delivery through the skin surface offers several advantages compared to systemic (oral) drug administration [71, 72]. For example, drugs administrated topically or transdermally provide the targeted delivery to the site of inflammation, which minimizes adverse side-effects associated with systemic delivery. The 'ideal' materials for topical delivery should be robust, light, and provide sustained drug release without drug solidification on the skin surface. Furthermore, the possibility of storing material in a dry state and its rehydration prior to usage are important factors for device transportation and shelf-life stability.

In this regard, film preparation using [C_2mim][OAc] IL is a quick and relatively simple process performed by casting of an IL-solution of regenerated chitin onto a glass substrate followed by coagulation in an anti-solvent (alcohol or water) coagulation bath [35, 73]. IL-regenerated chitin films are lightweight, relatively thin, robust, and flexible enough to be applied as skin patches for delivery of Active Pharmaceutical Ingredients (APIs). They meet the air and moisture permeability and moisture absorbency requirements to enter the wound care markets, in particular the wound care sector [35].

Despite these advantages, there are not many studies reporting drug delivery from chitin films prepared via the IL-regeneration process. Our group investigated the potential application of chitin films as drug delivery vehicles using caffeine, which is often administrated topically, as a model API [74]. Loading and successful retention of APIs into films is usually challenging and strongly dependent on material–drug interactions; on the other hand, active compounds can be loaded into materials through changing their solubility by solvent exchange.

Thus, to load caffeine into chitin films, two strategies were explored in parallel [35]: loading caffeine from supersaturated aqueous solution of caffeine and from ethanoic solutions of caffeine (solubilities of 2 g per 100 mL, and 1.5 g per 100 mL, respectively [75]). The first strategy resulted in caffeine crystallization on the film surface, and film washing and drying resulted in complete caffeine removal. In contrast, in the second approach, when the wet films were loaded with caffeine in ethanoic solution, caffeine was successfully incorporated and retained after supercritical CO_2 (sc-CO_2) drying.

The release from caffeine-loaded sc-CO_2 dried chitin films was studied through direct immersion of the films into phosphate buffer saline (PBS) for 48 h. The caffeine release profile had a quick so-called burst release with ~80% of the cargo released in the first 20 min. After that, caffeine was slowly released, with all the caffeine released after 36 h.

3.2 *Chitin Beads for Drug Delivery*

Systemic drug administration is the conventional route for drug delivery into human body [76]. To increase the therapeutic efficiency and ensure that active compounds are delivered into the body with minimum side effects, drug delivery systems are designed as an essential part of the drug delivery process [77]. Among the different delivery systems, polymeric microbeads feature a high surface area and simple preparation method and, because of their spherical shape, they are less prone to aggregation when administered [78, 79]. Although in the literature, microspheres

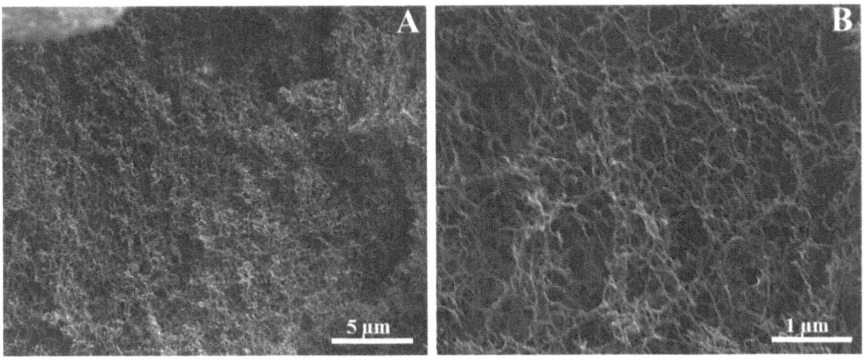

Fig. 6 SEM images of chitin beads interior. Magnifications are 5,000× (**a**) and 23,000× (**b**). (Au-sputter coated)

from chitin are generally prepared using corrosive solvents (dimethylacetamide/lithium chloride (DMAc/LiCl) or sodium hydroxide-urea (NaOH/urea) mixtures [80, 81], which may lead to chitin deacetylation, we developed an IL-based chitin processing method to synthesize chitin microbeads from high-MW chitin and investigated its drug delivery properties.

The microbeads were synthesized by coagulation of chitin dissolved in IL in hot polypropylene glycol (PPG) under an applied shear force. The formed beads exhibited relatively low size distribution with ~60% of the beads falling into the range 250–125 μm. The beads had highly porous interiors (Fig. 6) when dried with supercritical CO_2; all of them showed nanoscale microstructure with macropores of the chitinous beads of <100 nm and thin pore walls, which may be related to the acetylamino groups on chitin that increase the steric hindrance and thus decrease hydrogen bonding between polymer chains.

Drug delivery capabilities of synthesized chitin beads were investigated with two model compounds – salicylic acid and indigo dye – which were loaded into the beads from their 10 wt% solutions. Both were released with similar release rates – a fast initial release of about 70% observed during the first 4 h and release of about 90% in the full 7 h.

3.3 Chitin Nanomaterials

In more advanced bio-related applications, such as tissue engineering and scaffolding, fibers with nanometer diameters are in great demand [82, 83]. One of the easiest and most straightforward methods to prepare nanofibers in large quantities is electrospinning [84, 85]. Furthermore, electrospinning makes it possible to design materials with micro-, nano-, or a combination of sizes, which results in high porosity and large surface areas mimicking an extracellular matrix [86, 87]. In contrast to traditional electrospinning from organic solvents (VOCs), electrospinning of chitin

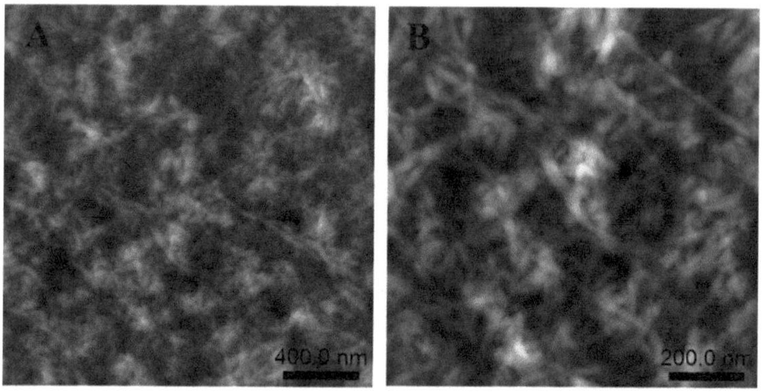

Fig. 7 Atomic force microscopy (AFM) images of air-dried electrospun chitin mats obtained after electrospinning of 0.4 wt% chitin solutions. Scan sizes 2 × 2 μm (**a**) and 1 × 1 μm (**b**) with the corresponding Z-heights of 75 and 45 nm

and other biopolymers from IL requires fiber coagulation using an anti-solvent (water or alcohol) coagulation bath, where biopolymers are solidified in a form of interconnected fiber network and IL is mixed (and then washed) with anti-solvent [88–92].

In an effort to obtain nanofibers and increase the output of a conventional single-needle electrospinning set-up, our group developed an electrospinning system equipped with a multi-needle spinneret head that resulted in a high throughput of 0.3 L [37–39]. In search of chitin nanofibers and an understanding of parameters regulating their formation, solutions of regenerated chitin obtained from different biomass sources were electrospun at different chitin concentrations. An example of nanofibers obtained after electrospinning and air-drying of chitin mats is shown in Fig. 7. The effect of various solution properties such as viscosity, concentration, conductivity, and surface tension on solution electrospinnability and fiber diameter was also determined.

Although electrospinning from IL was shown to be viable for chitin-nanofiber production, applications for these materials in the biomedical field have still not been found. In the future the focus should be not only on gaining better control over fiber formation but also on addressing the effect of fiber diameter and mechanical strength on the performance of these materials regarding application in tissue engineering.

4 Outlook

Multiple reviews of the biological properties of chitin can be found in the literature. Chitin is biodegradable; its rate of degradation depends on DA, subunit ordering, and MW. It is biocompatible, an effective immunomodulator, and controls various aspects of cell function because of structural similarity to GAGs, which form a major macromolecular component of the extracellular matrix, and is particularly abundant in cartilaginous tissues. Chitin promotes fibroblast proliferation, granulation, and vascularization.

Despite these promising properties, biomedical materials are primarily mass-manufactured from the chitin derivative chitosan for several reasons. First, biomedical data vary greatly with chitin MW and degree of deacetylation, and rather than going into details of understanding this phenomenon and finding a reliable chitin producer, people prefer to work with fully deacetylated chitosan polymer whose properties are known and consistent. As a result, many reviews in the field are biased and favor one polymer over the other. Second, chitin's inherent insolubility in common solvents and the need to use IL-based technologies present hurdles for scaling them up to a commercial level.

Our experience demonstrated, on the example of extraction of chitin from crustacean biomass, that the process could be scaled into a viable industrial process. In 2010 we demonstrated the dissolution and extraction of chitin directly from shrimp shells, with the resultant chitin polymer maintaining the high MW and providing a material with high strength, antimicrobial properties, and unprecedented control. As the technology progressed it went through several stages, each one successively larger in scale – bench, pilot, and demonstration scale. Thus, in 2014–2015, Rogers' Group (together with 525 Solutions, Inc.) built a scaled, highly automated, and customized 20-L early pilot stage system applicable to chitin extraction as part of the US Department of Energy Small Business Innovation Research Program (*DOE-SBIR Grant No. DE-SC0010152*). The project allowed for a demonstration of the operational viability of the scale-up and a generation of data for the full-scale operating plant design.

Currently this 20-L pilot unit serves as a basis for construction of a 3,000-L plant for a scaled-up chitin extraction by Mari Signum, Mid-Atlantic. Scaling up IL-based technologies permit isolation of chitin from nature by simply dissolving and extracting it directly from any crustacean biomass source, allowing the generation of sufficient supplies of chitin of quality unobtainable by any known chemical pulping processes. The ability to produce not only chitin itself but also products from chitin gives Mari Signum a competitive advantage to diversify the range of its products and to enter several profitable specialized markets. Our vision is that because of the tremendous promise of chitin as a raw biopolymer, it can soon be used for commercial production of chitin-based biomaterials including sutures, scaffolds, wound healing gauges, and drug delivery devices, utilizing the full potential of the IL-based platform.

Acknowledgements The authors would like to thank the 525 Solutions, and U.S. Department of Energy (DOE) SBIR Office of Science (DE-SC0010152) for support.

Notes RDR is an owner and president of 525 Solutions, Inc. and has a partial ownership of Mari Signum, Mid-Atlantic. JLS and OZ are employees of Mari Signum, Mid-Atlantic. RDR, JLS, and OZ are named inventors on related patent applications. The authors have no other relevant affiliations or financial involvement with any organization or entity with a financial interest in or financial conflict with the subject matter or materials discussed in this chapter apart from those disclosed.

References

1. Eriksen M, Lebreton LCM, Carson HS et al (2014) Plastic pollution in the world's oceans: more than 5 trillion plastic pieces weighing over 250,000 tons afloat at sea. PLoS One 9:e111913. https://doi.org/10.1371/journal.pone.0111913
2. Lunt J, Shafer AL (2000) Polylactic acid polymers from corn. Applications in the textiles industry. J Ind Text 29:191–205. https://doi.org/10.1177/152808370002900304
3. Rasal RM, Janorkar AV, Hirt DE (2010) Poly(lactic acid) modifications. Prog Polym Sci 35: 338–356. https://doi.org/10.1016/j.progpolymsci.2009.12.003
4. Rizvi R, Cochrane B, Naguib H et al (2011) Fabrication and characterization of melt-blended polylactide-chitin composites and their foams. J Cell Plast 47:283–300. https://doi.org/10.1177/0021955X11402549
5. Arulrajah A, Maghoolpilehrood F, Disfani MM et al (2014) Spent coffee grounds as a non-structural embankment fill material: engineering and environmental considerations. J Clean Prod 72:181–186. https://doi.org/10.1016/j.jclepro.2014.03.010
6. Marchessault RH (1984) Carbohydrate polymers: nature's high performance materials. In: Vandenberg EJ (ed) Contemporary topics in polymer science. Springer, Boston, pp 15–53
7. Parkes A (1857) Application for British patent 1856. In: Patents for inventions. UK Patent office, 255
8. Hyatt JW (1869) Billiard balls. US Patent 50359, Oct 1869
9. Plastics Market Analysis by Product (PE, PP, PVC, PET, Polystyrene, Engineering Thermoplastics), by Application (Film & Sheet, Injection Molding, Textiles, Packaging, Transportation, Construction) and Segment Forecasts to 2020. Grand View Research, Inc. 2015
10. Muzzarelli RAA (1983) Chitin and its derivatives: new trends of applied research. Carbohydr Polym 3:53–75. https://doi.org/10.1016/0144-8617(83)90012-7
11. Seoudi R, Nada AMA (2007) Molecular structure and dielectric properties studies of chitin and its treated by acid, base and hypochlorite. Carbohydr Polym 68:728–733. https://doi.org/10.1016/j.carbpol.2006.08.009
12. Kunike G (1926) Chitin and chitosan. J Soc Dyers Colorists 42:318–342
13. Von Weimarn PP (1927) Conversion of fibroin, chitin, casein, and similar substances into the ropy-plastic state and colloidal solution. Ind Eng Chem 19:109–110. https://doi.org/10.1021/ie50205a034
14. Brine CJ, Austin PR (1975) Renaturated chitin fibrils, films and filaments. In: Church TD (ed) Marine chemistry in coastal environment, ACS symposium series, vol 18. ACS Publications, Washington, pp 505–518. https://doi.org/10.1021/bk-1975-0018.ch031
15. Austin PR (1975) Solvents for and purification of chitin. US Patent 3,892,731, Jul 1975
16. Tamura H, Hamaguchi T, Tokura S (2004) Destruction of rigid crystalline structure to prepare chitin solution. In: Boucher I, Jamieson K, Retnakaran A (eds) Advances in chitin science, proceedings of the 9th international conference on Chitin and Chitosan. European Chitin Society, Montreal, pp 84–87
17. Carpozza RC (1976) Spinning and shaping poly-(N-acetyl-D-glucosamine). US Patent 3,988,411, Oct 1976

18. Welton T (1999) Room-temperature ionic liquids. Chem Rev 99:2071–2084. https://doi.org/10.1021/cr980032t
19. Swatloski RP, Spear SK, Holbrey JD et al (2002) Dissolution of cellulose with ionic liquids. J Am Chem Soc 124:4974–4975. https://doi.org/10.1021/ja025790m
20. Fort DA, Remsing RC, Swatloski RP et al (2007) Can ionic liquids dissolve wood? Processing and analysis of lignocellulosic materials with 1-*N*-butyl-3-methylimidazolium chloride. Green Chem 9:63–69. https://doi.org/10.1039/B607614A
21. Wang H, Gurau G, Rogers RD (2012) Ionic liquid processing of cellulose. Chem Soc Rev 41: 1519–1537. https://doi.org/10.1039/C2CS15311D
22. Sun N, Rahman M, Qin Y et al (2009) Complete dissolution and partial delignification of wood in the ionic liquid 1-ethyl-3-methylimidazolium acetate. Green Chem 11:646–655. https://doi.org/10.1039/B822702K
23. Li WY, Sun N, Stoner B et al (2011) Rapid dissolution of lignocellulosic biomass in ionic liquids using temperatures above the glass transition of lignin. Green Chem 13: 2038–2047. https://doi.org/10.1039/C1GC15522A
24. Sun N, Jiang XY, Maxim ML et al (2011) Use of polyoxometalate catalysts in ionic liquids to enhance the dissolution and delignification of woody biomass. ChemSusChem 4:65–73. https://doi.org/10.1002/cssc.201000272
25. Cheng F, Wang H, Chatel G et al (2014) Facile pulping of lignocellulosic biomass using choline acetate. Bioresour Technol 164:394–401. https://doi.org/10.1016/j.biortech.2014.05.016
26. Lan W, Liu C-F, Sun RC (2011) Fractionation of bagasse into cellulose, hemicelluloses, and lignin with ionic liquid treatment followed by alkaline extraction. J Agric Food Chem 59: 8691–8701. https://doi.org/10.1021/jf201508g
27. Qin Y, Lu X, Sun N et al (2010) Dissolution or extraction of crustacean shells using ionic liquids to obtain high molecular weight purified chitin and direct production of chitin films and fibers. Green Chem 12:968–971. https://doi.org/10.1039/C003583A
28. Tietze AA, Heimer P, Stark A et al (2012) Ionic liquid applications in peptide chemistry: synthesis, purification and analytical characterization processes. Molecules 17:4158–4185. https://doi.org/10.3390/molecules17044158
29. Schröder C (2017) Proteins in ionic liquids: current status of experiments and simulations. Top Curr Chem 375:25–51. https://doi.org/10.1007/s41061-017-0110-2
30. Cheng FC, Wang H, Rogers RD (2014) Oxygen enhances polyoxometalate-based catalytic dissolution and delignification of woody biomass in ionic liquids. ACS Sustain Chem Eng 2: 2859–2865. https://doi.org/10.1021/sc500614m
31. Mantz RA, Fox DM, Green MIII et al (2007) Dissolution of biopolymers using ionic liquids. Z Naturforsch A 62:275–280. https://doi.org/10.1515/zna-2007-5-60
32. Shamshina JL, Gurau G, Block LE et al (2014) Chitin–calcium alginate composite fibers for wound care dressings spun from ionic liquid solution. J Mat Chem B 2:3924–3936. https://doi.org/10.1039/C4TB00329B
33. Barber PS, Kelley SP, Griggs CS et al (2014) Surface modification of ionic liquid-spun chitin fibers for the extraction of uranium from seawater: seeking the strength of chitin and the chemical functionality of chitosan. Green Chem 16:1828–1836. https://doi.org/10.1039/C4GC00092G
34. Qin Y, Rogers RD, Daly DT (2010) Process for forming films, fibers, and beads from chitinous biomass, US 9096743 B2, Jun 2010
35. King C, Shamshina JL, Gurau G et al (2017) A platform for more sustainable chitin films from an ionic liquid process. Green Chem 19:117–126. https://doi.org/10.1039/C6GC02201D
36. King CA, Shamshina JL, Zavgorodnya O et al (2017) Porous chitin microbeads for more sustainable cosmetics. ACS Sustain Chem Eng 5:11660–11667. https://doi.org/10.1021/acssuschemeng.7b03053
37. Barber PS, Griggs CS, Bonner JR et al (2013) Electrospinning of chitin nanofibers directly from an ionic liquid extract of shrimp shells. Green Chem 15:601–607. https://doi.org/10.1039/C2GC36582K

38. Zavgorodnya O, Shamshina JL, Bonner JR et al (2017) Electrospinning biopolymers from ionic liquids requires control of different solution properties than volatile organic solvents. ACS Sustain Chem Eng 5:5512–5519. https://doi.org/10.1021/acssuschemeng.7b00863
39. Shamshina JL, Zavgorodnya O, Bonner JR et al (2016) "Practical" electrospinning of biopolymers in ionic liquids. ChemSusChem 10:106–111. https://doi.org/10.1002/cssc.201601372
40. Kimura M, Shinohara Y, Takizawa J et al (2015) Versatile molding process for tough cellulose hydrogel materials. Sci Rep 5:16266. https://doi.org/10.1038/srep16266
41. Gao X, Chen X, Zhang J et al (2016) Transformation of chitin and waste shrimp shells into acetic acid and pyrrole. ACS Sustain Chem Eng 4:3912–3920. https://doi.org/10.1021/acssuschemeng.6b00767
42. Hirano S, Nakahira T, Nakagawa M et al (1999) The preparation and application of functional fibers from crab shell chitin. J Biotechnol 70:373–377. https://doi.org/10.10116/S0079-6352(99)80130-1
43. Vázquez JA, Rodríguez-Amado I, Montemayor MI et al (2013) Chondroitin sulfate, hyaluronic acid and chitin/chitosan production using marine waste sources: characteristics, applications and eco-friendly processes: a review. Mar Drugs 11:747–774. https://doi.org/10.3390/md11030747
44. Peluso G (1994) Chitosan-mediated stimulation of macrophage function. Biomaterials 15: 1215–1220. https://doi.org/10.1016/0142-9612(94)90272-0
45. Prudden JF, Migel P, Hanson P et al (1970) The discovery of a potent pure chemical wound-healing accelerator. Am J Surg 119:560–564. https://doi.org/10.1016/0002-9610(70)90175-3
46. Jayakumar R, Prabaharan M, Kumar PTS et al (2011) Novel chitin and chitosan materials in wound dressing. In: Laskovski AN (ed) Biomedical engineering, trends in materials science. InTech, Rijeka, pp 3–24. https://doi.org/10.5772/13509
47. Mori T, Okumura M, Mastuura M et al (1997) Effects of chitin and its derivatives on the proliferation and cytokine production of fibroblasts in vitro. Biomaterials 18:947–951. https://doi.org/10.1016/S0142-9612(97)00017-3
48. Wan ACA, Tai BCU (2013) Chitin - a promising biomaterial for tissue engineering and stem cell technology. Biotech Adv 31:1776–1785. https://doi.org/10.1016/j.biotechadv.2013.09.007
49. Shigemasa Y, Minami S (1996) Applications of chitin and chitosan for biomaterials. Biotechnol Genet Eng Rev 13:383–420. https://doi.org/10.1080/02648725.1996.10647935
50. Yang T-L (2011) Chitin-based materials in tissue engineering: applications in soft tissue and epithelial organ. Int J Mol Sci 12:1936–1963. https://doi.org/10.3390/ijms12031936
51. Unitika's history creates its next history. https://www.unitika.co.jp/e/company/history.html. Accessed 14 Oct 2017
52. Unitika Ltd. http://www.unitika.co.jp/e/. Accessed 14 Oct 2017
53. Syvec. http://syvek.com/. Accessed 11 Oct 2017
54. Hemostasis LLC. http://www.hemostasisllc.com/excelarrest-techinfo.html. Accessed 11 Oct 2017
55. Nakajima M, Atsumi K, Kifune K et al (1986) Chitin is an effective material for sutures. Jpn J Surg 16:418–424. https://doi.org/10.1007/BF02470609
56. Chu C-C, von Fraunhofer JA, Greisler HP (eds) (1996) Wound closure biomaterials and devices. CRC Press, New York, p 416
57. Rinaudo M (2006) Chitin and chitosan: properties and applications. Prog Polym Sci 31: 603–632
58. Khoushab F, Yamabhai M (2010) Chitin research revisited. Mar Drugs 8:1988–2012
59. Shamshina JL, Barber PS, Gurau G et al (2017) Pulping of crustacean waste using ionic liquids: to extract or not to extract. ACS Sust Chem Eng 4:6072–6081. https://doi.org/10.1021/acssuschemeng.6b01434
60. King K, Stein RS, Shamshina JL (2017) Measuring the purity of chitin with a clean, quantitative solid-state NMR method. ACS Sust Chem Eng 5:8011–8016. https://doi.org/10.1021/acssuschemeng.7b01589. Unpublished results (2016)
61. Rogers RD (2016) Unpublished results

62. Sall B (2000–2013) Regulation of medical devices. In: Madame Curie bioscience database: landes bioscience. Available via NIH. Available via https://www.iso.org/standard/40884.html. Accessed 15 Oct 2017. https://www.ncbi.nlm.nih.gov/books/NBK6534/. Accessed 15 Oct 2017
63. Considerations for the biocompatibility evaluation of medical devices (2001) https://www.mddionline.com/considerations-biocompatibility-evaluation-medical-devices. Accessed 15 Oct 2017
64. ISO 10993-10:2010 (2010) Biological evaluation of medical devices - Part 10: Tests for irritation and skin sensitization. Available via https://www.iso.org/standard/40884.html. Accessed 15 Oct 2017
65. Silva SS, Duarte ARC, Carvalho AP et al (2011) Green processing of porous chitin structures for biomedical applications combining ionic liquids and supercritical fluid technology. Acta Biomater 7:1166–1172. https://doi.org/10.1016/j.actbio.2010.09.041
66. ISO/10993-5 (1992) Biological evaluation of medical devices- Part 5: test for cytotoxicity, *In Vitro* methods: 8.2 test on extracts
67. Wang L, Khor E, Wee A et al (2002) Chitosan-alginate PEC membrane as a wound dressing: assessment of incisional wound healing. J Biomed Mater Res 63:610–618. https://doi.org/10.1002/jbm.10382
68. Domard AA (2011) Perspective on 30 years research on chitin and chitosan. Carbohydr Polym 87:696–703. https://doi.org/10.1016/j.carbpol.2010.04.083
69. Li J, Du Y, Yang J et al (2005) Preparation and characterisation of low molecular weight chitosan and chito-oligomers by a commercial enzyme. Polym Degrad Stabil 87:441–448. https://doi.org/10.1016/j.polymdegradstab.2004.09.008
70. Rogers RD (2015) Unpublished results
71. Prausnitz MR, Langer R (2008) Transdermal drug delivery. Nat Biotechnol 26:1261–1268. https://doi.org/10.1038/nbt.1504
72. Paudel KS, Milewski M, Swadley CL et al (2010) Challenges and opportunities in dermal/transdermal delivery. Ther Deliv 1:109–131. https://doi.org/10.4155/tde.10.16
73. Wu Y, Sasaki T, Irie S et al (2008) Novel biomass-ionic liquid platform for the utilization of native chitin. Polymer 49:2321–2327. https://doi.org/10.1016/j.polymer.2008.03.027
74. Luo L, Lane ME (2015) Topical and transdermal delivery of caffeine. Int J Pharm 490:155–164. https://doi.org/10.1016/j.ijpharm.2015.05.050
75. Budavari S (1996) The Merck index, an encyclopedia of chemicals, drugs, and biologicals.12th edn. Merck & Co., Inc., Whitehouse Station, p 1674
76. Hassan BAR (2012) Overview on drug delivery system. Pharm Anal Acta 3:e137. https://doi.org/10.4172/2153-2435.1000e137
77. Neuse EW (2008) Synthetic polymers as drug-delivery vehicles in medicine. Met Based Drugs 2008:1–19. https://doi.org/10.1155/2008/469531
78. Mahdavi SA, Jafari SM, Ghorbani M et al (2014) Spray-drying microencapsulation of anthocyanins by natural biopolymers: a review. Dry Technol 32:509–518
79. Kawaguchi H (2000) Functional polymer microsphere. Prog Polym Sci 25:1171–1210. https://doi.org/10.1080/07373937.2013.839562
80. Yusof NL, Lim LY, Khor E (2001) Preparation and characterization of chitin beads as a wound dressing precursor. J Biomed Mater Res 54:59–68. https://doi.org/10.1002/1097-4636(200101)54:1<59::AID-JBM7>3.0.CO;2-U
81. Wang Y, Li Y, Liu S et al (2015) Fabrication of chitin microspheres and their multipurpose application as catalyst support and adsorbent. Carbohydr Polym 120:53–59. https://doi.org/10.1016/j.carbpol.2014.12.005
82. Sill TJ, von Recum HA (2008) Electrospinning: applications in drug delivery and tissue engineering. Biomaterials 29:1989–2006. https://doi.org/10.1016/j.biomaterials.2008.01.011
83. Singh N, Chen J, Koziol KK et al (2016) Chitin and carbon nanotube composites as biocompatible scaffolds for neuron growth. Nanoscale 8:8288–8299. https://doi.org/10.1039/c5nr06595

84. Schiffman JD, Schauer CL (2008) A review: electrospinning of biopolymer nanofibers and their applications. Polym Rev 48:317–352. https://doi.org/10.1080/15583720802022182
85. Li D, Xia Y (2004) Electrospinning of nanofibers: reinventing the wheel? Adv Mater 16: 1151–1170. https://doi.org/10.1002/adma.200400719
86. Lee KY, Jeong L, Kang YO et al (2009) Electrospinning of polysaccharides for regenerative medicine. Adv Drug Deliv Rev 61:1020–1032. https://doi.org/10.1016/j.addr.2009.07.006
87. Noh HK, Lee SW, Kim JM et al (2006) Electrospinning of chitin nanofibers: degradation behavior and cellular response to normal human keratinocytes and fibroblasts. Biomaterials 27: 3934–3944. https://doi.org/10.1016/j.biomaterials.2006.03.016
88. Xu S, Zhang J, He A et al (2008) Electrospinning of native cellulose from nonvolatile solvent system. Polymer 49:2911–2917. https://doi.org/10.1016/j.polymer.2008.04.046
89. Quan S-L, Kang S-G, Chin I-J (2006) Characterization of cellulose fibers electrospun using ionic liquid. Cellulose 17:223–230. https://doi.org/10.1007/s10570-009-9386-x
90. Viswanathan G, Murugesan S, Pushparaj V et al (2006) Preparation of biopolymer fibers by electrospinning from room temperature ionic liquids. Biomacromolecules 7:415–418. https://doi.org/10.1021/bm050837s
91. Freire MG, Teles ARR, Ferreira RAS et al (2011) Electrospun nanosized cellulose fibers using ionic liquids at room temperature. Green Chem 13:3173–3180. https://doi.org/10.1039/C1GC15930E
92. Zheng Y, Miao J, Maeda N et al (2014) Uniform nanoparticle coating of cellulose fibers during wet electrospinning. J Mater Chem A 2:15029–15034. https://doi.org/10.1039/C4TA03221G

Adv Biochem Eng Biotechnol (2019) 168: 199–214
DOI: 10.1007/10_2018_68

Published online: 10 August 2018

Synthesis of Ionic Liquids Originated from Natural Products

Hiroyuki Ohno

Contents

Abstract In this chapter, preparation, basic properties, and some applications of ionic liquids composed of naturally-derived ions are summarized. There are many candidate ions in nature suitable for ionic liquid preparation. Physicochemical properties and preparation of some ionic liquids based on carboxylate anions, cholinium cation, and even amino acids are mentioned. Some interesting applications based on these ionic liquids composed of naturally-derived ions are also briefly introduced.

H. Ohno (✉)
Tokyo University of Agriculture and Technology, Fuchu, Tokyo, Japan
e-mail: ohnoh@cc.tuat.ac.jp

Graphical Abstract

Keywords Amino acids, Biological membrane, Carboxylic acids, Hydrophobicity, Polarity, Proteins

1 Introduction

Because ionic liquids have been synthesized with suitably sized ions, there is no limit to the use of some ions derived from natural products. In nature there are many medium-sized ions that look suitable to be components of ionic liquids or organic salts with relatively low melting points. Compared to synthetic chemicals, naturally-derived materials are considered to be biocompatible. Naturally-derived organic ions are, however, believed to be expensive. Many of them are really expensive but there are also many less-expensive ions existing in nature, and there are a wide variety of ions and charged molecules derived from petroleum. However, taking various factors such as global environmental problems into consideration, society tends to use natural products. Accordingly, there are many opportunities to prepare ionic liquids using charged natural products. In this chapter, recent trends in the synthesis of ionic liquids using naturally-derived ions are summarized as well as their properties and applications. Because it is not easy to review all the bio-derived ionic liquids, the major ionic liquids are highlighted here. Discussion is not limited to ionic liquids composed of naturally-derived cations and anions. Some of ionic liquids containing naturally-derived cations or anions and petroleum-derived ions are also mentioned.

2 Why Bio-Derived?

Many people simply believe that naturally-derived materials (or molecules), especially from animals and plants, are safe. This is unfortunately not always true. We should remember bio-derived toxic molecules such as tetrodotoxin found in some pufferfish, so there is no guarantee that ionic liquids from naturally-derived ions are safe. However, because most such ions are frequently found in biological systems, there is strong expectation that we can prepare biologically safe (or friendly) ionic liquids by the combination of these ions. However, we should not forget the fact that ionic liquids are salts composed only of ions. Salts show osmotic pressure in water and this is a decisive disadvantage for safety. Safety is also a function of the dosage of materials for animals, but we do not want to discuss this here. It should also be mentioned that the largest merit of naturally-derived ions is their wide availability. This should open a new category of ionic liquids such as cheap and (relatively) safe ionic liquids. Considering recent trends in the global issue, it is important to use bio-derived materials. The policy of bio-economics leads the whole material-based society, so it is the perfect time to discuss bio-derived ionic liquids. I want to introduce here some bio-derived ionic liquids that are likely to be the major ionic liquids used in future society.

3 Carboxylate Anions as Potential Components of Ionic Liquids

There are many carboxylic acids in nature. They are generally known as "fatty acids," and most of them can be found as components of esters or salts in biological systems. Those with from 1 (formic acid) to 30 (triacontanoic acid) carbon atoms are often mentioned in biology textbooks. Some of these acids are useful in the design of ionic liquids, but it should be noted here that the properties of ionic liquids are a function of the number of carbon atoms in these carboxylic acids. A melting point (T_m) of a carboxylic acid generally increases with an increasing number of carbon atoms. For example, T_m of formic acid, propionic acid, lauric acid, and stearic acid is 8.4, −21, 44, and 69.9°C, respectively. These are good examples of the fact that the physicochemical properties of carboxylic acid are not solely governed by the alkyl chain length but also by other interactions, such as hydrogen bonding. This rough relationship between T_m and alkyl chain length of carboxylates is useful when considering the properties of ionic liquids containing these carboxylic acids.

Larger anions generally form ionic liquids showing lower T_m. A decreasing surface charge density of anions contributes to a lower T_m of the ionic liquids. However, in the case of anions containing the same anionic site and different alky chains, the charge density of the carboxylic acid residue is almost the same. Accordingly, the properties of the ionic liquids formed are a function of the structure of their alkyl chains. As mentioned above, there is not a linear relation because of the effects of conformation, hydrogen bonding, and other intermolecular interactions.

The effect of a double bond is also remarkable in lowering T_m, as typically seen in unsaturated fatty acids such as linoleic acid (T_m is −5°C). It should therefore be better to use such unsaturated carboxylic acids to prepare ionic liquids with low T_m. Again, the T_m of ionic liquids is not directly related to the T_m of the corresponding carboxylic acids, even when the same cation is used, because of the cancelation of some intermolecular interaction factors such as the hydrogen bond.

Viscosity is another important property of the ionic liquids. Carboxylic acids with longer alkyl chains generally give ionic liquids with higher viscosity. In some cases of ionic liquids composed of carboxylic acids with longer alkyl chains, they give some unique assembled structures even after forming the salts. Many of these salts show multi-phases including a liquid crystalline phase as a function of temperature. These assembled structures are quite attractive for designing ionic liquids as they have ordered structures. The introduction of long alkyl chain(s) onto a cation is also effective in forming an assembled ionic liquid structure in spite of the fact that ordinary anions containing no long alkyl chains may have been originally used [1]. The advantage of ionic liquids having ordered structures is typically seen in the design of ion conductive materials. In such materials, successive ion conduction pathways are formed to transport ions in the matrix. It should be noted that the ionic conductivity of these ionic liquids having ordered structures is greater than that of ionic liquids having polydomain (or randomly oriented domain) structures. Some other physicochemical properties of carboxylate-type ionic liquids are the function of alkyl chain length, so a suitable design of properties can be carried out by the selection of carboxylates with suitable alkyl chain lengths. The thermal stability of general ionic liquids is generally quite good, but that of carboxylate-type ionic liquids is not. These ionic liquids are therefore not so suitable for use at higher temperatures.

It is also possible to maintain other properties of component ions after forming ionic liquids. We should mention ionic liquids prepared by the combination of suitable cations and biologically active anions. Betulinic acid (Fig. 1, left) is known as a natural product with a few favorable biological properties such as anti-cancer, anti-malarial, anti-HIV, and so on. This cholesterol-like carboxylic acid was coupled with cholinium cation to improve its water solubility by more than 100 times [2]. The use of ionic liquids is also expected to improve the biological affinity of water-insoluble drugs such as this betulinic acid.

The polarity of ionic liquids has frequently drawn much attention. It is not easy to determine the polarity of salts and, accordingly, Kamlet–Taft parameters of ionic

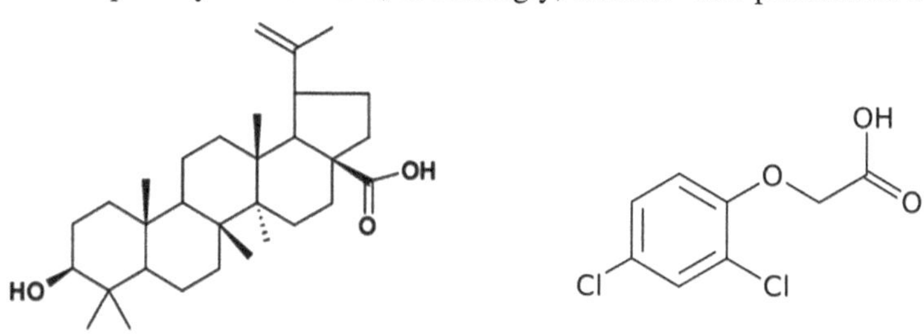

Fig. 1 Structure of betulinic acid (left) and 2,4-dichlorophenoxyacetic acid (right)

liquids are usually used to evaluate their polarity [3]. Proton donating ability (α value), proton accepting ability (β value), and dipolarity (π^* value) are determined by the solvatochromic effect of dye molecules in the ionic liquids [3]. A series of carboxylic acid-derived ionic liquids show very high hydrogen bond accepting ability, almost independent of alkyl chain length. These carboxylic acid-derived ionic liquids are useful in breaking hydrogen bonding of many systems, and sometimes these ionic liquids are powerful solvents to dissolve complex materials composed of intermolecular hydrogen bonding, such as cellulose. Some carboxylic acid-derived ionic liquids actually have the power to dissolve cellulose that cannot be solubilized in ordinary solvents under mild conditions [4].

On the other hand, pharmaceutical application is also in progress. For example, 2,4-dichlorophenoxyacetic acid (Fig. 1, right) was coupled with imidazolium, pyridinium, and ammonium cations to form ionic liquids. This acetic acid derivative shows auxin-like herbicide activity with good chemical and thermal stability [5, 6]. Pharmaceutical ionic liquids have been reviewed by some scientists [7] and these reviews make good reading for those who have been engaging in work in this pharmaceutical field.

A unique paper should be mentioned here. Some ants (*Solenopsis invicta*) spray alkaloid-based venom when fighting with other ant species. When ants (*Nylanderia fulva*) are attacked they detoxify the alkaloid-based venom by grooming with their own venom, formic acid. Davis et al. reported that this neutralized mixture was a viscous liquid to be classified as an ionic liquids [8]. This strongly suggests that naturally-derived ionic liquids can be found elsewhere.

There are some other naturally-derived carboxylic acids with specific properties, and it is quite interesting to design ionic liquids with these carboxylic acids. In other words, it is possible to design (or to introduce functional groups onto) the alkyl chain of carboxylic acids, and this is one potential tool to prepare functional ionic liquids. Many examples are found in the review on the functional design of ionic liquids [9]. One noteworthy application is the improvement of solubility in water for a series of hydrophobic molecules. In spite of the insolubility properties of some drugs, they can be made water-soluble after forming ionic liquids, a very useful improvement. The physicochemical properties of a series of ionic liquids are summarized in the literature [10].

4 Amino Acid Ionic Liquids

Most well-known bio-derived molecules are amino acids. We first reported ionic liquids derived from 20 amino acids in 2005 [11]. After this paper appeared, the number of papers on amino acid-derived ionic liquids has been increasing. Why so many? Because amino acids are typical bio-derived and charged materials. They are non-toxic, abundant in nature, biologically active, and raw materials of proteins and other biomolecules. As seen in Fig. 2, there are 20 essential amino acids known to be important in our body. Their properties are different and solely dependent on their

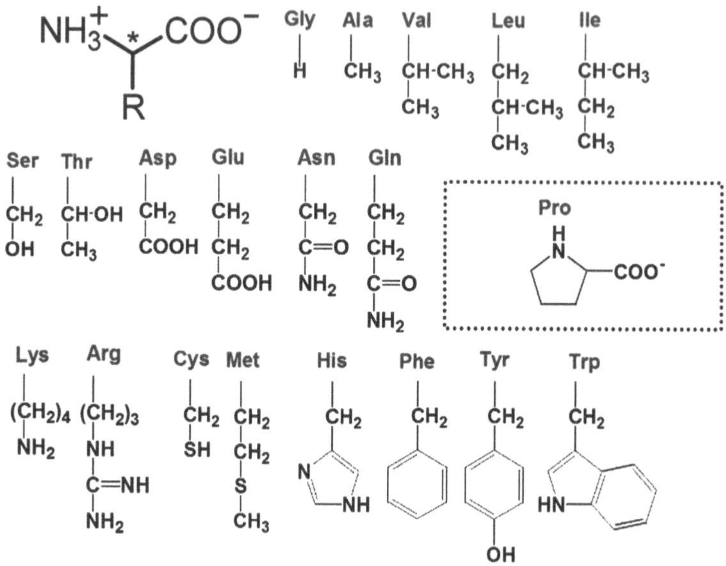

Fig. 2 Basic structure of 20 essential amino acids

side chain structure. It is quite natural to use these amino acids as components of ionic liquids. In 2007 we also published an account of amino acid-derived ionic liquids [12]. As the definition of amino acid is a molecule having both an amino group and a carboxylic acid group, there are many other amino acids. For examples, ornithine, creatine, and 4-aminobutanoic acid are known as not essential but naturally-derived amino acids. Taking the diverse structure of side groups (see R in Fig. 2) into account, there are many possibilities for obtaining unique and functional ionic liquids.

There are many papers on the preparation and properties of ionic liquids produced using amino acids as anions [13–18]. Also, because amino acids behave as cations after protonation (or quaternization) of amino groups, there is a large number of research papers on the preparation and properties of ionic liquids using amino acids as cations [19]. As shown in Fig. 3, amino acids can be used as either cations or anions, and ionic liquids have been prepared by coupling them with anions (for IL a) or cations (for IL b), respectively. Furthermore, ionic liquids are prepared by simple mixing of pristine amino acids with acids or bases through neutralization. This neutralization method is quite a simple way to produce ionic liquids [20]. For example, when an amino acid is neutralized with an acid (HX) to protonate the amino group, the acid residue was coupled with the ammonium cation formed (Fig. 3, IL c). A series of ionic liquids prepared by such neutralization are sometimes called "protic ionic liquids." They have the potential to show proton conductivity produced by labile protons [21]. Equimolar mixing of these components is essential, but this is quite an easy method to prepare one kind of ionic liquids. After basic evaluation of physicochemical properties of salts obtained by neutralization, one can prepare the "real" ionic liquids by coupling amino acids with suitable counterions.

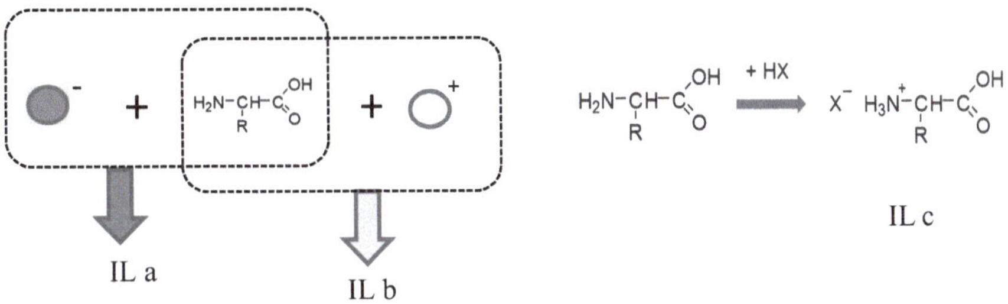

Fig. 3 Amino acids are coupled with acid or base to form ionic liquids (left). Ionic liquids can also be prepared by the neutralization of amino acids (right)

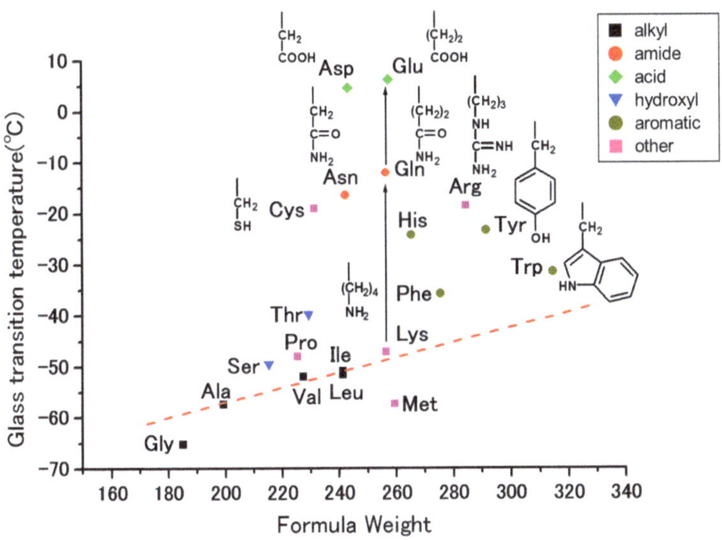

Fig. 4 Relationship between formula weight and glass transition temperature of a series of amino acid-derived ionic liquids. C2mim cation was used to prepare these ionic liquids

In one of studies on the properties of a series of amino acid-derived ionic liquids, we used a series of amino acids as anions to couple with 1-ethyl-3-methylimidazolium cation (C2mim). There is a strong correlation between glass transition temperature (T_g) and viscosity [15]. The T_g of a series of [C2mim] [amino acid] ionic liquids is also the function of their formula weight as shown in Fig. 4. It is quite understandable that the T_g increased with the formula weight of corresponding amino acids. Furthermore, it jumped when amino acids containing some polar groups were used. For example, the T_g of [C2mim][Lys], [C2mim][Gln], and [C2mim][Glu] is −47, −12, and 6°C, respectively. The formula weight of these three ionic liquids is almost the same, and the increased T_g is the result of interaction among ions, for example, through hydrogen bond and other interaction forces. The additional interaction between component ions is quite useful and effective in regulating the physicochemical properties of ionic liquids. Accordingly, [C2mim] [Gly] showed the lowest T_g because of no extra interaction between ions.

One of the serious weak points of amino acids is that they are all only soluble in water. It was impossible to dissolve 20 different amino acids in the same organic solvent. However, after forming the ionic liquids, all these amino acid-based ionic liquids are easily soluble in many organic solvents as well as water. Quite interestingly, pristine amino acids are also soluble in these amino acid-derived ionic liquids, considerably improving the chemistry of amino acids through the wide availability of solvents.

Furthermore, amino acids formed ionic liquids with excellent properties, such as high polarity [15]. Many amino acid-derived ionic liquids showed much higher proton accepting ability than those containing the chloride anion and other polar ionic liquids. Some amino acid-derived ionic liquids show higher proton accepting ability than carboxylate-derived ones, and are reported to dissolve cellulose [22]. Itoh et al. succeeded in regenerating cellulose just by treating cellulose with alanine-derived ionic liquids [22].

Because polar materials are hydrophilic and there are many intermolecular interactions, such as hydrogen bonding, polar materials are generally viscous. It was a challenge to design polar and low viscosity ionic liquids by mixing two distinct amino acid ionic liquids [23]. It is interesting that the mixture of tetrabutylphosphonium aspartate ([P_{4444}][Asp]) and tetrabutylphosphonium lysinate ([P_{4444}][Lys]) showed a strange polarity change as shown in Fig. 5. Dipolarity (π*), one of the Kamlet–Taft polarity parameters, of both pristine ionic liquids was around 1.05, but it was higher than 1.15 when the mixing ratio of [P_{4444}][Asp] and [P_{4444}][Lys] was 1:3. In spite of a very high π* value, viscosity was as low as that of [P_{4444}]

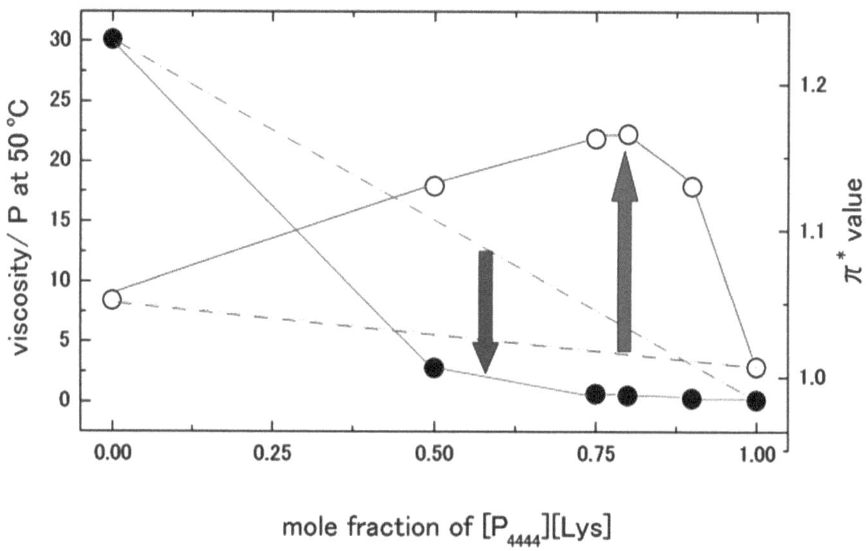

Fig. 5 Effect of [P_{4444}][Lys] fraction on the π* value (open circles) and viscosity (filled circles) at 50°C for the mixture of [P_{4444}][Asp] and [P_{4444}][Lys]. Dashed lines show the expected value where no particular interaction among these ionic liquids

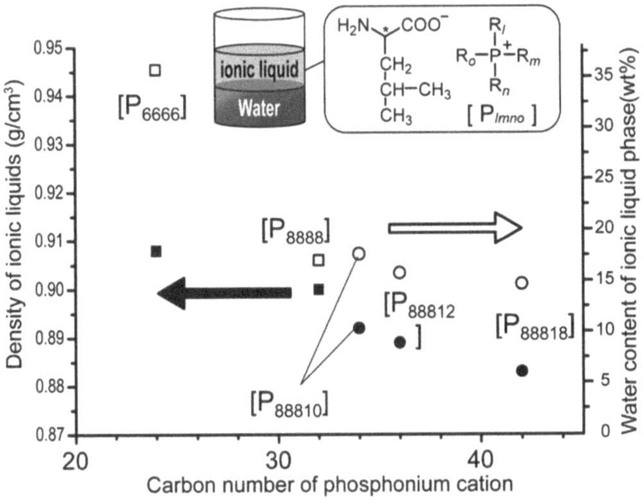

Fig. 6 Density (25°C) and water content (hydrophobicity) of a series of Ile-based ionic liquids [25]

[Lys] (see Fig. 5). Here it should be mentioned that mixing two different amino acid ionic liquids may give you unexpected properties from their individual natures.

Toxicity of amino acid-derived ionic liquids has been widely examined. Recently, ionic liquids composed of cholinium cation and some amino acids such as arginine, glutamine, glutamic acid, and cysteine were examined. They showed much less toxicity than those containing imidazolium and pyridinium cations [24]. There is a strong tendency to use cholinium cation, one of naturally-obtained ions, to design ionic liquids with different properties including biocompatibility.

Hydrophobic and low-density amino acid ionic liquids have also been proposed by controlling the hydrophobicity of phosphonium cations [25]. In the case of tetraalkylphosphonium isoleucinate salts, the density of the ionic liquids is decreased by increasing the alkyl chain length of the cation as shown in Fig. 6. Of course, water content in the ionic liquid phase decreased alongside this. For example, those with phosphonium cations having alkyl chains longer than hexyl were phase-separated with water to cover the surface of water as seen in Fig. 7. Hydrophobic and ionic liquids lighter than water are useful to cover the aqueous phase to suppress vaporization of water and to increase some functions based on the used ionic liquids. These properties were also affected by the amino acid species [25]. Other amino acid ionic liquids, however, showed similar phase separation when cations had enough hydrophobicity. Such polar and hydrophobic ionic liquids should also be useful in many solution-based chemistry fields.

It should be noted that the study on hydrophobic amino acid ionic liquids led to the dawning of the study on the dynamic phase change of ionic liquid/water mixtures. Ionic liquids are generally divided into two major groups such as water-soluble and water-insoluble. We have prepared many ionic liquids with different properties, and analyzed their miscibility with water. We then found some ionic liquids with suitable

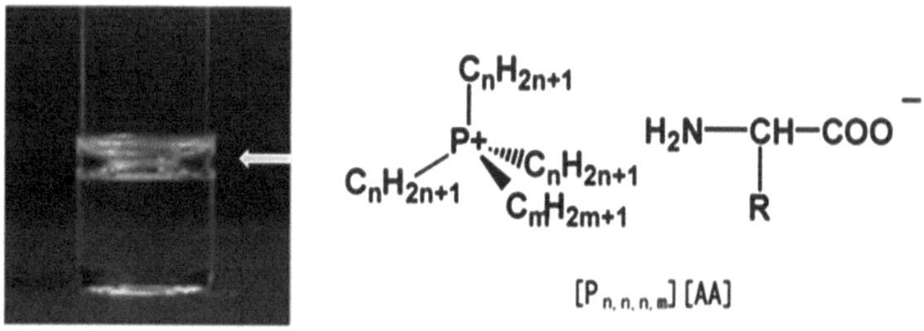

Fig. 7 Tetradecylphosphonium leucinate ($[P_{10,10,10,10}]$[Leu]) (upper phase with arrow) was phase separated with water (lower phase) at 25°C [25]

hydrophobicity that showed a lower critical solution temperature (LCST)-type dynamic phase transition after mixing them with water. The first was composed of amino acid-derived anion [26]. These were homogeneously mixed at temperatures below the LCST, but they were phase separated to show two phases at temperatures above the LCST. After this finding, we analyzed the required properties of component ions to show the LCST-type phase separation. A series of studies revealed that the hydrophobic ionic liquids having some bound water showed the LCST-type phase transition [27–30]. Some hydrophobic ionic liquids containing more than seven water molecules per ion pair showed LCST-type phase separation after mixing them with water. This requirement is effective even for ionic liquid mixtures, and some amino acid ionic liquid mixtures can be designed to show this LCST-type phase separation by selecting suitable cations as well as amino acid anions [31]. This dynamic phase change was applied to separate water soluble proteins within a very short period of time [32, 33]. There are many applications for this kind of dynamic phase change of ionic liquid/water mixtures, and bio-derived ionic liquids also have chances to be used as key ionic liquids for this purpose.

Chirality of amino acids is one of advantages for a variety of applications, the chirality being maintained after forming ionic liquids [11, 34]. However, formation of ionic liquids does not improve the thermal stability of the chiral properties of amino acids, and racemization occurs at temperatures of 120°C and above [11, 34]. Figure 8 shows the time and temperature dependence of the chirality of $[P_{4444}]$[L-Val]. The $[P_{4444}]$[L-Val] kept the chirality at 100°C for hours, but it turned racemic mixture after being kept at 150°C for 5 h. The ionic liquid formation was strongly suggested not to be good to retain the chiral properties of amino acids, especially at higher temperatures.

Because amino acid ionic liquids contain free amino groups, they are potential candidates for CO_2 capture [35–37]. There are many papers on the absorption/desorption of CO_2 with amino acid-based ionic liquids and their membranes. These are also useful for desulfurization of flue gas [38]. Gas separation and/or specific absorption with ionic liquids is one of the practical applications of ionic liquids.

There are other uses of amino acid ionic liquids. Papers on these subjects are still published every month and should be seen.

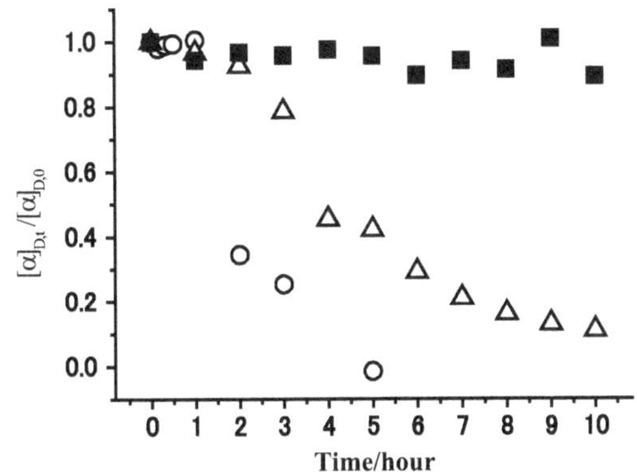

Fig. 8 Temperature dependence of chiral stability of $[P_{4444}]$[L-Val] [11]. The ionic liquids were kept at 100°C (filled square), 120°C (filled triange), or 150°C (open circle)

5 Choline-Based Salts

5.1 *Cholinium Cation*

Choline is a typical natural amine known as Vitamin B, and its acetylated ester is acetylcholine, one of the well-known neurotransmitters. Choline is a component of membrane lipid, phosphatidylcholine, and choline chloride is known to be a good partner to form deep eutectic mixtures. For example, a mixture of solid choline chloride and solid urea (1:2 by mol) forms a liquid known as a deep eutectic fluid [39]. There are many papers on the choline chloride-based deep eutectic solvents [40] and on the deep eutectic solvents based on natural products, and these natural deep eutectic solvents were reviewed by Paiva et al. [41, 42]. Among these, choline chloride is the major component.

Similarly, there are increasing numbers of papers in which the cholinium cation is used as a major component to prepare ionic liquids. Many different anions were coupled with the cholinium cation to form salts, and some anions were bio-derived. As mentioned above, because cholinium cation is frequently found in biological systems, there are many examples of wholly bio-derived ionic liquids in cholinium-based ionic liquids.

5.2 *Cholinium Amino Acid ILs*

It is well-known that proteins and enzymes are mainly composed of amino acids. Amino acids can also be found as components of other biomolecules, such as cell membrane lipids and neurotransmitters. As mentioned above, amino acids are used as components to prepare polar and functionalized ionic liquids. It is quite natural to use these amino acids and cholinium cations for the preparation of whole natural ionic liquids. The number of published papers on cholinium amino acid-type ionic liquids is greater than those on other bio-derived ionic liquids.

Because amino acid-derived ionic liquids are known to have a very strong proton accepting ability, cholinium amino acid-type ionic liquids are well-used as solvents for biomass treatments. The effect of amino acid species on the dissolution of cellulose, hemicellulose, and lignin has been evaluated [43]. The interaction between component polymers of biomass and ionic liquids depends on the amino acids used. Pre-treated microcrystalline cellulose with cholinium glycinate ([Ch][Gly]) showed better results on enzymatic hydrolysis because of looser cellulose fibrils [43]. Similarly, this [Ch][Gly] was used to stabilize aged cellulose paper through moderate interaction with it [44].

There are some effects on cell growth found in the presence of [Ch][Lys] and [Ch][Ser], but it was inhibited by further addition of cholinium acetate ([Ch][OAc]) [45]. They also affected microbial lipid production by *Trichosporon fermentans*. This study strongly suggested that the anion species deeply affected cell growth and lipid production.

5.3 Cholinium Dihydrogen Phosphate

Phosphoric acid is a typical bio-based molecule, and it is one of the important components in biological systems. It is well-known that phosphoric acid is a component of DNA, RNA, ATP, and so on in our living bodies. Dihydrogen phosphate is one of the analogues of phosphoric acid. Accordingly, cholinium dihydrogen phosphate is also a typical bio-derived salt. It is a solid salt at room temperature, and its melting point is 119°C. So it is not a "real" ionic liquid, but this salt has a great power as a solvent when mixed with a small amount of water. Aqueous buffer solutions are the best solvents for biomolecules such as proteins, because of similar ion concentrations and osmotic pressures to our human body. As mentioned in Fig. 9, pure water is not a good solvent for biomolecules such as enzymes, although an aqueous salt solution, a so-called buffer solution, is known to be a good solvent for them. Further addition of salt to the buffer solution does not help to dissolve them, and most biomolecules are denatured. So everybody believed there were no good solvents for biomolecules when further salts were added to buffer solutions. It is impossible to add extremely large amounts of salt into an aqueous solution because of the limited solubility of the salts in water. However, starting with ionic liquids, it is easy to prepare an aqueous salt solution with very high salt concentration. Pure ionic liquids are mostly not good solvents for proteins. Their higher ordered structure is changed because of the very high charge density. Because of electrostatic shielding of charges on polypeptide chains, their compact structure was loosened. However, some ionic liquids mixed with a small amount of water are good solvents for proteins. Most ionic liquid/water mixtures can dissolve proteins, keeping their native higher ordered structure. This region (marked with "**?**" in Fig. 9) cannot be found from an approach adding salt to aqueous salt solutions. Cholinium dihydrogen phosphate (or choline dihydrogen phosphate) ([Ch][DHP]) was found to be an excellent solvent for proteins.

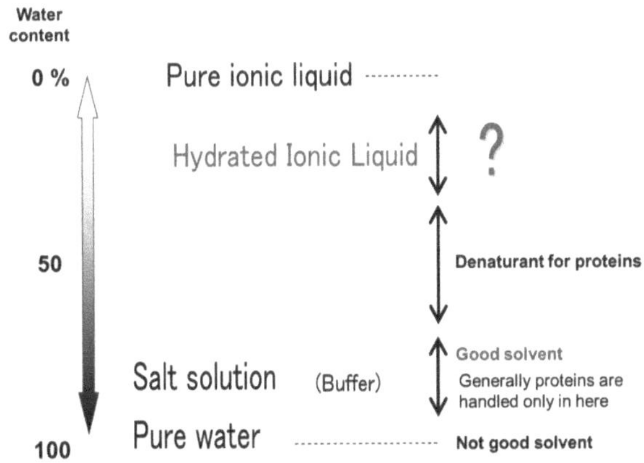

Fig. 9 Relation between salt content (water content) of salt/water mixtures and biological activity of proteins

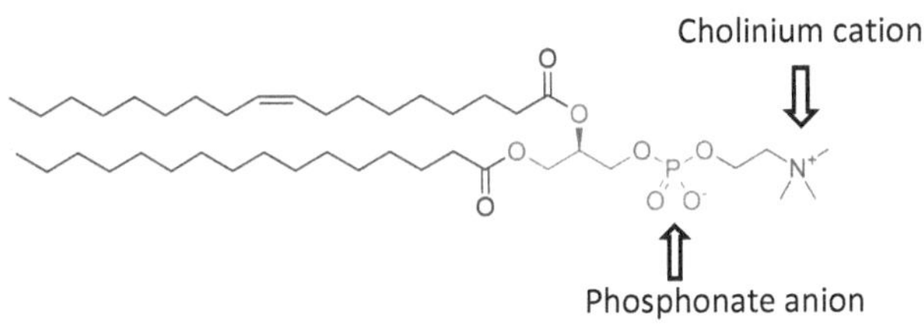

Fig. 10 Structural similarity of the hydrophilic part of the phospholipid and [Ch][DHP]

Stabilization of enzymes and other biological materials in hydrated ionic liquids is mentioned in detail in [46].

It is possible to discuss the biocompatibility of ([Ch][DHP]/water mixtures by comparing their structure with the polar part of phospholipids. It is known that phospholipids are a major component of cell membranes. The hydrophilic part of the phospholipid is phosphorylcholine, a pair of cholinium cations and a phosphonate anion. A mixture of [Ch][DHP] and a small amount of water can be regarded as a hydrophilic part of the cell membrane. As shown in Fig. 10, they are quite similar in structure.

In the case of cell membrane surfaces, all the water molecules are bound to ions. This chemical environment is quite similar to that of [Ch][DHP] containing a small amount of water. This should contribute to the dissolution and stable preservation of proteins and other biological molecules. As mentioned above, phosphorylcholine is the hydrophilic part of the phospholipid, and this phosphorylcholine is regarded as the tethered ion pair of [Ch][DHP], a so-called zwitterion. Tethering cation with anion generally raises the melting point, and, accordingly, this is not a good method to prepare low melting point salts. However, addition of a small amount of water solves this. The phosphorylcholine/water mixture is liquid and this should also be biocompatible. Please refer to [46] for further discussion.

6 Future Aspects

The future for ionic liquids composed of naturally-derived ions is bright. The number of ionic liquids used practically should increase not only in industry but in the whole of future society. Considering society based on the bioeconomy, it is quite reasonable to use naturally-derived ions as components for functional ionic liquids. There should be other naturally-derived ionic liquids newly found to have unique properties, because there is still a great possibility for coupling many different ions in nature. Further studies on ionic liquids composed of naturally-derived ions should gather more and more interest, and not only industry but also society require these bio-based ionic liquids for our future.

7 Conclusions

Preparation, basic properties, and some applications of ionic liquids composed of naturally-derived ions are summarized in this chapter. There are many candidate ions in nature suitable for ionic liquid preparation. Many applications are waiting for these ionic liquids in a wide variety of fields, including life science. It should be mentioned that ionic liquids composed of naturally-derived ions are not safe enough because they are salts. In an aqueous medium they dissociated into ions to generate osmotic pressure. Again, these ionic liquids are quite useful in many situations, and continuous study on the functionalization of ionic liquids should be pursued.

Acknowledgements Some of the results introduced in this chapter were obtained by the considerable efforts of laboratory members. The following scientists should be acknowledged: the late Dr. Kenta Fukumoto, Dr. Kyoko Fujita, Dr. Junko Kagimoto, Dr. Yukinobu Fukaya, Dr. Yuki Kohno, and Dr. Satomi Taguchi. I also acknowledge the financial support of a Grant-in-Aid for Scientific Research from the Japan Society for the Promotion of Science (KAKENHI).

References

1. Yoshio M, Mukai T, Kanie K, Yoshizawa M, Ohno H, Kato T (2002) Liquid-crystalline assemblies containing ionic liquids: an approach to anisotropic ionic materials. Chem Lett 31:320–321
2. Moriel P, García-Suárez EJ, Martínez M, García AB, Montes-Morán MA, Calvino-Casilda V, Bañares MA (2010) Synthesis, characterization, and catalytic activity of ionic liquids based on biosources. Tetrahedron Lett 51:4877–4881
3. Kamlet MJ, Abboud JLM, Abraham MH, Taft RW (1983) Linear solvation energy relationships. 23. A comprehensive collection of the solvatochromic parameters, π^*, α and β, and some methods for simplifying the generalized solvatochromic equation. J Org Chem 48:2877–2887
4. Fukaya Y, Hayashi K, Wada M, Ohno H (2008) Cellulose dissolution with polar ionic liquids under mild conditions: required factors for anions. Green Chem 10:44–46

5. Syguda A, Janiszewska D, Materna K, Praczyk T (2011) Ionic liquids with herbicidal anions. Tetrahedron 67:4838–4844
6. Pernak J, Syguda A, Materna K, Janus E, Kardasz P, Praczyk T (2012) 2,4-D based herbicidal ionic liquids. Tetrahedron 68:4267–4273
7. Marrucho IM, Branco LC, Rebelo LP (2014) Ionic liquids in pharmaceutical applications. Ann Rev Chem Biomol Eng 5:527–546
8. Davis JJ et al (2014) On the formation of a protic ionic liquid in nature. Angew Chem Int Ed 53:11762–11765
9. Ohno H (2006) Functional design of ionic liquids. Bull Chem Soc Jpn 79:1665–1680
10. Ohno H (2017) Physical properties of ionic liquids for electrochemical applications. Chapter 3. In: Endres F et al (eds) Electrodeposition from ionic liquids2nd edn. Wiley VCH, Weinheim, pp 55–94. ISBN: 978-3-527-33602-9
11. Fukumoto K, Yoshizawa M, Ohno H (2005) Room temperature ionic liquids from 20 natural amino acids. J Amer Chem Soc 127:2398–2399
12. Ohno H, Fukumoto K (2007) Amino acid ionic liquids. Acc Chem Res 40:1122–1129
13. Zhao H, Jackson L, Song Z, Olubajo O (2006) Enhancing protease enantioselectivity by ionic liquids based on chiral- or ω-amino acids. Tetrahedron Asymm 17:1549–1553
14. Allen CR, Richard PL, Ward AJ, van de Water LGA, Masters AF, Maschmeyer T (2006) Facile synthesis of ionic liquids possessing chiral carboxylates. Tetrahedron Lett 47:7367–7370
15. Kagimoto J, Fukumoto K, Ohno H (2006) Effect of tetrabutylphosphonium cation on the physico-chemical properties of amino-acid ionic liquids. Chem Commun 21:2254–2256
16. Tao G, He L, Liu W, Xu L, Xiang W, Welton T, Kou Y (2006) Preparation, characterization and application of amino acid-based green ionic liquids. Green Chem 8:639–646
17. Plaquevent JC, Levillain J, Guillen F, Malhiac C, Gaumont AC (2008) Ionic liquids: new targets and media for alpha-amino acid and peptide chemistry. Chem Rev 108:5035–5060
18. Gonzalez L, Altava B, Bolte M, Burguete MI, Garcia-Verdugo E, Luis SV (2012) Synthesis of chiral room temperature ionic liquids from amino acids - application in chiral molecular recognition. Eur J Org Chem 26:4996–5009
19. Tao G-H, He L, Sun N, Kou Y (2005) New generation ionic liquids: cations derived from amino acids. Chem Commun 28:3562–3564
20. Hirao M, Sugimoto H, Ohno H (2000) Preparation of novel room temperature molten salts by neutralization of amines. J Electrochem Soc 147:4168–4172
21. Martinelli A, Matic A, Jacobsson P, Borjesson L, Fernicola A, Panero S, Scrosati B, Ohno H (2007) Physical properties of proton conducting membranes based on a protic ionic liquid. J. Phys Chem B 111:12462–12467
22. Ohira K, Abe Y, Kawatsura M, Suzuki K, Mizuno M, Amano Y, Itoh T (2012) Design of cellulose dissolving ionic liquids inspired by nature. ChemSusChem 5:388–391
23. Kagimoto J, Noguchi K, Murata K, Fukumoto K, Nakamura N, Ohno H (2008) Polar and low viscosity ionic liquid mixtures from amino acids. Chem Lett 37:1026–1027
24. Gouveia W, Jorge TF, Martins S, Meireles M, Carolino M, Cruz C, Almeida TV, Araújo MEM (2014) Toxicity of ionic liquids prepared from biomaterials. Chemosphere 104:51–56
25. Kagimoto J, Taguchi S, Fukumoto K, Ohno H (2010) Hydrophobic and low-density amino acid ionic liquids. J Molecular Liq 153:133–138
26. Fukumoto K, Ohno H (2007) LCST type phase changes of a mixture of water and ionic liquids derived from amino acids. Angew Chem Int Ed 46:1852–1855
27. Kohno Y, Arai H, Saita S, Ohno H (2011) Material design of ionic liquids to show temperature-sensitive LCST-type phase transition after mixing with water. Australian J Chem 64:1560–1567
28. Kohno Y, Ohno H (2012) Key factors to prepare polyelectrolytes showing temperature-sensitive LCST-type phase transition in water. Aust J Chem 65:91–94
29. Kohno Y, Ohno H (2012) Temperature-responsive ionic liquid/water interfaces: relation between hydrophilicity of ions and dynamic phase change. Phys Chem Chem Phys 14:5063–5070

30. Kohno Y, Ohno H (2012) Ionic liquid/water mixtures: from hostility to conciliation. Chem Commun 48:7119–7130
31. Saita S, Kohno Y, Nakamura N, Ohno H (2013) Ionic liquids showing phase separation with water prepared by mixing hydrophilic and polar amino acid ionic liquids. Chem Comm 49:8988–8990
32. Kohno Y, Saita S, Murata K, Nakamura N, Ohno H (2011) Extraction of proteins with temperature sensitive and reversible phase change of ionic liquid/water mixture. Polym Chem 2:862–867
33. Kohno Y, Nakamura N, Ohno H (2012) Selective transport of water-soluble proteins from aqueous to ionic liquid phase *via* a temperature-sensitive phase change of these mixtures. Australian J Chem 65:1548–1553
34. Fukumoto K, Kohno Y, Ohno H (2006) Chiral stability of phosphonium-type amino acid ionic liquids. Chem Lett 35:1252–1253
35. Zhang J, Zhang S, Dong K, Zhang Y, Shen Y, Lv X (2006) Supported absorption of CO_2 by tetrabutylphosphonium amino acid ionic liquids. Chem Eur J 12:4021–4026
36. Goodrich BF, de la Fuente JC, Gurkan BE, Lopez ZK, Price EA, Huang Y, Brennecke JF (2011) Effect of water and temperature on absorption of CO_2 by amine-functionalized anion-tethered ionic liquids. J Phys Chem B 115:9140–9150
37. Zhang Y, Yu P, Luo Y (2013) Absorption of CO_2 by amino acid-functionalized and traditional dicationic ionic liquids: properties Henry's law constants and mechanisms. Chem Eng J 214:355–363
38. Wu W, Han B, Gao H, Liu Z, Jiang T, Huang J (2004) Desulfurization of flue gas: SO_2 absorption by an ionic liquid. Angew Chem Int Ed 43:2415–2417
39. Liao JH, Wu PC, Bai YH (2005) Eutectic mixture of choline chloride/urea as a green solvent in synthesis of a coordination polymer:$[Zn(O_3PCH_2CO_2)]\cdot NH_4$. Inorg Chem Commun 8:390–392
40. Smith EL, Abbott AP, Ryder KS (2014) Deep eutectic solvents (DESs) and their applications. Chem Rev 114:11060–11082
41. Paiva A, Craveiro R, Aroso I, Martins M, Reis RL, Duarte ARC (2014) Natural deep eutectic solvents – solvents for the 21st century. ACS Sus Chem Eng 2:1063–1071
42. Espino M, Fernández MA, Gomez FJV, Silva MF (2016) Natural designer solvents for greening analytical chemistry. Trends Anal Chem 76:126–136
43. Liu Q-P, Hou XD, Li N, Zong MH (2012) Ionic liquids from renewable biomaterials: synthesis, characterization and application in the pretreatment of biomass. Green Chem 14:304–307
44. Scarpellini E, Ortolani M, Nucara A, Baldassarre L, Missori M, Fastampa R, Caminiti R (2016) Stabilization of the tensile strength of aged cellulose paper by cholinium-amino acid ionic liquid treatment. J Phys Chem C 120:24088–24097
45. Liu L, Hu Y, Wen P, Li N, Zong M, Ou-Yang B, Wu H (2015) Evaluating the effects of biocompatible cholinium ionic liquids on microbial lipid production by *Trichosporon fermentans*. Biotech Biofuels 8:119
46. Fujita K (2018) Ionic liquids as stabilization and refolding additives and solvents for proteins. Adv Biochem Eng Biotechnol. https://doi.org/10.1007/10_2018_65

Adv Biochem Eng Biotechnol (2019) 168: 215–226
DOI: 10.1007/10_2018_65

Published online: 13 July 2018

Ionic Liquids as Stabilization and Refolding Additives and Solvents for Proteins

Kyoko Fujita

Contents

Abstract This chapter focuses on recent advances in the use of ionic liquids as additives and solvents in protein applications. The solvent properties of ionic liquids can be tuned by the appropriate selection of cation and anion. The effects of different kinds of ionic liquids on protein stability and refolding behavior have been investigated and reported. The ionic liquid properties affect the intermolecular interactions of proteins, inducing different formations and folding behavior. These effects also vary with the concentration of ionic liquids. Although many of the associated mechanisms are not completely clear, some of this behavior may be attributed to the kosmotropicity of the ions and their Hofmeister effects.

K. Fujita (✉)
Department of Pathophysiology, School of Pharmacy, Tokyo University of Pharmacy and Life Sciences, Hachioji, Tokyo, Japan
e-mail: kyokof@toyaku.ac.jp

Graphical Abstract

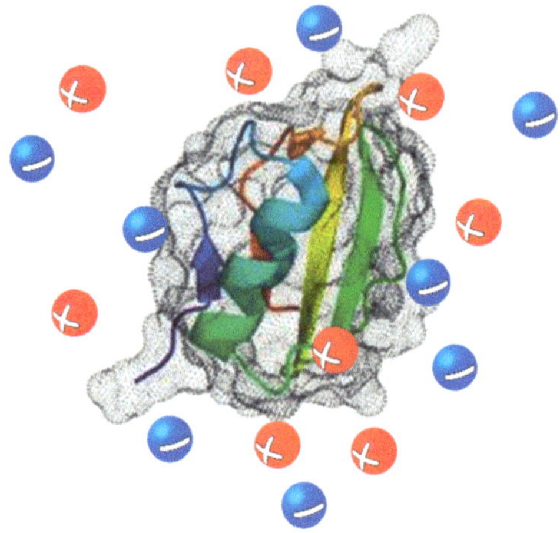

Keywords Additive, Aggregation, Component ion, Protein refolding, Protein stability, Solvation

1 Introduction

Ionic liquids (ILs) have emerged as novel solvents possessing attractive features that are difficult to achieve in ordinary molecular solvents [1, 2]. Hence, ILs are extensively used in a wide range of fields, including chemistry, chemical engineering, biotechnology, and pharmaceuticals [3]. ILs have attracted particular interest as solvents and/or additives for biotechnology and biochemistry applications, namely biocatalysis [4], biotransformation [5], biopreservation [6], and bioseparation [7]. Recently, several reviews focusing on advances in the use of ILs in applications involving biomolecules have been published [8, 9].

Although ILs have been used as novel solvents for biomolecules, it is difficult to dissolve biomolecules in neat ILs and still retain their native structure and activity, regardless of the component ions of the IL. However, some enzymes that show activity as biocatalysts under anhydrous conditions, for example, lipase, show catalytic reactivity when dispersed in neat ILs. To preserve the higher-order structure of biomolecules dissolved in neat ILs, additional procedures are required, such as chemical modification of the protein surface [10] and the addition of specific molecules to the solvent [11]. However, surface modification involves tedious procedures and additional molecules in the IL complicate the system. Thus, the widespread application of neat ILs as protein solvents remains impractical until a significant breakthrough is achieved. For this reason, most studies to date have used mixtures of ILs and conventional molecular solvents for biomolecular applications.

A recent interesting application of such IL/solvent mixtures is their use as additives for protein refolding [12]. By choosing component ions with suitable properties and optimizing the IL concentration, the solubility and the stability of proteins can be improved, apparently through reduced self-aggregation. This chapter discusses the effects of the component ions and the concentration of the IL in the solvent on the stability and structure of dissolved proteins. Specifically, it focuses on accomplishments in the field of protein stability and folding behavior in IL/water mixtures.

2 Effect of ILs on the Intermolecular Interactions of Proteins

Protein aggregation is one of the most important problems in industrial production and storage processes. Protein aggregation and structural stability also play important roles in protein misfolding diseases such as Alzheimer's and Parkinson's diseases. Moreover, self-assembled structures, including amyloid fibrils and nonamyloid fibrillar aggregates of globular proteins such as bovine serum albumin (BSA), human serum albumin (HSA), and ovalbumin, are of great importance in several scientific areas [13]. ILs have shown great potential in controlling protein structures and morphologies. By understanding the effects of the functional groups of ILs on protein structures, it may be possible to achieve deeper insight into IL-induced structural modification of proteins.

Singh and Kang [14] reported the self-assembly of BSA into microparticles, microrods, and long helical fibers mediated by three different imidazolium-based ILs with long alkyl chains, which are called "IL surfactants (ILSs)." Functionalization of the alkyl chain of the ILs with amide and ester moieties, which are capable of hydrogen bonding, had a remarkable effect on the size and shape of the resulting self-assembled BSA structures compared to those obtained using the non-functionalized IL. The formation of self-assembled structures at different IL concentrations was monitored by dynamic light scattering (DLS), zeta potential, fluorescence, and circular dichroism measurements. Figure 1 shows the variation of count rate monitored by DLS of aqueous solutions of BSA as a function of the concentration of various ILs. The ILs interacted with BSA as monomers at low IL concentrations; in contrast, at higher IL concentrations, the monomers underwent hierarchical self-assembly to form morphologically distinct aggregated complexes. The differences in the nature of the prevailing interactions between BSA and ILs in the different IL concentration regimes were demonstrated by the variations in size, surface charge, and degree of unfolding or refolding of BSA. The IL functionalized with a flexible, hydrogen-bonding-prone ester moiety induced refolding of BSA, although the refolded structure was different from the native structure. The ester-type IL also

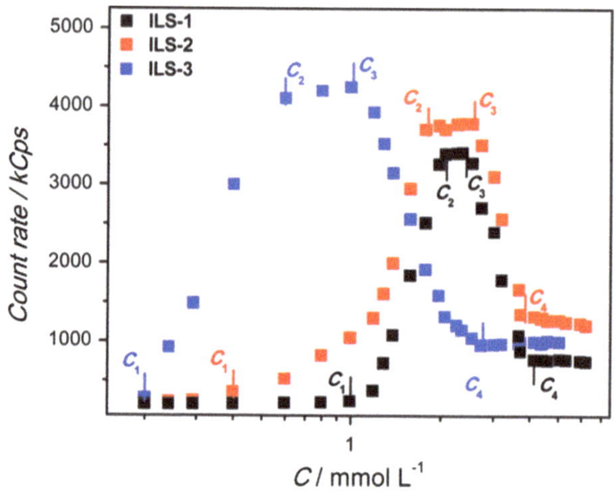

Fig. 1 Count rate observed by dynamic light scattering in aqueous solutions of BSA as a function of concentration of ILSs (Singh and Kang [14])

showed a distinct set of interactions with BSA compared to the other systems, affording long, right-hand-twisted amyloid fibers of BSA in a certain concentration range (Fig. 2).

Similarly, Mangialardo et al. [15] reported the effect of IL cations on the structure of hen egg white lysozyme (HEWL). HEWL, suspended in the ammonium-based ILs 2-methyl ethyl ammonium nitrate, ethyl ammonium nitrate, propyl ammonium nitrate, and butyl ammonium nitrate, was subsequently washed and then analyzed using Raman spectroscopy. The refolding enhancement properties of the ILs were found to depend on the cation structure. Cations with long alkyl chains influenced the proportion of ordered β-aggregates in the fibril conformation and prevented the conversion of β-sheets into α-helices. In contrast, ethyl ammonium nitrate, which has a short alkyl chain, induced both fibril melting and refolding of the secondary structure.

3 ILs as Refolding Additives

Hofmeister reported the effects of coexisting salts on the solubility of proteins in 1888. In the so-called "Hofmeister series," ions are ranked on the basis of their salting-in (protein-solubilizing) and salting-out (protein-desolubilizing) effects. Ions that tend to solubilize and denature proteins are classified as chaotropes. Conversely, ions that induce the structuring of water and hydrogen bonding, accelerate the aggregation of proteins, and stabilize the protein structure are categorized as kosmotropes [16]. Ammonium sulfate, which is one of the most widely used salting-out agents, is employed in protein folding methods as a stabilizer and is categorized as a kosmotrope. Similarly, sugars, polyols, betains, and hydrophilic polymers have also been used to stabilize proteins [17]. The effect of IL addition on the stability and folding behavior of dissolved proteins has also been investigated [18], and in this section the use of ILs as protein refolding additives is summarized.

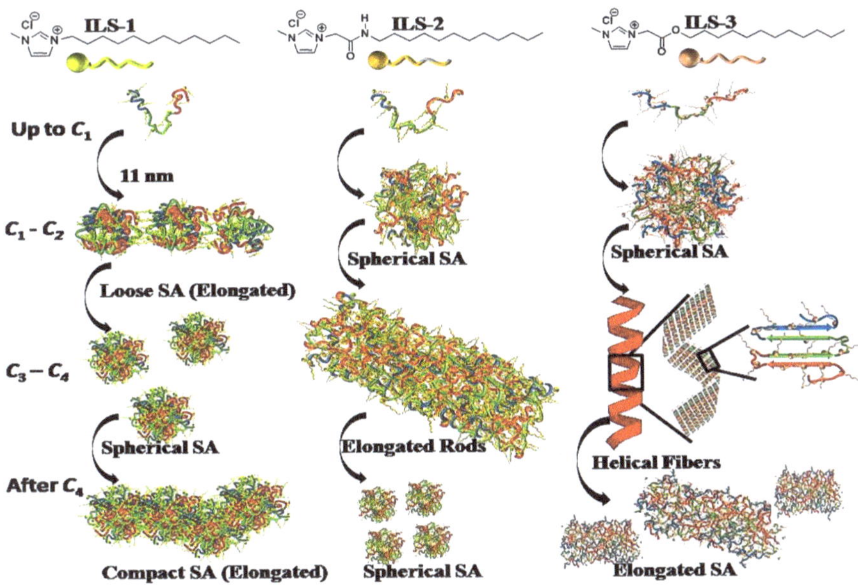

Fig. 2 Schematic view showing various self-assembled structures of BSA formed by mediation of different ILSs in different concentration regimes (Singh and Kang [14]). SA refers to self-assembly

3.1 *Cosolute Systems with Urea-Induced Solubilization and Refolding*

To solubilize protein aggregates, a relatively low concentration (0.5–2.0 M) of a chaotropic additive, such as urea or guanidine hydrochloride, is generally used to dissolve the aggregates, and then the additives are gradually removed during the protein refolding procedure. However, this method results in a limited refolding rate and low efficiency caused by reaggregation of the protein. Some ILs have been reported to be effective for protein refolding when used in conjunction with chaotropic additives. Attri et al. [19] reported that the addition of 100 μL of the protic IL triethyl ammonium phosphate induced structural refolding of the enzymes α-chymotrypsin and succinylated Con A from the urea-induced chemically denatured states (3 or 4 M urea in 2 mg/mL enzyme solution). Bae et al. [20] investigated the effect of additives, including metal cofactors, organic cosolvents, and imidazolium-based ILs, on the refolding of horseradish peroxidase (HRP) that had undergone urea-induced chemical denaturing. Refolding of the denatured sample was initiated by tenfold dilution of the sample with a refolding buffer containing the additives. Among the tested ILs, the highest HRP activity was observed with 1-ethyl-3-methylimidazolium chloride (1.0 vol%). The effect of the ILs on HRP refolding was attributed to Hofmeister effects. The HRP refolding yield increased significantly, up to 75%, with the addition of ILs, whereas yields of 45% and 10%

were obtained in refolding buffers containing hemin and without hemin, respectively. The effect of temperature on the refolding of HRP was also studied, and the highest and most stable refolding yield was obtained at 4°C.

Bisht et al. [21] reported the refolding behavior of urea-denatured lysozyme in the presence of hydrophobic ILs. Ammonium-based ILs with the trifluromethylsulfonyl imide anion (1 vol%) were added to a lysozyme solution pretreated with urea. Structural refolding was observed for some hydrophobic ammonium-based ILs using several optical methods, although it was unclear how the refolding and the activity of the lysozyme in ILs were related.

Molecular dynamics (MD) studies have been used to provide insight into the detailed mechanisms by which urea and ILs affect protein refolding dynamics, many of which remain unclear. For example, an MD simulation conducted by Ghosh et al. [22] showed that triethylammonium acetate efficiently counteracts the urea-induced denaturation of a small S-peptide analogue, even when the preservative IL is present in a very low molar ratio with respect to the denaturant urea. Further studies might help to improve the understanding of the refolding mechanism related to various ILs.

3.2 *Aggregation Inhibition*

Small molecular additives are frequently employed to inhibit aggregation during the refolding procedure [23]. Summers and Flowers [24] were the first to explore the ILs tetraethyl and tetrabutyl ammonium nitrate as additives for protein refolding. These ILs effectively suppressed the aggregation of HEWL and led to a significant increase in refolding yields during oxidative refolding at concentrations up to 5% (0.5 M). However, neat tetraalkyl ammonium nitrates were found to denature the proteins.

Systematic trends in the effects of imidazolium-based ILs on protein refolding were investigated by Lange et al. [25]. The renaturation of two model proteins, HEWL and an anti-oxazolone single-chain antibody fragment, were investigated in the presence of a series of *N*′-alkyl- and *N*′-(hydroxyalkyl)-*N*-methylimidazolium chlorides with alkyl chain lengths of two to six carbon atoms. All the investigated ILs acted as refolding enhancers, and in some instances performed even better than L-arginine hydrochloride, which is one of the most widely used additives for protein renaturation. The most favorable effect on refolding was observed at a concentration of about 1 M, although the optimal concentration varied slightly depending on the protein and IL used. The ability of the ILs to suppress protein aggregation and increase stability was determined to be the key feature responsible for renaturation. Hydrophobic imidazolium cations with longer alkyl chains increased destabilization, whereas cations with hydroxyl-terminated alkyl chains improved protein stability. Similar results in terms of alkyl chain length, IL concentration, and prevention of aggregation during protein refolding were also reported for *N*-alkylpyridinium and *N*-alkyl-*N*-methylpyrrolidinium ILs by Yamamoto et al. [26] (Fig. 3).

Variation of the anion was found to have a profound effect on the renaturation of the recombinant plasminogen activator (rPA) [12], with anions other than chloride leading to a reduction in refolding yields. 1-Ethyl-3-methylimidazolium ([C_2mim]) salts with different anions were screened as additives for in vitro refolding of rPA. The renaturation yields of rPA in the presence of 2-(2-methoxyethoxy)ethyl sulfate ($MDEGSO_4^-$), ethyl sulfate ($EtSO_4^-$), and hexyl sulfate ($HexSO_4^-$) were qualitatively correlated with the hydrophobicity of the three salts. The influence of the anion on the efficacy of the IL as a refolding enhancer for rPA decreased in the following order: $Cl^- > MDEGSO_4^- >$ acetate > tosylate > diethyl phosphate > $HexSO_4^-$. In vitro refolding of rPA was promoted more effectively by [C_2mim][Cl] (refolding yield of 10% at 1.2 M) than by L-arginine hydrochloride (refolding yield of 14% at 3 M), which has been possibly the most widely used additive for in vitro refolding over the last two decades.

These reports reveal that IL additives can act as direct enhancers of protein refolding. Moreover, by preferentially binding to and being slightly or moderately chaotropic for proteins, IL additives prevent protein aggregation.

3.3 *Use of Vesicle Systems*

Bharmoria et al. [27] reported a vesicle-forming biamphiphilic IL that induced significant folding alterations in protein structures. The binding behavior of 3-methyl-1-octylimidazolium dodecylsulfate ([C_8mim][$C_{12}OSO_3$]) with BSA in an aqueous medium at pH 7.0 was investigated at different concentrations. Both the cation and anion of this IL are amphiphilic in nature, and the IL acts as a catanionic surfactant. At low concentrations, [C_8mim][$C_{12}OSO_3$] induced a small amount of unfolding of BSA because of the formation of monomer complexes via exothermic

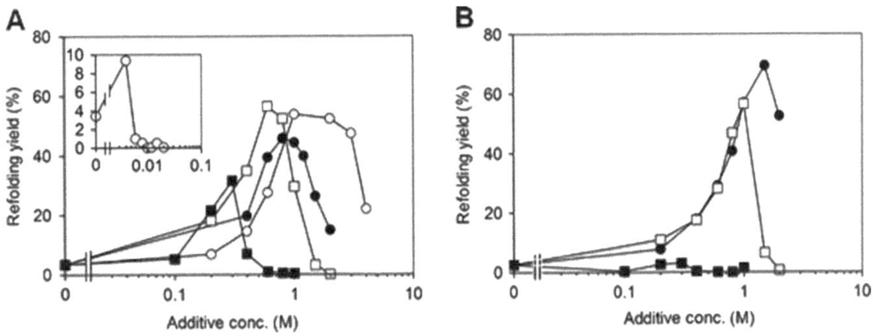

Fig. 3 Refolding yields (recovered activities) of denatured and reduced lysozyme obtained by dilution with refolding buffers containing carious *N*-alkylpyridinium chlorides and *N*-alkyl-*N*-methylpyrrolidinium chlorides as a function of concentrations. (**a**) *N*-Alkylpyridinium chlorides: *N*-ethyl- (open circles), *N*-butyl- (closed circles), *N*-hexyl- (open squares), *N*-octyl- (closed squares), and *N*-dodecyl- (inset, open circles) pyridinium chlorides. (**b**) *N*-Alkyl-*N*-methylpyrrolidinium chlorides: *N*-butyl- (closed circles), *N*-hexyl- (open squares), *N*-hexyl- (open squares) *N*-methylpyrrolidinium chlorides (Yamamoto et al. [26])

electrostatic interactions with the charged amino acid residues on the surface of the protein. As the concentration was increased, protein refolding was observed, until the critical aggregation concentration of $[C_8mim][C_{12}OSO_3]$ was reached. This aggregation resulted from crosslinking of $[C_8mim][C_{12}OSO_3]$ via electrostatic interactions at one end and hydrophobic interactions at the other. Above the critical aggregation concentration, a small amount of BSA unfolding was again observed until the critical vesicular concentration was reached, which suggested that aggregation of $[C_8mim][C_{12}OSO_3]$ at the BSA surface occurred via cooperative electrostatic and hydrophobic interactions in this concentration regime. BSA remained stable against folding alterations in the vesicular and postvesicular regimes because of the strong cooperative hydrophobic and electrostatic interactions among $[C_8mim][C_{12}OSO_3]$ ions (Fig. 4). BSA in the postvesicular concentration regime of $[C_8mim][C_{12}OSO_3]$ showed for a month high stability against aggregation, which is the major cause of protein destabilization. Similar results have been reported for the stability and activity of the enzyme cellulase [28]. This work provided insights into the design of surface-active ILs as artificial chaperones for protein stabilization.

4 Hydrated ILs as Refolding Fields

Hydrated ILs are prepared by mixing ILs with a small amount of water, such that there is almost no free water. The threshold for the existence of free water was suggested to be seven or more water molecules per one ion pair, regardless of the ion structure [29]. Depending on the IL, the IL concentration in hydrated ILs can be greater than 4 M; hence, it is predicted that their effects on proteins are different from those of ILs used as additives in solution. By selecting an optimized ion structure and

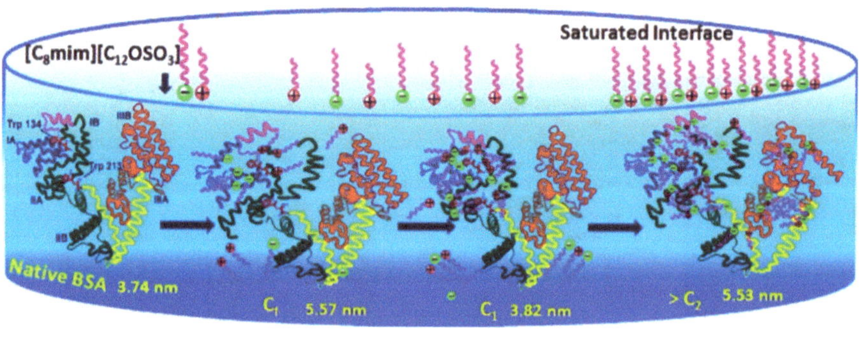

Fig. 4 Folding alterations in BSA at variable concentration of $[C_8mim][C_{12}OSO_3]$ (Bharmoria et al. [28]). C_f signifies the maximum unfolding concentration resulting from association of biamphiphilic IL monomers to BSA; C_1 corresponds to the critical aggregation concentration up to which refolding of BSA occurs as a result of cross-linking of IL ion on BSA, forming BSA – IL aggregate complexes; $>C_2$ shows the post-vesicular regime where BSA – IL aggregates and BSA adsorbed on vesicles exist

water content, hydrated ILs can achieve the dissolution of some proteins, at the same time maintaining their higher-order structure. One of the most effective hydrated ILs as a biomolecular solvent that dissolves biomolecules but maintains the higher-order structure was the pair composed of the cholinium cation and the dihydrogen phosphate anion ([ch][dhp]) [30]. Improvements of the long-term stability and thermal stability of proteins after dissolution were observed in hydrated [ch][dhp] [31]. The refolding of aggregated cellulase, which is a protein heterologously expressed in *Escherichia coli*, was recently reported to occur in hydrated [ch][dhp] without the use of denaturants or additives [32]. A common method for regenerating aggregated proteins that are abundantly expressed in *E. coli* (inclusion bodies) is the use of a high concentration of denaturant to dissolve aggregates after cell breakage by sonication. However, excess amounts of denaturant inhibit protein refolding and recovery, even after dialysis or dilution, and reaggregation of the dissolved proteins occurs during this step. Furthermore, problems, such as the generation of large amounts of wastewater and the long time required to remove the denaturant, make this procedure impractical. A more efficient and easily performed method is needed to regenerate aggregated proteins. Notably, ILs and hydrated ILs have potential as novel refolding solvents that are able to dissolve aggregated protein and/or induce refolding behavior directly. The solubility of aggregated cellulase was not high, but dissolved cellulase was observed to have a folded state similar to that of its native state in hydrated [ch][dhp]. The dissolution and restructuring patterns of aggregated cellulase in hydrated ILs differed depending on the component ions of the IL. When other imidazolium-type ILs (1-ethyl-3-methylimidazolium methylphosphate and 1-ethyl-3-methylimidazolium tetrafluoroborate) were used, unsatisfactory results were obtained, which demonstrated the difficulty of dissolving aggregated cellulase, even in the hydrated state.

The number of water molecules per [ch][dhp] pair clearly influenced the renaturation procedure. The spectral intensity decreased when the number of water molecules per ion pair in hydrated [ch][dhp] was increased. This behavior indicated that the dissolution ability of hydrated [ch][dhp] for aggregated cellulase decreased with increasing water content, even though the viscosity of hydrated [ch][dhp] decreased with increasing water content. Furthermore, the position of the fluorescence spectral maxima, which indicates the folding state of the protein, was affected by the water content (Fig. 5). Fluorescence spectra with a spectral maximum around 340 nm were observed for both native cellulase and cellulase in hydrated [ch][dhp] with four, seven, or ten water molecules per ion pair, which suggests that the aggregated cellulase dissolved in hydrated [ch][dhp] had a folding state similar to that of its native structure in buffer. In contrast, the fluorescence maximum showed a blue shift to 300 nm when cellulase was dissolved in [ch][dhp] with around 2,800 water molecules per ion pair, that is, a 20 mM solution. This spectral shift indicated the formation of an aggregated state similar to that observed in buffer and in water. In [ch][dhp] with 50 water molecules per ion pair, the spectrum had two peaks at around 300 and 340 nm, suggesting that cellulase was present in two different states: the native folded state and the aggregated state. Renaturation and correct refolding of the aggregated cellulase dissolved in hydrated [ch][dhp] was observed. Furthermore,

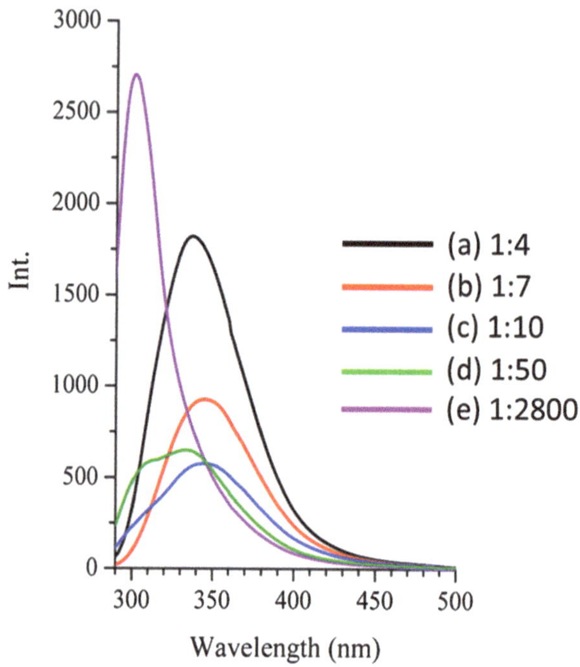

Fig. 5 Fluorescence spectra of supernatants mixed with white precipitate in [ch][dhp] containing (a) 4, (b) 7, (c) 10, (d) 50, and 2,800 water molecules per ion pair (Fujita et al. [32])

cellobiose was produced by the refolded cellulase after desalination of [ch][dhp]. Understanding and controlling the water state in hydrated ILs may facilitate the development of improved and convenient methods for the renaturation of aggregated proteins.

5 Concluding Remarks

ILs have diverse solvation properties that can be tuned by the proper choice of cation and anion. For applications in biotechnology and bioscience, many kinds of ILs have been investigated at different concentrations. ILs have been reported to have notable effects on protein activity, solubility, stability, separation, crystallization, morphology, refolding, and unfolding. Some of this behavior can be attributed to the kosmotropicity of the ions and their Hofmeister effects [33]. Other effects, however, are difficult to explain because of the complexity and variety of IL moieties, as well as their distinct behavior depending on the concentration of water molecules in the solution. The potential of neat or hydrated ILs for bioapplications has not yet been fulfilled. Studies on the mechanisms of biocompatible hydrated ILs and the interactions among biomolecules, ions, and water may reveal useful information for understanding intercellular phenomena.

Acknowledgments This research was supported by a Grant-in-Aid for Scientific Research from the Japan Society for the Promotion of Science (KAKENHI, No. 17H01225). K.F. is grateful for the fund from the Asahi Glass Foundation. Reproduced from Fujita et al. 2016 with permission from The Royal Society of Chemistry.

References

1. Welton T (1999) Room-temperature ionic liquids. Solvents for synthesis and catalysis. Chem Rev 99:2071–2083
2. Seddon KR (1997) Ionic liquids for clean technology. J Chem Technol Biotechnol 68:351–356
3. Armand M, Endres F, MacFarlane DR, Ohno H, Scrosati B (2009) Ionic-liquid materials for the electrochemical challenges of the future. Nat Mater 8:621–629
4. Itoh T, Han SH, Matsushita Y, Hayase S (2004) Enhanced enantioselectivity and remarkable acceleration on the lipase-catalyzed transesterification using novel ionic liquids. Green Chem 6:437–439
5. Moniruzzaman M, Kamiya N, Goto M (2010) Activation and stabilization of enzymes in ionic liquids. Org Biomol Chem 8:2887–2899
6. Fujita K, MacFarlane DR, Forsyth M (2005) Protein solubilising and stabilising ionic liquids. Chem Commun 38:4804–4806
7. Oppermann S, Stein F, Kragl U (2011) Ionic liquids for two-phase systems and their application for purification, extraction and biocatalysis. Appl Microbiol Biotechnol 89:493–499
8. Kumar A, Bisht M, Venkatesu P (2017) Biocompatibility of ionic liquids towards protein stability: a comprehensive overview on the current understanding and their implications. Int J Biol Macromol 96:611–651
9. Sivapragasam M, Moniruzzaman M, Goto M (2016) Recent advances in exploiting ionic liquids for biomolecules: solubility, stability and applications. Biotechnol J 11:1000–1013
10. Ohno H, Suzuki C, Fukumoto K, Yoshizawa M, Fujita K (2003) Electron transfer process of poly(ethylene oxide)-modified cytochrome c in imidazolium type ionic liquid. Chem Lett 32:450–451
11. Shimojo K, Nakashima K, Kamiya N, Goto M (2006) Crown ether-mediated extraction and functional conversion of cytochrome c in ionic liquids. Biomacromolecules 7:2–5
12. Buchfink R, Tischer A, Patil G, Rudolph R, Lange C (2010) Ionic liquids as refolding additives: variation of the anion. J Biotechnol 150:64–72
13. Lara C, Reynolds NP, Berryman JT, Xu A, Zhang A, Mezzenga R (2014) ILQINS hexapeptide, identified in lysozyme left-handed helical ribbons and nanotubes, forms right-handed helical ribbons and crystals. J Am Chem Soc 136:4732–4739
14. Singh G, Kang TS (2015) Ionic liquid surfactant mediated structural transitions and self-assembly of bovine serum albumin in aqueous media: effect of functionalization of ionic liquid surfactants. J Phys Chem B 119:10573–10585
15. Mangialardo S, Gontrani L, Leonelli F, Caminiti R, Postorino P (2012) Role of ionic liquids in protein refolding: native/fibrillar versus treated lysozyme. RSC Adv 2:12329–12336
16. Collins KD (2004) Ions from the Hofmeister series and osmolytes: effects on proteins in solution and in the crystallization process. Methods 34:300–311
17. Kohyama K, Matsumoto T, Imoto T (2010) Refolding of an unstable lysozyme by gradient removal of a solubilizer and gradient addition of a stabilizer. J Biochem 147:427–431
18. Yamaguchi S, Yamamoto E, Mannen T, Nagamune T, Nagamune T (2013) Protein refolding using chemical refolding additives. Biotechnol J 8:17–31
19. Attri P, Venkatesu P, Kumar A (2012) Water and a protic ionic liquid acted as refolding additives for chemically denatured enzymes. Org Biomol Chem 10:7475–7478

20. Bae SW, Eom D, Mai NL, Koo YM (2016) Refolding of horseradish peroxidase is enhanced in presence of metal cofactors and ionic liquids. Biotechnol J 11:464–472
21. Bisht M, Kumar A, Venkatesu P (2016) Refolding effects of partially immiscible ammonium-based ionic liquids on the urea-induced unfolded lysozyme structure. Phys Chem Chem Phys 18:12419–12422
22. Ghosh S, Dey S, Patel M, Chakrabarti R (2017) Can an ammonium-based room temperature ionic liquid counteract the urea-induced denaturation of a small peptide? Phys Chem Chem Phys 19:7772–7787
23. Middelberg AP (2002) Preparative protein refolding. Trends Biotechnol 20:437–443
24. Summers CA, Flowers 2nd. RA (2000) Protein renaturation by the liquid organic salt ethylammonium nitrate. Protein Sci 9:2001–2008
25. Lange C, Patil G, Rudolph R (2005) Ionic liquids as refolding additives: N′-alkyl and N′-(omega-hydroxyalkyl) N-methylimidazolium chlorides. Protein Sci 14:2693–2701
26. Yamamoto E, Yamaguchi S, Nagamune T (2011) Protein refolding by N-alkylpyridinium and N-alkyl-N-methylpyrrolidinium ionic liquids. Appl Biochem Biotechnol 164:957–967
27. Bharmoria P, Rao KS, Trivedi TJ, Kumar A (2014) Biamphiphilic ionic liquid induced folding alterations in the structure of bovine serum albumin in aqueous medium. J Phys Chem B 118:115–124
28. Bharmoria P, Mehta MJ, Pancha I, Kumar A (2014) Structural and functional stability of cellulase in aqueous-biamphiphilic ionic liquid surfactant solution. J Phys Chem B 118:9890–9899
29. Ohno H, Fujita K, Kohno Y (2015) Is seven the minimum number of water molecules per ion pair for assured biological activity in ionic liquid-water mixtures? Phys Chem Chem Phys 17:14454–14460
30. Fujita K, MacFarlane DR, Forsyth M, Yoshizawa-Fujita M, Murata K, Nakamura N, Ohno H (2007) Solubility and stability of cytochrome c in hydrated ionic liquids: effect of oxo acid residues and kosmotropicity. Biomacromolecules 8:2080–2086
31. Fujita K, Ohno H (2012) Stable G-quadruplex structure in a hydrated ion pair: cholinium cation and dihydrogen phosphate anion. Chem Commun 48:5751–5753
32. Fujita K, Kajiyama M, Liu Y, Nakamura N, Ohno H (2016) Hydrated ionic liquids as a liquid chaperon for refolding of aggregated recombinant protein expressed in *Escherichia coli*. Chem Commun 52:13491–13494
33. Zhao H (2016) Protein stabilization and enzyme activation in ionic liquids: specific ion effects. J Chem Technol Biotechnol 91:25–50

Adv Biochem Eng Biotechnol (2019) 168: 227–240
DOI: 10.1007/10_2018_80

Published online: 16 November 2018

Extraction and Isolation of Natural Organic Compounds from Plant Leaves Using Ionic Liquids

Toyonobu Usuki and Masahiro Yoshizawa-Fujita

Contents

Abstract Plants contain many kinds of natural organic compounds, and their compounds possess many useful properties. Natural organic compounds are important for the development of medicines, pesticides, fragrances, cosmetics, and synthetic chemicals. In this chapter, we introduce efficient methods for extraction and isolation of valuable natural organic compounds from various plant leaves by using cellulose-dissolving ionic liquids. High-polarity ionic liquids, which can dissolve cellulose, contribute to the extraction of natural organic compounds from plant leaves probably by breaking down plant cell walls, which are composed of cellulose, hemicellulose, and lignin. Extraction and isolation of shikimic acid from ginkgo leaves, caffeoylquinic acids from sweet potato leaves, and neral and geranial (which combine to form citral) from lemon myrtle leaves were performed. Ionic liquids can achieve extraction rates greater than those achieved with water and other organic solvents.

T. Usuki (✉) and M. Yoshizawa-Fujita (✉)
Department of Materials and Life Sciences, Faculty of Science and Technology, Sophia University, Tokyo, Japan
e-mail: t-usuki@sophia.ac.jp; masahi-f@sophia.ac.jp

Graphical Abstract

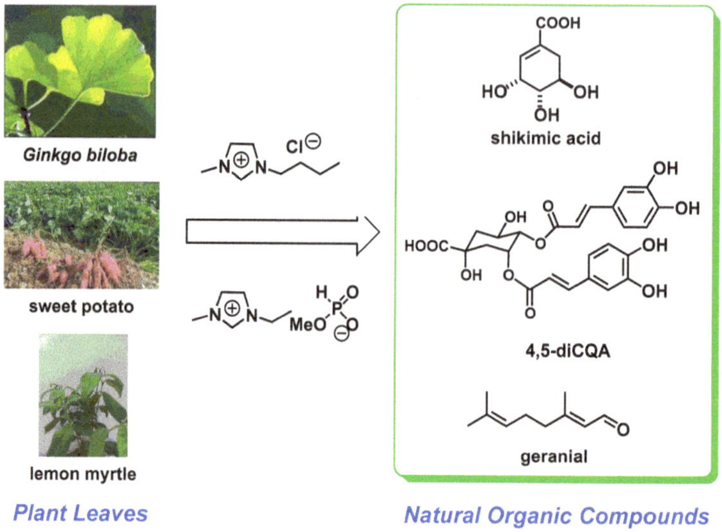

Keywords Extraction, Ionic liquids, Isolation, Natural products, Plant leaves

1 Introduction

Biomass is a resource that can be converted into energy, providing a possible replacement for depleted fossil fuels. Cellulose is an example of biomass and represents the most plentiful type of biomass on earth. Since cellulose can be converted into useful substances such as glucose, it has attracted attention as a source of energy. Cellulose is a polysaccharide composed of glucose polymerized with β(1,4) bonds [1] and involves various types of intramolecular and intermolecular hydrogen bonds. Therefore, cellulose is insoluble in water and most organic solvents [2]. The demand for biomass, however, has resulted in the development of processes that require strong acids, high temperatures, and high pressures. These processes also require large amounts of energy, which means that cellulose is not presently a viable source of energy.

In 2002, Rogers and coworkers reported that cellulose was soluble in an ionic liquid [3], which prompted additional research on the utilization of ionic liquids to process biomass. According to published reports, the anionic structure of the ionic liquid is the key to the solubility of cellulose. The stronger the Lewis basicity (electron-donating property) of the anions in an ionic liquid, the greater the solubility of cellulose [4]. After determining the solubility of cellulose in various ionic liquids,

Rogers et al. reported that chloride ions were best for dissolving cellulose. In subsequent studies, ionic liquids formed from acetate anions [5, 6] and formate anions [7] displayed the same cellulose-dissolving properties as chloride-based ionic liquids. Ohno and coworkers discovered that phosphorous acids were also viable anions for this purpose [8]. Typical structures of cellulose-dissolving ionic liquids are shown in Fig. 1.

Over the years, extraction methodologies have been used to carry out the extraction of target natural organic compounds from plants by using water and volatile organic solvents. In general, the current extraction efficiencies are low. Recently, researchers have been studying the development of alternative extraction processes with ionic liquids that can dissolve cellulose. A few studies have examined the extraction of natural organic compounds from plants using ionic liquids [9–13]. However, only a very small number of these studies have reported isolation of natural products, mainly because of the nonvolatility of the ionic liquids, which makes isolating organic compounds extracted from plant material into the ionic liquid very difficult. Bica and coworkers, however, succeeded in utilizing an ionic liquid to extract limonene, the main component of orange essential oil, from orange peel [14]. First, orange skin was dissolved in an ionic liquid and then stirred at a set temperature. After vacuum distillation, isolation of the volatile limonene was achieved. This is a good example of the nonvolatile nature of an ionic liquid.

This background prompted an investigation into the effectiveness of isolating bioactive natural compounds, which are stored as secondary metabolites in the vacuoles of plant leaf cells, using ionic liquids that can dissolve cellulose – the primary component of plant cell walls. This report describes developments in the extraction and isolation of natural organic compounds from plants by using ionic liquids, focusing on the extraction and isolation of shikimic acid from ginkgo leaves [15], caffeoylquinic acids (CQAs) from sweet potato leaves [16], and neral and geranial (which combine to form citral) from lemon myrtle leaves [17] (Fig. 2).

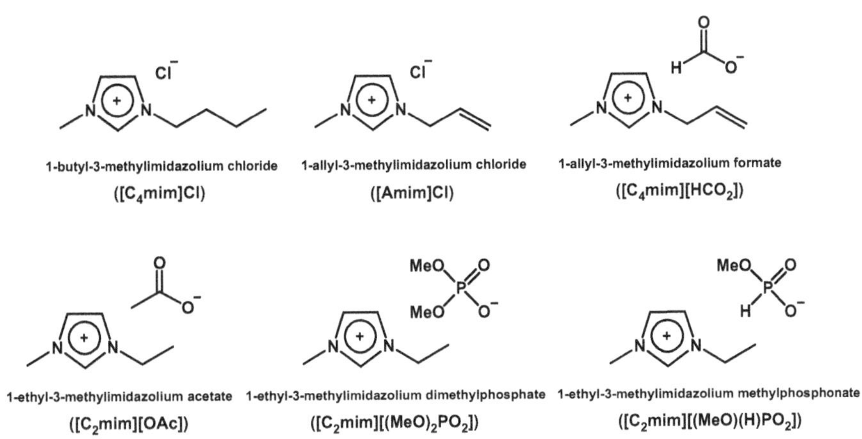

Fig. 1 Typical structures of cellulose-dissolving ionic liquids

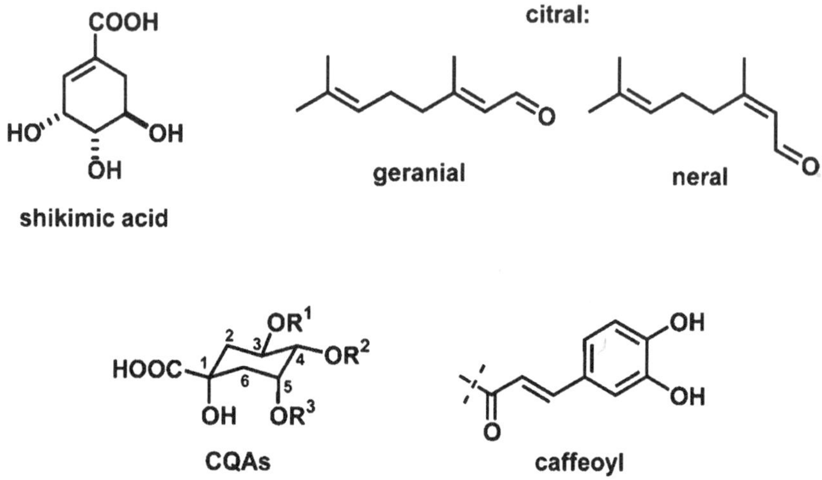

caffeoylquinic acids (CQAs)	R1	R2	R3
3,4-dicaffeoylquinic acid (3,4-diCQA)	caffeoyl	caffeoyl	H
3,5-dicaffeoylquinic acid (3,5-diCQA)	caffeoyl	H	caffeoyl
4,5-dicaffeoylquinic acid (4,5-diCQA)	H	caffeoyl	caffeoyl
3,4,5-tricaffeoylquinic acid (3,4,5-triCQA)	caffeoyl	caffeoyl	caffeoyl

Fig. 2 Chemical structures of shikimic acid, citral, and caffeoylquinic acids

Fig. 3 Molecular structure of oseltamivir phosphate (Tamiflu®)

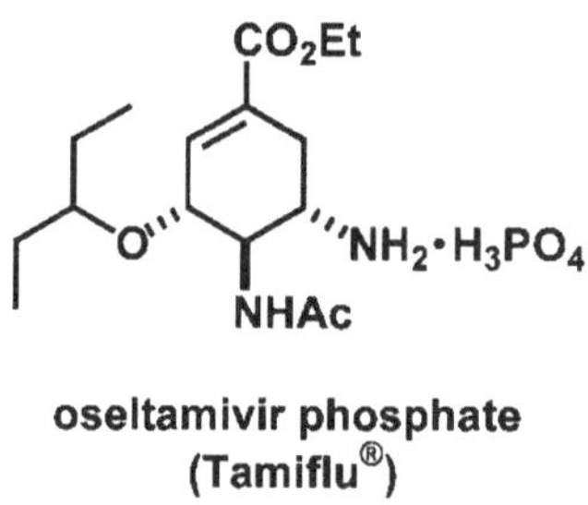

2 Shikimic Acid from Ginkgo Leaves

Shikimic acid – a nonvolatile natural organic compound first isolated from the fruit of Japanese star anise (*shikimi* in Japanese) by Eijkman in 1885 [18] – is the starting material for oseltamivir (also known as Tamiflu®; see Fig. 3), which is used to treat influenza. The pharmaceutical company Roche produces oseltamivir by a 12-step process with shikimic acid as the starting material. Most of the shikimic acid (70%) is extracted from star anise (also called *hakkaku* or Chinese star anise) growing on in southwest China [19]. However, overharvesting of star anise has depleted the resources, leading to supply problems. Considering the widespread occurrence of influenza, a steady supply of shikimic acid is needed. The cosmetics company Kosé recently announced a new whitening cosmetic that uses shikimic acid as its active ingredient [20], indicating that shikimic acid can be expected to have various potential applications beyond the production of oseltamivir.

Securing natural sources of shikimic acid other than star anise is the key to resolving the supply problems. Therefore, the present research focused on ginkgo (*Ginkgo biloba*) because it contains shikimic acid and displays a low level of regional specificity. Ginkgo's worldwide distribution indicates that its main requirement is a minimum average yearly temperature of 0°C [21]. Thus, it can live in a wide variety of places and is easy to acquire.

Shikimic acid was extracted from the green leaves of female ginkgo trees growing on the campus of Sophia University, by freeze drying them with liquid nitrogen and grinding them to produce the experimental material. The protocol for extraction of shikimic acid is shown in Fig. 4. The powdered ginkgo was stirred into 1-*n*-butyl-3-methylimidazolium chloride ([C_4mim]Cl) for 1 h at 100–150°C, and methanol (MeOH) was then added to decrease the viscosity and induce precipitation of cellulose. Stirring was continued until the cellulose was removed by filtration, resulting in collection of [C_4mim]Cl extract. Meanwhile, as a control, the experimental material was stirred in MeOH for 1 h under reflux conditions to obtain a MeOH extract after filtration and removal of cellulose. Extracts were also obtained using ethyl alcohol (EtOH), H_2O, and dimethylformamide (DMF). Then,

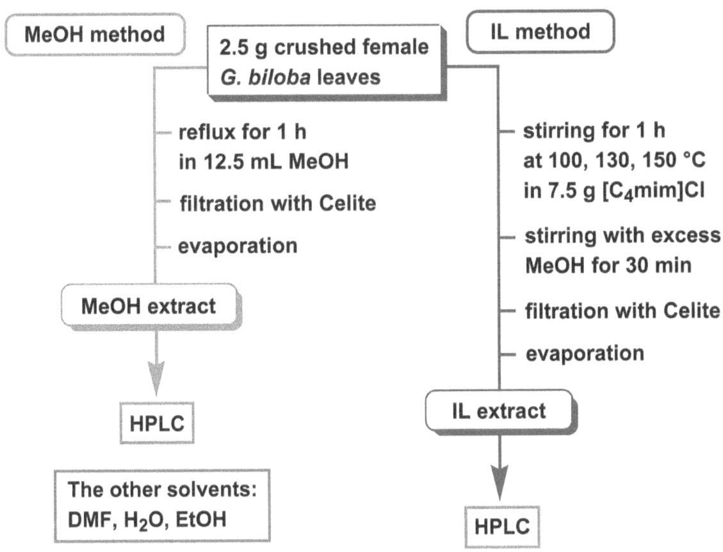

Fig. 4 Protocol for extraction of shikimic acid from ginkgo leaves. *[C₄mim]Cl* 1-*n*-butyl-3-methylimidazolium chloride, *DMF* dimethylformamide, *EtOH* ethyl alcohol, *HPLC* high-performance liquid chromatography, *IL* ionic liquid, *MeOH* methanol

quantitative analysis of each extract was performed using reverse-phase high-performance liquid chromatography (HPLC). Using a calibration curve, the shikimic acid extraction rate was determined against the ginkgo leaf weight for each extraction solvent (MeOH, EtOH, H_2O, DMF, and [C_4mim]Cl) and at different temperatures (Fig. 5).

When using [C_4mim]Cl, the extraction rate varied greatly with temperature. The extraction rate was 1.1% at 100°C, 1.7% at 130°C, and 2.3% at 150°C; thus, the extraction rate increased with temperature. The extraction rate using MeOH was 0.93%, a value slightly lower than that using [C_4mim]Cl, even when stirred at 100°C. Similar extraction rates were obtained with the other solvents: 0.86% with EtOH and 0.86% with H_2O. The use of DMF resulted in an extraction rate of 1.1%, even at 150°C. Thus, the 2.3% extraction rate achieved with [C_4mim]Cl at 150°C indicated that efficient dissolution of cellulose is strongly favored in ionic liquid. The extraction rate with [C_4mim]Cl at 150°C (2.3%) was twice that achieved with methanol (0.93%). Thus, an ionic liquid is more effective than an organic solvent for the extraction of shikimic acid.

Bica and coworkers used ionic liquids that included sulfonate groups to induce esterification reactions, and performed extractions with the derivatives of these processes [22]. After dissolving star anise in an acidic ionic liquid, esterification resulted in the production of shikimic acid ethyl ester and shikimic acid ketal ester, indicating that acidic ionic liquids can act as esterification catalysts, as well as solvents. The prospect of this methodology, which is able to derive target natural organic compounds and extracts, is fascinating.

The postextraction ginkgo leaves were examined using scanning electron microscopy (SEM) (Fig. 6). The image on the left shows the leaf surface before extraction, compared with the leaf residue after MeOH extraction (center image), in which the

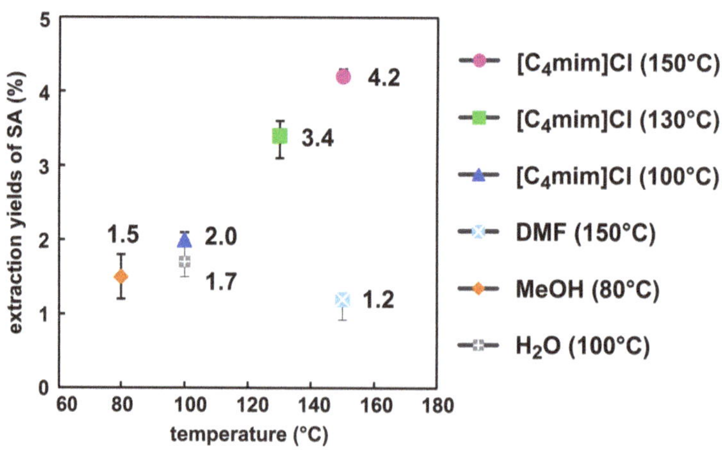

Fig. 5 Relation between shikimic acid (*SA*) extraction rate and stirring temperature for ginkgo leaves. *[C_4mim]Cl* 1-*n*-butyl-3-methylimidazolium chloride, *DMF* dimethylformamide, *MeOH* methanol

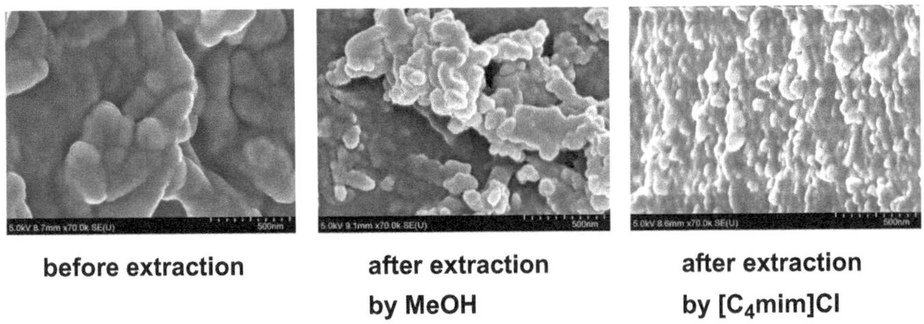

Fig. 6 Scanning electron microscopy images of a ginkgo leaf surface. *[C₄mim]Cl* 1-*n*-butyl-3-methylimidazolium chloride, *MeOH* methanol, *SE(U)* upper secondary electron detector

cell wall structures are broken down. The image on the right, which shows the leaf surface after extraction with [C_4mim]Cl at 150°C, indicates that the leaf surface is broken down more than with MeOH extraction. The ionic liquids must break down the leaves more effectively to enhance the extraction of shikimic acid. Bica and coworkers compared shikimic acid extraction rates using ionic liquids with different types of anions [23]. They combined the 1-ethyl-3-methylimidazolium cation with six different anions – acetate (OAc), chloride (Cl), bis(trifluoromethanesulfonyl) azanide (TFSA), triflate (OTf), tetrafluoroborate (BF_4), and hexafluorophosphate (PF_6) – and then chose ionic liquids with different rates of cellulose solubility. As predicted, anions that resulted in greater cellulose solubility led to greater rates of shikimic acid extraction, with OAc showing the highest value. These results demonstrate that ionic liquids with cellulose-dissolving properties are important for efficient extraction of shikimic acid. Furthermore, shikimic acid could be extracted efficiently from conifer needles using an ionic liquid, indicating that the methodology outlined above can be applied to a variety of plant species.

Although studies related to the harvesting of natural organic compounds from plant leaves using ionic liquids are increasing in number, only a few examples have reported isolation of the target compounds. Perhaps the nonvolatile nature of ionic liquids makes isolation difficult to achieve. A method for isolating shikimic acid from [C_4mim]Cl extract using an ion-exchange resin was investigated [15]. The [C_4mim]Cl extract was supported on the strong basic anion-exchange resin Amberlite IRA 400 (Cl), allowing absorption into the resin, followed by removal of the [C_4mim]Cl using a bath of deionized water. Then, the shikimic acid was obtained in a 25% acetic acid solution. The chemical structure of shikimic acid after isolation by Amberlite was confirmed by ^{1}H nuclear magnetic resonance (NMR) (Fig. 7). For comparison, the spectrum of commercially available shikimic acid is shown in Fig. 7. The shikimic acid isolated using Amberlite was less pure than the commercially available one. Purification using HPLC resulted in isolation of pure shikimic acid, as can be seen in Fig. 7. Many ionic liquids are known to be toxic to bacteria, yeasts, and other organisms [24]. In particular, cellulose-dissolving ionic

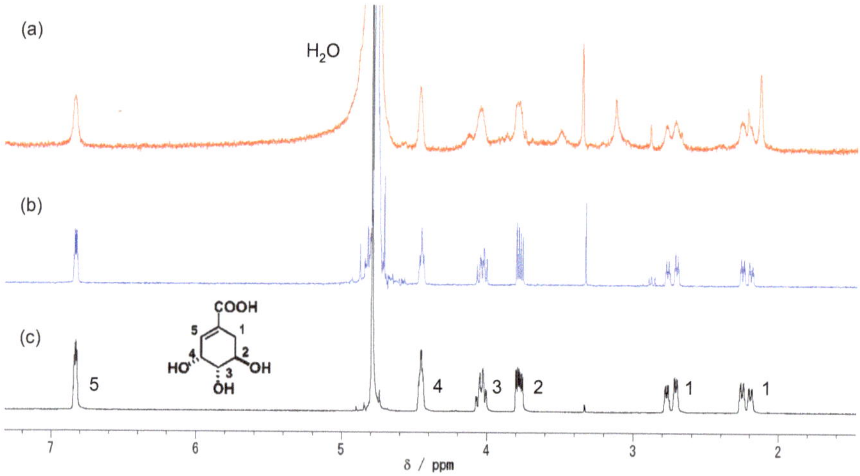

Fig. 7 ^{1}H NMR spectra of shikimic acid: (**a**) after isolation by Amberlite, (**b**) after isolation by high-performance liquid chromatography, and (**c**) as a commercially available reagent. *ppm* parts per million

liquids are highly toxic because they change the structures of proteins dissolved in such ionic liquids [25–27]. Purification of isolated natural organic compounds is also important.

Thus, an efficient method was developed for extracting shikimic acid from ginkgo leaves by using the ionic liquid [C_4mim]Cl as an extraction solvent with cellulose-dissolving ability. In addition, shikimic acid was successfully isolated from the [C_4mim]Cl extract using an ion-exchange resin. The naturally occurring organic compound bilobalide – a terpene lactone found only in ginkgo leaves – was also isolated 1.4 times as efficiently using [C_4mim]Cl as with the previously reported method using MeOH [28].

3 Caffeoylquinic Acid Derivatives from Sweet Potato Leaves

The sweet potato (*Ipomoea batatas*) has been used as a food source since ancient times; it was first cultivated mainly in China but is now grown throughout the world. Its root is eaten, but the leaves are generally discarded. However, these leaves are rich in caffeoylquinic acid derivatives (CQAs) – polyphenols that make the leaves attractive as suppliers of natural biomolecules. The CQAs contain ester bonds between hydroxyl groups of quinic acids and carboxyl groups of caffeic acids. Many CQA analogs exist, categorized by the number and position of the caffeoyl groups – for example, 3-CQA, 3,4-diCQA, 4,5-diCQA, 3,5-diCQA, and 3,4,5-triCQA (Fig. 2) [29]. CQAs have antioxidant properties and cancer-fighting cytostatic effects. CQAs have also been reported to possess aggregation-inhibiting effects on amyloid β (Aβ42), a possible cause of Alzheimer's disease; 4,5-diCQA has the highest rate of effectiveness (half-maximal inhibitory concentration (IC_{50}) 0.1 μM) in this application [30]. Considering the promising health benefits of CQAs, the present study attempted to establish a method using ionic liquids with cellulose-dissolving capabilities to extract and isolate them efficiently from sweet potato leaves that otherwise would be discarded.

The sweet potato leaves used in this experiment were from Suiō potatoes, which have been selectively bred as food potatoes. The [C_4mim]Cl ionic liquid was used to extract the CQAs by adding powdered sweet potato leaves to either a [C_4mim]Cl or [C_4mim]Cl/MeOH (1:1) mixed-solvent solution at 100°C, stirring it for 1 h, and then performing Celite filtration to produce a [C_4mim]Cl extract. That extract was subsequently dissolved in an H_2O/MeOH (7:3) solution to obtain a [C_4mim]Cl layer that contained the CQAs after washing with hexane. The CQA extraction rate was determined using HPLC by comparing the amount of CQAs isolated with the original weight of the leaves. As a control, extraction was performed as was previously done by adding MeOH to powdered sweet potato leaves and stirring the experimental material into MeOH for 1 h under reflux conditions. The MeOH extract was obtained after Celite filtration and removal. The extract was dissolved into an H_2O/MeOH (7:3) mixed-solvent solution and washed with hexane. Quantitative analysis was performed on the crude extract using HPLC, and the CQA extraction rate was calculated (Fig. 8).

The results showed that the [C_4mim]Cl/MeOH (1:1) mixed-solvent solution enabled extraction of 3,5-diCQA 1.9 times as efficiently as use of MeOH. Furthermore, the use of [C_4mim]Cl as the only extraction solvent increased the comparative efficiency of the extraction rate to 2.2 times for 3,4-diCQA and 3.2 times for 3,4,5-

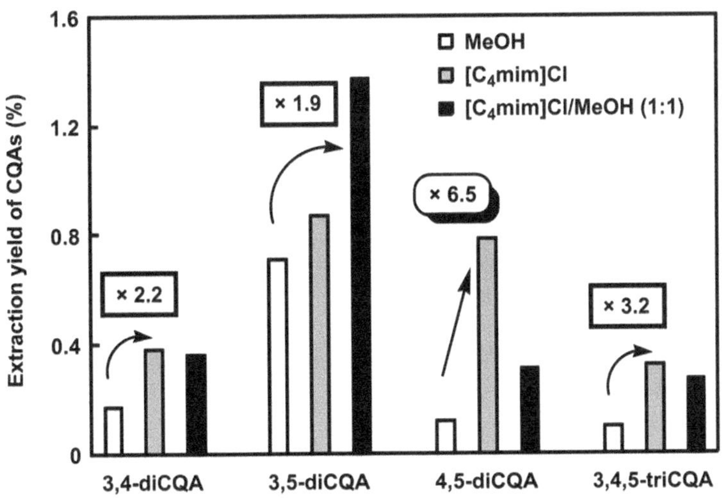

Fig. 8 Extraction yields of caffeoylquinic acids (*CQAs*) from sweet potato leaves. *[C_4mim]Cl* 1-*n*-butyl-3-methylimidazolium chloride, *MeOH* methanol

triCQA. For 4,5-diCQA, which has a high aggregation-inhibition effect on Aβ42, the extraction became 6.5 times as efficient. Although significant differences in extraction rates were observed among CQAs, depending on the number and position of the caffeoyl group, use of [C_4mim]Cl as an extraction solvent was able to increase the extraction rate for every CQA extracted from the sweet potato leaves. Recently, Tan and coworkers dissolved Flos Lonicerae Japonicae in an ionic liquid and, using ultrasonic waves, observed increases in CQA extraction rates [31]. These results indicate that combining ionic liquids and ultrasonic waves is a very interesting approach for increasing the efficiency of extracting natural materials.

Next, isolation of the CQAs contained within the [C_4mim]Cl was investigated. A saturated saline solution was added to the previously obtained [C_4mim]Cl layer and stirred. This was followed by addition of a mixed-solvent solution of ethyl acetate (EtOAc), MeOH, and 90% formic acid (100:20:1). The mixture was then decanted four times to isolate the CQA mixture. A qualitative analysis was then performed using HPLC. A quantitative analysis showed that 90% 3,4-diCQA, 85% 3,5-diCQA, 100% 4,5-diCQA, and 100% 3,4,5-triCQA were isolated. This method enabled simpler separation of the target organic compounds from the ionic liquid than use of supercritical carbon dioxide or recrystallization.

4 Citral from Lemon Myrtle Leaves

The next target was a plant-derived natural compound with volatile essential oils, because previous studies had used ionic liquids to extract essential oils from orange peel and rosemary. The starting material used was the leaf of the lemon myrtle (*Backhousia citriodora*) [32], a plant native to Australia. Lemon myrtle possesses the most citral (a lemon-scented mixture of geranial and neral) among plants and has been used in aromatherapy and herbal teas. The volatility of the citral prompted the use of the ionic liquid 1-ethyl-3-methylimidazolium methylphosphonate ([C_2mim][(MeO)(H)PO_2]) because of its ability to dissolve cellulose at room temperature.

The lemon myrtle leaves were harvested from Tomy Green Farm in Shizuoka Prefecture. As a control study, extraction was attempted using an organic solvent. Lemon myrtle leaves were placed into EtOH, EtOAc, and hexane, and stirred under reflux conditions for 1 h, followed by filtration and qualitative analysis using gas chromatography–mass spectrometry (GC-MS) with methyl decanoate as the internal standard. Then, extraction was performed using the ionic liquid by adding lemon myrtle leaves to [C_2mim][(MeO)(H)PO_2] at 45°C and at 80°C, as well as to a mixed-solvent solution combining [C_2mim][(MeO)(H)PO_2] with an organic solvent at 80°C, and stirring each mixture for 1 h. To reduce the viscosity of the [C_2mim][(MeO)(H)PO_2], water was added with additional stirring, followed by filtration. Extraction was done with water and EtOAc (which does not dissolve in [C_2mim][(MeO)(H)PO_2]), and the extract was subjected to GC-MS quantitative analysis. The extraction rate was calculated by comparing the weights of both geranial and neral with the weight of the lemon myrtle leaves (Fig. 9).

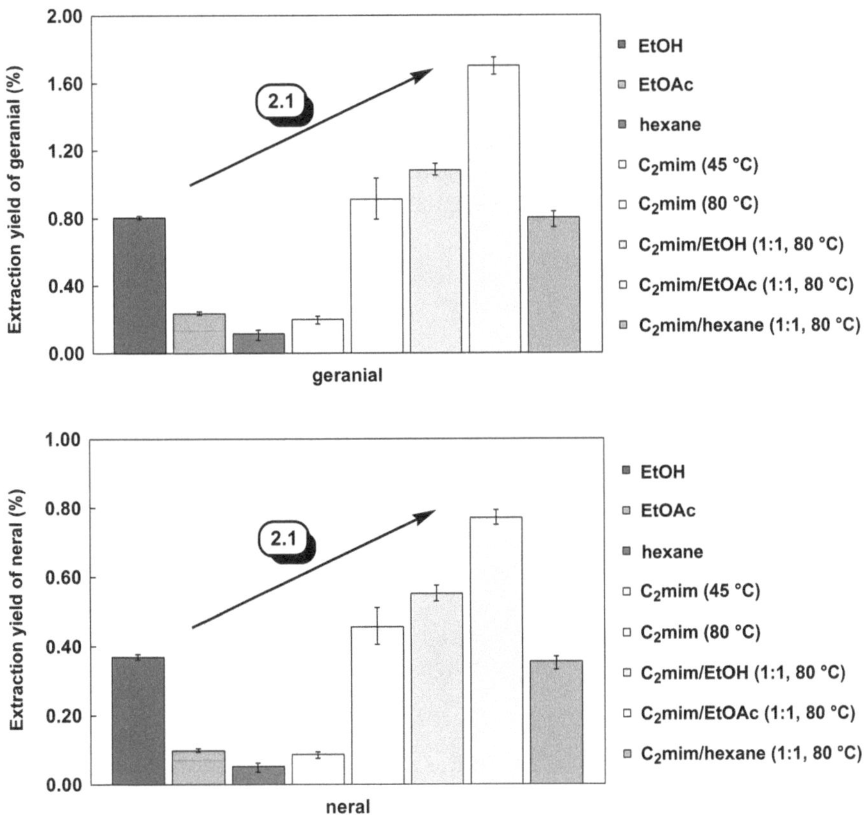

Fig. 9 Extraction yields of geranial and neral from lemon myrtle leaves. *C_2mim* 1-ethyl-3-methylimidazolium methylphosphonate ([C_2mim][(MeO)(H)PO_2]), *EtOAc* ethyl acetate, *EtOH* ethyl alcohol

When the extraction was performed with organic solvents, the greatest extraction rates for both geranial (0.80%) and neral (0.37%) were achieved when using EtOH. In contrast, an extremely low extraction rate was obtained when using only [C_2mim][(MeO)(H)PO_2] at 45°C. However, when the mixture was heated to 80°C, the extraction rate improved. In addition, the use of mixed-solvent solutions of [C_2mim][(MeO)(H)PO_2] combined with either EtOH or EtOAc increased the extraction rate. A mixed-solvent solution of [C_2mim][(MeO)(H)PO_2]/EtOAc (1:1) resulted in extraction rates for geranial and neral of 1.70% and 0.77%, respectively – 2.1 times as efficient as extraction using EtOH.

The viscosity of the [C_2mim][(MeO)(H)PO_2] was determined to assess the effectiveness of stirring. Increasing the temperature from 45°C to 80°C reduced the viscosity of the ionic liquid (from 39.6 mPa·s at 45°C to 13.4 mPa·s at 80°C). These results indicated that increasing the temperature during stirring and combining [C_2mim][(MeO)(H)PO_2] with organic solvents to create a mixed-solvent solution increased the effectiveness of stirring because of the lower viscosity, which increased the extraction rates.

Thus, a method using the cellulose-dissolving ionic liquid [C_2mim][(MeO)(H)PO_2] for extracting the fragrant substance citral from lemon myrtle leaves was developed that was 2.1 times as efficient as extraction using EtOH [17]. The results also confirmed that the ionic liquid could be used for efficient extraction of the fragrant substance geraniol from lemongrass [33].

5 Conclusions

Efficient methods for extracting bioactive natural organic compounds have been developed by using cellulose-dissolving ionic liquids to break down plant cell walls. The results demonstrated ionic liquid extraction and isolation of shikimic acid, caffeoylquinic acids, and citral from the leaves of ginkgo, sweet potato, and lemon myrtle, respectively. The use of ionic liquids that can dissolve cellulose allowed isolation of greater amounts of the natural organic compounds from plant leaves than previous extraction methods using organic solvents or water. This approach to the extraction of natural organic compounds by using ionic liquids can also be applied to other plants; a wide range of ionic liquids and technologies are available for extraction and isolation.

Acknowledgements This work was supported by a Sophia University Special Grant for Academic Research. We thank Prof. Masahiro Rikukawa (Sophia University) for his support and suggestions.

References

1. Smith HD (1937) Structure of cellulose. Ind Eng Chem 29:1081–1084
2. Brandt A, Gräsvik J, Hallett JP, Welton T (2013) Deconstruction of lignocellulosic biomass with ionic liquids. Green Chem 15:550–583
3. Swatloski RP, Spear SK, Holbrey JD, Rogers RD (2002) Dissolution of cellose with ionic liquids. J Am Chem Soc 124:4974–4975
4. Ohno H, Fukaya Y (2009) Task specific ionic liquids for cellulose technology. Chem Lett 38: 2–7
5. Zhao H, Baker GA, Song Z, Olubajo O, Crittle T, Peters D (2008) Designing enzyme-compatible ionic liquids that can dissolve carbohydrates. Green Chem 10:696–705
6. Vitz J, Erdmenger T, Haensch C, Schubert US (2009) Extended dissolution studies of cellulose in imidazolium based ionic liquids. Green Chem 11:417–424
7. Fukaya Y, Sugimoto A, Ohno H (2006) Superior solubility of polysaccharides in low viscosity, polar, and halogen-free 1,3-dialkylimidazolium formates. Biomacromolecules 7:3295–3297
8. Fukaya Y, Hayashi K, Wada M, Ohno H (2008) Cellulose dissolution with polar ionic liquids under mild conditions: required factors for anions. Green Chem 10:44–46
9. Du FY, Xiao XH, Li GK (2007) Application of ionic liquids in the microwave-assisted extraction of trans-resveratrol from Rhizoma Polygoni Cuspidati. J Chromatogr A 1140:56–62

10. Du FY, Xiao XH, Luo XJ, Li GK (2009) Application of ionic liquids in the microwave-assisted extraction of polyphenolic compounds from medicinal plants. Talanta 78:1177–1184
11. Lu Y, Ma W, Hu R, Dai X, Pan YJ (2008) Ionic liquid-based microwave-assisted extraction of phenolic alkaloids from the medicinal plant *Nelumbo nucifera* Gaertn. J Chromatogr A 1208: 42–46
12. Ma W, Lu Y, Hu R, Chen J, Zhang Z, Pan Y (2010) Application of ionic liquids based microwave-assisted extraction of three alkaloids N-nornuciferine, O-nornuciferine, and nuciferine from lotus leaf. Talanta 80:1292–1297
13. Zeng H, Wang Y, Kong J, Nie C, Yuan Y (2010) Ionic liquid-based microwave-assisted extraction of rutin from Chinese medicinal plants. Talanta 83:582–590
14. Bica K, Gaertner P, Rogers RD (2011) Ionic liquids and fragrances—direct isolation of orange essential oil. Green Chem 13:1997–1999
15. Usuki T, Yasuda N, Yoshizawa-Fujita M, Rikukawa M (2011) Extraction and isolation of shikimic acid from *Ginkgo biloba* leaves utilizing an ionic liquid that dissolves cellulose. Chem Commun 47:10560–10562
16. Usuki T, Onda S, Yoshizawa-Fujita M, Rikikawa M (2017) Use of [C_4mim]Cl for efficient extraction of caffeoylquinic acids from sweet potato leaves. Sci Rep 7:6890
17. Munakata K, Yoshizawa-Fujita M, Rikukawa M, Usuki T (2017) Improved extraction yield of citral from lemon myrtle using a cellulose-dissolving ionic liquid. Aust J Chem 70:699–704
18. Eijkman JF (1885) Sur les principes constituants de l'*Illicium religiosum* (Sieb.) (Shikimi-no-Ki en Japonais). Rec Trav Chim Pays-Bas 4:32–54
19. Farina V, Brown JD (2006) Tamiflu: the supply problem. Angew Chem Int Ed 45:7330–7334
20. KOSÉ Corporation, Maniwa Fumio, Itagura Laboratory. Whitening agent with shikimic acid as the active ingredient. Japanese patent nos. 2008-063308 and 2009-215266
21. Togame Y (1984) The worldwide distribution of the ginkgo. Koshien Jr Univ J 4:1–14
22. Ressmann AK, Gaertner P, Bica K (2011) From plant to drug: ionic liquids for the reactive dissolution of biomass. Green Chem 13:1442–1447
23. Zirbs R, Strassl K, Gaertner P, Schroder C, Bica K (2013) Exploring ionic liquid–biomass interactions: towards the improved isolation of shikimic acid from star anise pods. RSC Adv 3: 26010–26016
24. Pham TPT, Cho C-W, Yun Y-S (2010) Environmental fate and toxicity of ionic liquids: a review. Water Res 44:352–372
25. Erbeldinger M, Mesiano AJ, Russell AJ (2000) Enzymatic catalysis of formation of Z-aspartame in ionic liquid—an alternative to enzymatic catalysis in organic solvents. Biotechnol Prog 16:1129–1131
26. Lau RM, Sorgedrager MJ, Carrea G, van Rantwijk F, Secundo F, Sheldon RA (2004) Dissolution of *Candida antarctica* lipase B in ionic liquids: effects on structure and activity. Green Chem 6:483–487
27. De Diego TD, Lozano P, Gmouh S, Vaultier M, Iborra JL (2005) Understanding structure–stability relationships of *Candida antarctica* lipase B in ionic liquids. Biomacromolecules 6:1457–1464
28. Onda S, Usuki T, Yoshizawa-Fujita M, Rikukawa M (2015) Ionic liquid-mediated extraction of bilobalide from *Ginkgo biloba* leaves. Chem Lett 44:1461–1463
29. Islam MS, Yoshimoto M, Yahara S, Okuno S, Ishiguro K, Yamakawa O (2002) Identification and characterization of foliar polyphenolic composition in sweet potato (*Ipomoea batatas* L.) genotypes. J Agric Food Chem 50:3718–3722
30. Miyamae Y, Kurisu M, Murakami K, Han J, Isoda H, Irie K, Shigemori H (2012) Protective effects of caffeoylquinic acids on the aggregation and neurotoxicity of the 42-residue amyloid β-protein. Bioorg Med Chem 20:5844–5849

31. Tan T, Lai CJS, OuYang H, He MZ, Feng Y (2016) Ionic liquid-based ultrasound-assisted extraction and aqueous two-phase system for analysis of caffeoylquinic acids from Flos Lonicerae Japonicae. J Pharm Biomed Anal 120:134–141
32. Lassak EV. Revision of *Backhousia citriodora* essential oil standard. RIRDC publication no. 11/137. Rural Industries Research and Development Corporation, Canberra, 2012
33. Murata C, Yoshizawa-Fujita M, Rikikawa M, Usuki T (2017) Extraction of essential oils from lemongrass using ionic liquid. Asian J Chem 29:309–312

Adv Biochem Eng Biotechnol (2019) 168: 241–328
DOI: 10.1007/10_2018_79

Published online: 2 February 2019

Environmental Concerns Regarding Ionic Liquids in Biotechnological Applications

Chul-Woong Cho, Myung-Hee Song, Thi Phuong Thuy Pham, and Yeoung-Sang Yun

Contents

C.-W. Cho
School of Chemical Engineering, Chonbuk National University, Jeonju, South Korea

Department of Bioenergy Science and Technology, Chonnam National University, Gwangju, Korea

M.-H. Song and Y.-S. Yun (✉)
School of Chemical Engineering, Chonbuk National University, Jeonju, South Korea
e-mail: ysyun@jbnu.ac.kr

T. P. T. Pham
Faculty of Biotechnology, HoChiMihn University of Food Industry, Go Chi Minh City, Vietnam

Abstract Ionic liquids provide challenges and opportunities for sustainable industrial developments. However, the toxic impacts of ionic liquids reported by many researchers cannot be overlooked. Therefore, in this chapter, we introduce the antimicrobial activities of ionic liquids in bioprocesses and, in greater detail, we discuss their environmental impacts, including the toxicity, biodegradability, bioaccumulation, and mobility of ionic liquids. We believe that this presented information will support colleagues engaged in ionic liquid–related fields.

Graphical Abstract

Keywords Adsorption, Bioaccumulation, Biodegradation, Ionic liquids, Mobility, Quantitative structure–activity relationship (QSAR), Removal, Toxicity

Abbreviations

$[IM01]^+$	1-Methylimidazolium
$[IM0\text{-}12]^+$	1-Dodecylimidazolium
$[IM01O\text{-}4]^+$	1-(Butoxymethyl)imidazolium
$[IM04]^+$	1-Butylimidazolium
$[IM11O2]^+$	1-(Ethoxymethyl)-3-methylimidazolium

$[Ch]^+$	Cholinium
$[IM11COO1]^+$	1-(2-Methoxy-2-oxoethyl)-3-methylimidazolium
$[IM11COOH]^+$	3-Methyl-1-(carboxymethyl)imidazolium
$[IM12]^+$	1-Ethyl-3-methylimidazolium
$[IM12{=}1]^+$	1-Methyl-3-(2-propenyl)imidazolium
$[IM14]^+$	1-Butyl-3-methylimidazolium
$[IM16]^+$	1-Hexyl-3-methylimidazolium
$[IM18]^+$	1-Methyl-3-octylimidazolium
$[N1888]^+$	Trioctylmethylammonium
$[P4444]^+$	Tetrabutylphosphonium
[P1444]	Tributylmethylphosphonium
$[P666\text{-}14]^+$	Trihexyltetradecylphosphonium
$[P3(OH)3(OH)3(OH)\text{-}10]^+$	Decyltriproxylphosphonium
$[Py4]^+$	1-Butylpyridinium
$[Py4\text{-}3Me]^+$	1-Butyl-3-methylimidazolium
$[Pyr14]^+$	1-Butyl-1-methylpyrrolidinium
$[(2\text{-}OPhO)_2B]^-$	Bis[1,2-benzenediolato(2-)]-borate
$[4MePhSO_3]^-$	4-Methylbenzenesulfonate
$[1COO]^-$	Acetate
$[(244Me3Pen)_2PO_2]^-$	Bis(2,4,4-trimethylpentyl)phosphinate
$[(CF_3)_2N]^-$	Bis(trifluoromethyl)amide
$[(CF_3SO_2)_2N]^-$	Bis(trifluoromethylsulfonyl)amide
$[AC]^-$	6-Methyl-2,2-dioxo-1,2,3-oxathiazin-4-onate
$[Lac]^-$	(2S)-2-Hydroxypropanoate
$[Ala]^-$	L-Alaninate
$[Arg]^-$	L-Argininate
$[Asp]^-$	L-Asparaginate
$[Glu]^-$	L-Glutaminate
$[His]^-$	L-Histidinate
$[Iso]^-$	L-Isoleucinate
$[Leu]^-$	L-Leucinate
$[Lys]^-$	L-Lysinate
$[Met]^-$	L-Methioninate
$[Phe]^-$	L-Phenylanine
$[Pro]^-$	L-Prolinate
$[Ser]^-$	L-Serinate
$[Thr]^-$	L-Threoninate
$[Try]^-$	L-Tryptophan
$[Val]^-$	L-Valinate
$[2OSO_3]^-$	Ethylsulfate
$[2SO_3]^-$	Ethylsulfonate
$[SbF_6]^-$	Hexafluoroantimonate
$[PF_6]^-$	Hexafluorophosphate
$[HSO_4]^-$	Hydrogen sulfate
$[1OSO_3]^-$	Methylsulfate
$[N(CN)_2]^-$	*N*-Cyanocyanamide

$[NO_3]^-$	Nitrate
$[1O_2O_2OSO_3]^-$	*O*-2-(2-Methoxyethoxy)ethyl sulfate
$[8OSO_3]^-$	Octylsulfate
$[B(CN)_4]^-$	Tetracyanidoboranuide
$[BF_4]^-$	Tetrafluoroborate
$[(C_2F_5)_3PF_3]^-$	Trifluorotris(pentafluoroethyl)phosphate
$[CF_3COO]^-$	Trifluoroacetate
$[CF_3SO_3]^-$	Trifluoromethansulfonate
$[SbF_6]^-$	Hexafluoridoantimonate
$[Bic]^-$	Bicarbonate
$[Bit]^-$	Bitartrate
$[DHPhosp]^-$	Dihydrogenophosphate
$[DHCit]^-$	Dihydrogenocitrate
$[Sal]^-$	Salicylate
$[Prop]^-$	Propanoate
$[But]^-$	Butanoate
$[AgBr_2]^-$	Silver dibromide
$[CuCl_2]^-$	Copper dichloride
$[FeCl_4]^-$	Tetrachloroferrate
$[MnCl_4]^-$	Manganese(IV) chloride
$[CoCl_4]^-$	Cobalt tetrachloride
$[GdCl_6]^-$	Gadolinium hexachloride
$[(CH_3)_2PO_4]^-$	Ethylphosphoate
$[Ibu]^-$	Ibuprogenate
$[Doc]^-$	Docusate
$[dicamba]^-$	3,6-Dichloro-2-methoxybenzoic acid
$[MCPP]^-$	(RS)-2-(4-Chloro-2-methylphenoxy)propanoic acid
$[TFo]^-$	Triflate
$[MDEGSO_4]^-$	2-(2-Methoxyethoxy) ethylsulfate
MDA	Malondialdehyde
SOD	Superoxide dismutase
POD	Peroxidase
CAT	Catalase

1 Introduction

Ionic liquids (ILs) are organic molten salts of which the melting points are normally less than 100°C. An IL consists of a bulky asymmetrical cation and a weakly coordinating anion. Features of ILs such as negligible vapor pressure, nonflammability, high thermostability, and highly solvating yet noncoordinating properties mean that ILs are known as "green solvents" [1]. The properties of ILs can even be tailored by numerous combinations of cations and anions, so task-specific ILs can be designed for a specific application. Because of the almost unlimited possibility of structural combinations, ILs are called "designer solvents" [2–4]. These characteristics of ILs have widened their application areas, which include organic synthesis, catalysis, and biotechnological fields.

Although ILs are considered green solvents, ILs are not always green. Some ILs are reported to be toxic toward bacteria, yeasts, and other indicator organisms [5]. An interesting example is ILs with fluorine-containing anions. ILs with imidazolium cations and fluorine-containing anions, such as hexafluorophosphate ($[PF_6]^-$) and tetrafluoroborate ($[BF_4]^-$), are most widely studied in biotechnological processes [6]. However, $[PF_6]^-$ and $[BF_4]^-$ are more toxic toward *Pseudokirchneriella subcapitata* than other anion moieties [7, 8]. In addition, it has been found that the half-maximal effective concentration (EC_{50}) value of a $[BF_4]^-$-containing IL, $[IM14][BF_4]$, decreases significantly with time. The increase in the toxicity of F-containing ILs has been shown to result from hydrolysis of anions to fluoride ions, which are more toxic [9]. This indicates that this kind of IL can become more toxic after discharge into aqueous environments. It is a matter of concern that although ILs with fluorine-containing anions are toxic and become more toxic after discharge, they are being widely used in biotechnological studies.

This chapter covers the following scope. First, ILs can have toxic or inhibitory effects on bioprocesses. For instance, in enzyme reactions or microbial cultivations, ILs may cause adverse effects in spite of their purposed functions. After being used in bioprocesses and discharged into aqueous environments, ILs can be toxic toward aqueous living things. In addition to describing the toxic effects of ILs, this chapter discusses toxicity prediction models, with some successful examples, and provides information regarding biodegradation, bioaccumulation, adsorption, and mobility of ILs, which are of great importance for understanding their fate in natural systems. Finally, strategies for selecting ILs, especially for biotechnological applications, are suggested.

2 Inhibitory and Toxic Effects of ILs on Bioprocesses

Ionic liquids have been widely applied as solvents for whole-cell and enzymatic reactions in biotechnological processes. ILs can help improve the yield and efficiency of these bioprocess. However, several studies have reported that ILs could inhibit the activities of enzymes and microorganisms used in bioprocesses. In this section, the toxic effects of ILs on enzyme activity or growth of microorganisms and their toxicity mechanisms in bioprocesses are discussed. Moreover, to minimize or solve problems, some possible approaches are introduced.

2.1 Indirect Effects of ILs

Extractive fermentation of lactic acid using traditional organic solvents (e.g., 1-butanol, 1-pentanol, and hexane) has been extensively studied. However, the traditional organic solvents cause serious problems for environmental organisms, including humans, because of their high toxicity and diffusibility. Thus, to solve these problems and achieve better process efficiency, ILs have been considered as alternatives. Indeed, some publications [10, 11] have reported that ILs enhance selectivity and yield in enzyme reactions. However, for more appropriate application of ILs in biological processes, toxic effects of ILs on microbe or enzyme activity, especially in whole-cell processes, should be avoided. Therefore, the biological impacts of ILs need to be understood.

Matsumoto et al. [12] examined the toxic effects of ILs on nine lactic acid–producing bacteria (*Lactobacillus rhamnosus* NBRC 3863, *Lactobacillus homohiochi* NRIC 0119, *Lactobacillus homohiochi* NRIC 1815, *Lactobacillus fructivorans* NRIC 0224, *Lactobacillus fructivorans* NRIC 1814, *Lactobacillus delbruekii* subsp. *lactis* NRIC 1683, *Pediococcus pentosaceus* NRIC 0099, *Leuconostoc fallax* NRIC 0210, and *Bacillus coagulans* NBRC 12583). In that study, some typical ILs ([IM14][PF_6], [IM16][PF_6], and [IM18][PF_6]) were tested. For the experiment, 5% (v/v) of each IL or 1% (v/v) of each traditional organic solvent were incubated with each bacterium, and then microbial activities in lactic production and the microbial growth rate were measured. The results for microbial growth showed that the tested ILs were less toxic than toluene; however, all microbial activities were inhibited by ILs. The toxic effects of ILs differed according to the kind of bacterium. Among the tested bacteria, *L. delbruekii* had the highest tolerance of the ILs while the growth of the rest was greatly inhibited in a range of 20–90%. In that study, it was also reported that the toxicity of ILs was dependent on the alkyl chain length. In contrast to the general toxicity trend associated with the alkyl chain effect (see Sect. 3), the increase in the alkyl chain length decreased the antimicrobial activities. This may be why the molecular toxicity of ILs is related to their solubility in a solution.

Matsumoto et al. [13] also applied the same ILs in an in situ extraction process using organic acids (e.g., butyric, glycolic, propionic, lactic, pyruvic, and acetic acids) from microbial fermentation culture. The ILs were used as one phase of a binary phase. The results showed that the extractability depended on the degree of hydrophobicity of the organic acids and could be ordered as [IM18] < [IM14] < [IM16], although there was no remarkable difference. To improve the extraction efficiency of the solvents, Matsumoto et al. [13] added an extractant of tri-n-butylphosphate into the IL phase. This helped to achieve an extraction efficiency similar to that of traditional organic solvents. The researchers again tested the inhibition of physiological activities (e.g., glucose consumption, lethality, and growth of colony-forming units of *Lactobacillus rhamnosus*) in the presence of ILs and toluene. Although the ILs inhibited the microbial activities and led to slow growth (around 60% of the control growth rate), they were less toxic than toluene, as was also shown in the previous study [12].

Santos et al. [14] investigated the antimicrobial activities of ILs with microorganisms used in a fermentation process for food production. The tested ILs were cholinium-based ILs such as [Ch][1COO], [Ch]Cl, and [Ch][$(CF_3SO_2)_2N$]; [IM12]-based ILs with Cl, [2OSO_3], [2SO_3], [$(CF_3SO_2)_2N$], and [P4441][1OSO_3]; and [IM14] [$(CF_3SO_2)_2N$]. The tested microorganisms were Gram-positive bacteria (*Bacillus subtilis*, *Lactobacillus delbrueckii*), actinobacteria (*Streptomyces drozdowikzii*), yeasts (*Saccharomyces cerevisiae*, *Yarrowia lipolytica*, and *Kluyveromyces marxianus*), and filamentous fungi (*Aspergillus brasiliensis* and *Rhizopus oryzae*). The results showed that the growth inhibition rate was dependent on the kinds of microorganisms and ILs. In general, the Gram-negative bacterium *P. aeruginosa* had higher resistance than other microorganisms in the presence of cholinium-, phosphonium-, and [$(CF_3SO_2)_2N$]-based ILs, and all of the fungi had tolerance even at a high concentration. Meanwhile, only one yeast (*Y. lipolytica*) could grow in the presence of all ILs. The authors reported that cholinium-based ILs can be considered biocompatible cations.

2.2 Direct Effects of ILs

Agricultural residues such as rice straw, corn stover, and sugarcane bagasse are used for biofuel production. Before enzymatic hydrolysis (in which polysaccharides are hydrolyzed to monomeric sugars), pretreatment techniques such as treatment with dilute acid, ammonia fiber expansion, lime, hydrothermolysis, and organosolv are used to increase the rate of the hydrolysis reaction. Recently, IL pretreatment has emerged as another alternative technique. Thermal treatment of whole biomass with certain ILs has been found to enhance the rate of enzymatic hydrolysis of lignocellulose, removes a portion of the lignin, and is broadly applicable to both woody and herbaceous feedstocks. Among these ILs, [IM12][1COO] has been found to be most effective. To assess its toxic effects on fermentative microorganisms, Ouellet et al. [15] investigated the toxic impact of [IM12][1COO] on *Saccharomyces cerevisiae* BY4742, which is used for ethanol production. For the test, biomass resources of microcrystalline cellulose (MCC), corn stover (CS) and switchgrass (SG) were pretreated with different concentrations of [IM12][1COO]. Samples of MCC1 and CS1 with a high content of [IM12][1COO] showed significant decreases in *S. cerevisiae* growth and ethanol production, demonstrating its strong inhibition effect. To determine whether residual [IM12] was the primary cause of the growth inhibition, *S. cerevisiae* synthetic defined medium cultures with 20 g/L glucose were supplemented with [IM12][1COO] and grown under aerobic and anaerobic conditions. The inhibition of growth and ethanol production that was observed could be primarily attributed to the amount of residual [IM12] in the medium. Aerobic growth of *S. cerevisiae* in 20 g/L glucose in the presence of [IM12][1COO], [IM12]Cl, Na [1COO], and NaCl was investigated. The growth inhibition caused by a given concentration of [IM12][1COO] was far greater than that seen with the equivalent concentration of acetate alone. [IM12][1COO] showed greater levels of inhibition than [IM12]Cl, indicating that IM12 and [1COO] together have a synergistic effect. It is clear that the main cause of inhibition is the carry-over of IL when hydrolysates derived from biomass treated with [IM12][1COO] are used.

Similarly, IL pretreatment can help bioconverter Clostridia to more easily ferment carbon sugars. However, the residue of ILs after pretreatment could inhibit their growth and activities. Indeed, Venkata Nancharaiah and Francis [16] reported the effects of residual ILs on the growth and fermentative metabolism of *Clostridium sporogenes* at different IL concentrations. The results showed that the growth and fermentative metabolism of *Clostridium* sp. was affected by selected ILs such as [IM12][1COO], [IM12][DMF], and [IM11][DMP] when the bacteria were exposed to IL concentrations higher than 2.5, 4, and 4 g/L, respectively. In the case of the metabolism inhibition test, when the microorganisms were exposed to an IL concentration of >2.5 g/L, their fermentative activity was inhibited.

Lovejoy et al. [17] examined the efficiency and biocompatibility of ILs in extracting branched unsaturated (isoprenoid) hydrocarbons from two phototrophic microbes (*Synechocystis* sp. PCC6803 and *Botryococcus braunii*) and also studied the cytotoxic effects of ILs on the phototrophic microbes. The tested ILs were [Pyr14][$(CF_3SO_2)_2N$], [Ch]Cl, [Ch][$(CF_3SO_2)_2N$], [P666-14][$(CF_3SO_2)_2N$], [P666-14]Cl, [P4444]Cl, and [P3(OH)3(OH)3(OH)-10]Br. With a 24-h incubation time and under standard growth conditions, *Synechocystis* sp. PCC6803 demonstrated strong resilience to high concentrations of each IL ($\leq$50% w/v). When the concentrations were in excess of 50% w/v, a sharp decline in cell viability was observed in all ILs except for Cyphos 101©. The more hydrophobic ILs – including those based on [P666-14][$(CF_3SO_2)_2N$], [P666-14]Cl, and the water-immiscible [Pyr14][$(CF_3SO_2)_2N$] – had relatively low toxicities in relation to the microbes. These less cytotoxic ILs were said to undergo phase separation with water such that their deleterious interactions with the microbes were minimized. The larger phosphonium ILs trended toward complete depigmentation and cell death of *B. braunii*; however, the effect of [Pyr14][$(CF_3SO_2)_2N$] on culture health was comparable to the effects of standard organic solvents on *B. braunii*, with similar extraction efficiencies. On the other hand, the more toxic ILs, including [P3(OH)3(OH)3(OH)-10]Br and [Ch][$(CF_3SO_2)_2N$], were more miscible with water and partitioned into water to sufficient degrees that allowed for a single phase to be observed at ~70°C. The observations of toxicity were thus attributed to the amphiphilic nature of those compounds, which introduced detergent-like components into the aqueous layers. A positive association between the anion and the toxicity of ILs and, to a lesser extent, the cationic structure, was allegedly observed; however, further studies were suggested to substantiate this allegation. It can be summarized that the ILs were found to be effective in extracting high-value chemicals from the microbes, with significantly low toxicity, which could be further minimized by considering the structure and hydrophobicity of the ILs.

2.3 Suggested Mechanisms of IL Toxicity

In a bioprocess, the toxic effects of solvents, including ILs, can be explained by two scenarios: one is molecular toxicity and the other is phase toxicity [18]. Molecular toxicity can be generated when the solvent permeates the cell membrane and accumulates in the cell membrane. These activities result in direct inhibition of enzyme activity, protein denaturation, and modification or expansion of the cell membrane [18]. Phase toxicity can be generated when solvent covers the interface of the cells. In this case, although the chemicals cannot directly affect the microorganisms, they can interfere with nutrient access and consequently inhibit the growth rate.

A more detailed mechanism investigation of molecular toxicity was performed by Dickinson et al. [19], who identified a potential mechanism for toxicity of imidazolium ILs (IILs) in *Saccharomyces cerevisiae* through chemical genomic and proteomic profiling. The part damaged by IILs is mitochondria – specifically, ion transport across the mitochondrial membrane – because their chemical genomic profiles closely resemble that of the mitochondrial membrane–disrupting agent valinomycin. IILs induced hyperpolarization of the mitochondrial membrane, damaging mitochondrial function. Further, several deletions of genes encoding mitochondrial proteins (e.g., *ARG2*, *COQ2*, *HMI1*, *IMG2*, *QCR2*, *RIM1*, *SHE9*, and *YPT7*) resulted in increased sensitivity to IILs. The IILs displayed a dose-dependent effect on the mitochondrial structure – which could induce abnormal mitochondrial morphology, as well as altered polarization of the mitochondrial membrane potential – similar to that of valinomycin. Deletion of putative serine/threonine kinase PTK2 can increase IIL tolerance due to the decrease of the cation influx. Conversely, increased expression of PMA1 confers sensitivity to IILs, suggesting that the proton efflux may be coupled to the influx of the toxic imidazolium cation. The toxicity of most or all IILs (such as [IM12]Cl, [IM14]Cl, and [IM12][1COO]) can be reduced by deletion of PTK2. The pH has a strong effect on IIL toxicity; at a near-neutral pH (pH 6.5), [IM12]Cl displayed greater growth inhibition of xylose-converting *S. cerevisiae* engineered for xylose-fermentation (Y133) than a *ptk2Δ* mutant (Y133-IIL), whereas at a lower pH (pH 5.0), there was no obvious difference in growth between Y133 and Y133-IIL, perhaps because the lower pH decreased the proton efflux by mass action. Under the greatest IIL toxicity conditions of a near-neutral pH and aerobic conditions, the Y133-IIL strain consumed glucose and xylose faster and produced more ethanol in the presence of 1% [IM14]Cl than the wild-type PTK2 strain. At a low pH (pH 5.0), the Y133-IIL strain still converted significantly more xylose to ethanol even under anaerobic conditions.

2.4 Suggested Methods to Overcome IL Toxicity

As mentioned above, ILs can have several advantages and simultaneously inhibit the growth and activities of microorganisms in bioprocesses. To overcome these problems, some methods (e.g., selection of less toxic and highly functionalized ILs, finding of highly IL-tolerant microorganisms, and gene modification) have been suggested.

Xu et al. [20] isolated IL-tolerant *Fusarium oxysporum* BN producing cellulase from IL-contaminated microhabitats. The contaminated conditions were created by adding a medium and 10% (w/v) of [IM12][H_2PO_4]. In the tolerance test of the isolated fungus with exposure to 1–5% IL concentrations, the inhibition rate increased as the IL concentration increased. Actually, *F. oxysporum* BN can grow in a 10% concentration of [IM12][H_2PO_4], indicating its high resistance. However, in the presence of some ILs (e.g., [IM12][H_2PO_4] and [IM12][1OSO_3]), the growth rate was slightly inhibited. Meanwhile, the strain was more sensitive to Cl-based ILs (e.g., [IM12=1]Cl and [IM14]Cl) and thus its growth was remarkably inhibited. In the activity test of the cellulase produced by *F. oxysporum* BN in the presence of ILs, the results also showed that the enzyme inhibition rate was dependent on the types of ILs. Similar results were observed in fungal growth test of ILs with phosphate- or sulfate-based anions, which were less toxic to the fungal strain. Consequently, although the presence of ILs inhibits the activity of a cellulase solution, by selecting less toxic ILs and isolating an IL-tolerant fungus, ILs can be used for developing a simple and efficient bioethanol production process.

ILs used for dissolving cellulose can inhibit microbial growth and reduce the efficiency of the entire bioprocess, as mentioned above. To overcome the toxic effects of ILs, Ruegg et al. [21] studied IL-resistant genes of *Enterobacter lignolyticus*, which is a rain forest soil bacterium with resistance to imidazolium-based ILs. They identified key genetic elements against [IM12]Cl by using a targeted functional screening approach and then inserted target genes into *E. coli* hosts. The results showed that enhanced production of terpene-based biofuels could be observed in the presence of the IL. With gene modification of the bacteria, the production of biofuel precursors could be improved even in the presence of low concentrations of ILs. This finding could be instrumental for developing effective and sustainable production of biofuels and chemicals by improving the bottleneck of biomass pretreatment methods.

3 Concerns About Spent and Discharged ILs

3.1 Concerns About Toxicity

Discharge of ILs into the environment can be generated through human error, industrial system error, or inappropriate treatment of spent ILs. Spills of ILs into the environment can lead to serious adverse effects. This has been verified by toxicological studies on ILs, performed by various researchers. Therefore, it is necessary to understand the toxic effects of ILs to cope with unexpected accidents related to ILs or to design/select green IL structures. In this section, previously presented impacts of IL toxicity are summarized. To better understand the toxicological information, readers are advised to consult previous reviews on this topic [5, 22–29].

3.1.1 Toxicity in Enzymatic and Microbial Activities

Toxic Effects of ILs on Enzyme Activity

Because there are many possibilities for enzymes and ILs to come in contact with each other in several industrial processes, and released ILs can inhibit the enzyme activity of environmental organisms, enzyme inhibition tests for ILs are needed. Thus, several researchers have examined these properties, using some of the predominant enzymes (acetylcholinesterase, chloroperoxidase, catalase, carboxylase, glutathione reductase, trypsin, luciferin, etc.) that have important roles in maintaining the environment.

Since toxicity tests using the activities of enzymes are sensitive and quick to perform, they have been used for toxicity assessment of ILs. Stock et al. [30] first applied a method using acetylcholinesterase to investigate the alkyl side chain effects of imidazolium- and pyridinium-based ILs. As expected, as the alkyl side chain length increased, the inhibition of enzyme activity increased. Among the tested ILs, trihexyltetradecylphosphonium-based ILs were identified as the most toxic structures [30]. The alkyl side chain effect and the head group effect of the cation were investigated in detailed by Arning et al. [31]. In their study, they confirmed that elongation of the alkyl chain effect caused a negative effect, and they reported that IL toxicity was caused mainly by the cationic part; there was no clear observation of an anion effect. However, for further investigations on ILs, comparison of some toxic anionic structures ([$(CF_3SO_2)_2N$] and [$(2\text{-}OPhO)_2B$]) with less toxic anions (Cl, Br, and [$8OSO_3$]) were suggested. To develop more environmentally friendly IL structures, less toxic ILs combined with biocompatible anions and cations, and oxygen-functionalized groups, have been introduced. With regard to biocompatible anions, acesulfamate and saccharinates can be considered [32], and for biocompatible cations, cholinium [33], protic [34], and multicharged cations [35] have been introduced. Studies on the inhibitory effects of ILs on other enzymes are listed in Table 1.

Toxic Effects of ILs on Bacteria, Yeasts, and Fungi

ILs can affect the activities and growth rates of microbes. The properties of ILs can be used for pharmaceutical developments but also indicate that ILs play a role as environmental toxicants. Therefore, several researchers have investigated the antimicrobial activities of ILs. In the experimental conditions, the antimicrobial properties were measured in high concentrations of ILs because ILs can be used in IL-condensed systems of bioprocesses and large amounts of ILs can be released. In general, to measure the antimicrobial properties of chemicals, their minimal inhibitory concentration (MIC) and minimal biocidal concentration (MBC) are estimated. If the MIC and MBC values are high, it means that the antimicrobial activity is low, whereas if the MIC and MBC values are low, the antimicrobial activity is high. In another measurement method, the bacterial growth rate at a lower concentration than those used in the methods above is measured by the incubation time. In this case, the end point used is the EC_{50}. The physiological behaviors of *Vibrio* luminescent marine bacteria species are frequently used for toxicity measurement using the rate of inhibition of the light-emitting ability of the bacteria.

Before the concept of ILs was introduced, several researchers started making antimicrobials using current IL structures. Initially, Pernak et al. [54] designed pyridinium-based ILs with a long alkyl chain, such as dodecylthiomethyl and octylthiomethyl, combined with Cl as a counterion, and examined their

Table 1 Studies on enzyme inhibition by ionic liquids

Enzymes	Cations	Side chains	Anions	Reference
Acetylcholinesterase	Imidazolium, pyridinium, phosphonium	Alkyl, aryl	$[BF_4]$, $[1O2O2OSO_3]$, $[8OSO_3]$, Br, $[CF_3SO_3]$, Cl, $[N(CN)_2]$, $[NO_3]$, $[PF_6]$, $[(244Me3Pen)_2PO_2]$	Stock et al. [30]
Acetylcholinesterase	Imidazolium, pyridinium	Ether, alkyl	$[BF_4]$	Jastorff et al. [36]
Acetylcholinesterase	10 head groups	Alkyl, aryl, cyano, ether, hydroxyl, carboxyl, ester, sulfonyl	37 anions	UFT [37]
Acetylcholinesterase	Imidazolium	Alkyl	$[(2\text{-}OPhO)_2B]$, $[(CF_3SO_2)_2N]$, $[8OSO_3]$, Cl, $[(CF_3)_2N]$	Matzke et al. [38]
Acetylcholinesterase	Piperidinium, imidazolium, pyrrolidinium, pyridinium, quinolinium, morpholinium, ammonium, phosphonium	Alkyl, ether, aryl, ketone, cyano, hydroxyl	Br, I, Cl, $[BF_4]$	Arning et al. [39]
Acetylcholinesterase	Pyridinium	Alkyl, alkoxyl, hydroxyl	Cl, saccharinate, acesulfamate	Stasiewicz et al. [32]
Antioxidant enzymes	Imidazolium	Alkyl	Br	Yu et al. [40]
Chloroperoxidaze	Pyridinium, imidazolium	Alkyl	$[BF_4]$, $[CF_3SO_3]$, Br, $[PF_6]$, $[1OSO_3]$	Boskin et al. [41]
Acetylcholinesterase, cellulase	Imidazolium	Alkyl	Br	Luo et al. [42]
Acetylcholinesterase	Ammonium	Alkyl, aryl	*N,N,N*-trialkylammoniododecabotates	Schaffran et al. [43]
Penicillium expansum lipase, mushroom tyrosinase	Cholinium, ammonium, imidazolium	Alkyl	$[1OSO_3]$, $[H_2PO_4]$, $[1COO]$, $[NO_3]$	Lai et al. [33]
Catalase	Imidazolium, phosphonium, pyrrolidinium	Alkyl	$[BF_4]$, $[CF_3SO_3]$, Cl, $[1SO_3]$	Pinto et al. [44]

Acetylcholinesterase	Imidazolium	Alkyl	$[(C_2F_5)_3PF_3]$, $[(CF_3SO_2)_2N]$, $[B(CN)_4]$, $[C(CN)_3]$, $[N(CN)_2]$, Cl	Steudte et al. [45]
Acetylcholinesterase	Imidazolium, cholinium	Alkyl, hydroxyalkyl	$[BF_4]$, Cl, 18 amino acids	Hou et al. [46]
Acetylcholinesterase	Ammonium, imidazolium, pyridinium	Hydroxyalkyl	Alkanoate	Peric et al. [34]
Acetylcholinesterase	Imidazolium	Aryl, ester, alkyl	Br, I	Stolte et al. [47]
Human carboxylase	Imidazolium, pyridinium, benzethonium, phosphonium	Alkyl	Bistriflimide, salicylate, docusate, $[(CF_3SO_2)_2N]$, Cl	Costa et al. [48]
Acetylcholinesterase	Imidazolium-based dications	Pentacyclo, alkyl	Cl, $[1OSO_3]$, Br, I	Steudte et al. [35]
Luciferase	Imidazolium, pyridinium	Alkyl	$[(CF_3SO_2)_2N]$, Cl, Br, $[8OSO_3]$, $[BF_4]$	Ge et al. [49]
Glutathione reductase	Imidazolium, phosphonium, pyrrolidinium	Alkyl	$[BF_4]$, Cl, $[1OSO_3]$, $[BF_4]$, $[CF_3SO_3]$	Cunha et al. [50]
Trypsin	Imidazolium	Alkyl	Br, Cl, $[BF_4]$, [1COO], $[CF_3SO_3]$, $[NO_3]$	Fan et al. [51]
Lactic dehydrogenase	Imidazolium	Alkyl	Br, Cl, $[BF_4]$, [1COO], $[CF_3SO_3]$	Dong et al. [52]
Xylanase, endoglucanase	Imidazolium	Alkyl	[1COO]	Wu et al. [53]

antimicrobial effects on several microorganisms (e.g., Gram-negative bacteria, Gram-positive bacteria, yeasts, and fungi). The results showed that the prepared ILs had higher antimicrobial activities than some reference compounds (cetyltrimethylammonium Br and cetylpyridinium Br). Strictly speaking, those reference compounds are also included in the class of ILs. Later, Prof. Pernak and his coworkers introduced several types of new ILs as antimicrobial agents [55]. For designing high-antimicrobial agents, they designed IL structures by selection of a cationic moiety with different long alkyl and/or alkoxymethyl chains and several types of anions.

One of the greatest advantages of ILs is the fact that their physical and environmental properties can be designed. Therefore, the toxic effects of ILs can be interpreted using the concept of the structure-–activity relationship. Based on an imidazolium core as a cationic moiety, numerous ILs were designed and tested for antimicrobial activities. Pernak et al. [56] synthesized 1-alkylimidazolium and 1-alkoxymethylimidazolium with different alkyl chain lengths and different anions (L-lactate and DL-lactate). It was shown that when a short alkyl chain was substituted, the antimicrobial activities were not active, but in the case of a long alkyl chain, the properties were considerable. This is because the hydrophobicity of ILs is a main factor in determining their toxic effects. Indeed, this fact was proved by correlation with the cationic lipophilicity value (k) as a second-order model: $a + b \log k + c \log k^2$ [55]. The same phenomena were observed with elongation of a long alkyl chain in several studies [57–59]. An alkyl chain effect was observed in the anion. Petkovic et al. [60] synthesized cholinium-based ILs with an alkanoate anion with different alkyl chain lengths. As expected, as the chain length increased, the antimicrobial activity rose. However, elongation on the anion was less toxic than that on the cation. This might be due to the fact that the anion has less adsorptive interaction with microbes than the cation does. However, the effect of elongation of alkyl chains is not always observed. For example, if there is a long side chain of approximately 14, a cutoff can be observed for several reasons (e.g., steric hindrance, formation of a critical micelle concentration, or low water solubility).

As mentioned above, the toxicity of ILs can be controlled by structural modification of ILs. To increase the antimicrobial toxicity of ILs, several researchers have substituted high-antibioactive-structure molecules. Bussetti et al. [61] used a quinolinium moiety with different alkyl chain lengths and showed that the prepared ILs had higher antimicrobial activity than general ILs such as imidazolium, pyridinium, and phosphonium. Cole et al. [62] and Coleman et al. [63] developed β-lactam antibiotic–based ILs and ester- and dipeptidyl-functioned ILs, respectively, and observed enhanced antimicrobial properties. Moreover, theophylline- [64], menthol- [65], and herbicide-based ILs [66], which had toxicity similar to or greater than that of general types of ILs, were observed. On the other hand, more environmentally friendly IL structures can be designed. To reduce the toxic effect of ILs, some functionalization (e.g., using ether, ester, and hydroxyl groups) can be applied. Ventura et al. [67] and Prydderch et al. [68] developed mendelic acid–based ILs using a bioresource and reported that the prepared ILs had low antimicrobial activity with several types of bacteria and fungi. Another way to achieve low-toxicity ILs is to employ a low-toxicity IL structure. Actually, among the various IL cations, cholinium can be considered a low-toxicity structure. Thus, Hou et al. [46] prepared cholinium-based ILs combined with amino acid–based anions and confirmed that these types of ILs were only one tenth as toxic as [IM14][BF_4] in terms of the MIC and MBC values for four bacteria. Similarly, Gouveia et al. [69] and Jordan et al. [70] used amino acids for IL anions and designed low-toxicity ILs.

In the case of a toxicity test of ILs with the luminescent marine bacteria *V. fischeri*, the experimentally measured results followed the trend in which elongation of the alkyl chain increased IL toxicity [64, 71–77]. However, there was an exception in that guanidinium-based ILs did not follow this trend. The toxic effects of ILs on bacteria could be reduced by hydrophilic functionalization (e.g., using ether [78, 79], ester [79], nitrile [80], and hydroxyl groups [80]). Moreover, dications [80] and double-bonded cationic moieties [81] have low toxicity. By modification or selection of low-toxicity cation and anions, IL toxicity can be reduced. Ventura et al. [82] synthesized cholinium-based ILs with amino acid anions and showed that they have low toxicity. They also showed that even among these low-toxicity ILs, less toxic ILs structures could be selected as follows: [Ch][Bic] $\ll$ [Ch]Cl $\ll$ [Ch][But] $<$ [Ch][1COO] $<$ [Ch]Cl $<$ [Ch][Prop] $<$ [Ch][DH] $<$ [Ch][Sal] $<$ [Ch][Bit] ~ [Ch][DHCit]. Similarly, Ben Ghanem et al. [83] studied low-toxicity amino acid–based anions (glycinate, serinate, prolinate, and analinate). On the other hand, metal-containing anions were moderately toxic to the bacteria [84]. Studies on the antimicrobial activities of ILs are listed in Table 2.

Toxic Effects of ILs on Animal Cell Lines

Animal cell culture has been commonly used for toxicity estimation of chemicals because this method can aid understanding of the basal toxic effect of a chemical. For experimental evaluation of ILs, several types of animal cells (e.g., a leukemia rat cell line, C6 glioma, CaCO-2, HeLa, MCF, and channel catfish ovary (CCO) cells) were selected. Using these cell lines, the toxicity of ILs was tested by measuring the growth rate and viability of the cells.

The cellular viability in the presence of ILs was examined for the first time by Ranke et al. [77], who described the toxicity according to IL structures. Ranke et al. [77] observed the effect of the side chain and anions using alkylimidazolium-based ILs with different side chain lengths and anions, and the results showed that the inhibition of leukemia and C6 rat cell line viability was strongly related to IL hydrophobicity. However, the toxic effect of the anions not sufficiently proved. Similar observations were made in viability testing using HeLa cells [116]. Later, a clear difference in toxic effects on cell viability according to the types of anions was demonstrated by a toxicity comparison of [IM14][BF_4] and [IM14][$(CF_3SO_2)_2N$] [36]. In this comparison, [IM14][$(CF_3SO_2)_2N$] had much higher toxicity than [IM14][BF_4] [36]. The toxic effects of [$(CF_3SO_2)_2N$] on the viabilities of HT-29 and Caco-2 cells was described in other studies [117]. More detailed cell viability in the presence of IL anions was examined, using 27 anions, by Stolte et al. [118], who

Table 2 Studies on antimicrobial activities of ionic liquids

Test systems	End points	Head groups	Side chains	Anions	Reference
6 cocci, 5 rods, 1 fungus, 1 yeast-like fungus, 2 bacilli	MIC	Pyridinium	Dodecylthiomethyl, octylthiomethyl	Cl	Pernak et al. [54]
22 bacteria and yeasts	MBC, MFC	Ammonium	Alkanoyl	Br	Ahlstrom et al. [85]
3 cocci, 5 rods, 2 fungi, 1 bacillus	MIC, MBC	Pyridinium, benzimidazolium	Alkoxymethyl, nicotinylaminomethyl	Cl	Pernak et al. [86]
4 cocci, 4 rods, 1 bacil-lus, 2 fungi	MIC	Pyridinium	Alkoxymethyl	Halides and anions containing metals (Co, Cu, Zn, Mg, Fe)	Pernak et al. [55]
4 cocci, 4 rods, 2 fungi	MIC, MBC	Imidazolium	Alkyl, alkoxymethyl	L-Lactate, DL-lactate	Pernak et al. [56]
5 cocci, 4 rods, 2 fungi	MIC, MBC	Phosphonium	Alkyl	12 anions	Cieniecka-Roslonkiewicz et al. [87]
E. coli, *P. pastoris*, *B. cereus*	MIC, growth	Imidazolium	Alkyl	[PF_6], [BF_4]	Ganske and Bornscheuer [88]
5 cocci, 5 rods, 2 fungi	MIC, MBC	Ammonium	Alkyl, alkoxy, hydroxyl, cyclo	Cl, [AC], [$(CF_3SO_2)_2N$]	Pernak et al. [59]
10 fungi	Growth	Imidazolium, pyridinium, cholinium	Alkyl	Cl, [$2OSO_3$], [SCN], [Lac], [1COO], [$(CF_3SO_2)_2N$]	Petkovic et al. [89]
4 fungi	MIC, MFC	Cholinium	–	9 alkanoates	Petkovic et al. [60]
10 bacteria	MIC, MBEC	Imidazolium	Alkyl	Cl	Carson et al. [90]
10 bacteria	MIC, MBEC	Quinolinium	Alkyl	Br	Busetti et al. [61]

12 bacteria	MIC	Imidazolium	Alkyl	Cl, $[BF_4]$, $[1OSO_3]$, $[8OSO_3]$, $[(CF_3SO_2)_2N]$, $[CF_3SO_3]$, $[4MePhSO_3]$	Luczak et al. [58]
Aspergillus sp.	Growth	Imidazolium	Alkyl	[1COO]	Singer et al. [91]
4 bacteria, 1 yeast	MIC	Imidazolium	Ester, ether, hydroxyl	$[8OSO_3]$, $[(CF_3SO_2)_2N]$	Deng et al. [92]
3 cocci, 3 rods, 1 fungus, 1 bacillus	MIC	Imidazolium, pyridinium	Alkyl	Br	Cornellas et al. [57]
E. coli, *E. faecium*, *K. pneumoniae*, *Staphylococcus aureus*	MIC	Imidazolium, pyridinium	Alkyl	Br, ampicillin	Cole et al. [62]
Clostridium sp.	Growth	Imidazolium	Alkoxy	$[BF_4]$, $[PF_6]$, $[CF_3COO]$, $[(CF_3SO_2)_2N]$, $[CF_3SO_3]$	Wang et al. [93]
Clostridium sp.	Growth	Imidazolium	Alkyl	$[(CH3)_2PO_4]$, [1COO], $[(CH3CH_2)_2PO_4]$	Nancharaiah and Francis [16]
Pseudomonas fluorescens	Growth	Imidazolium, pyridinium	Alkyl	$[PF_6]$, $[BF_4]$, $[CF_3COO]$	Zhang et al. [94]
E. coli, *S. aureus*, *P. aeruginosa*, *C. albicans*	MIC	Imidazolium	Selenium + aryl, alkyl	$[PF_6]$, $[BF_4]$, Cl	Alberto et al. [95]
E. coli, *B. subtilis*, *S. aureus*, *S. epidermidis*	MIC	Phosphonium	Sulfophenyl, sulfobutyl, ether, hydroxyl, chloride	$[HSO_4]$	Banothu and Bavanthula [96]
E. coli	Growth (diffusion disk method)	Ammonium (dicationic structure) l	Ether, alkyl, hydroxy	Br	Li et al. [97]
29 bacteria	MIC, MFC	Ammonium	Pt complexes	Cl, [1COO]	Petkovic et al. [98]
8 bacteria, 12 fungi	MIC	Imidazolium	Amino acid ester	Br	Coleman et al. [63]

(continued)

Table 2 (continued)

Test systems	End points	Head groups	Side chains	Anions	Reference
E. coli, *S. aureus*, *Fusarium* sp., *C. albicans*	Agar diffusion test, growth	Phosphonium, imidazolium	Alkyl	[N(CN)$_2$], [4MePhSO$_3$], [Phosp], Cl, [PF$_6$], [HSO$_4$], [CF$_3$SO$_3$], [BF$_4$], [(CF$_3$SO$_2$)$_2$N]	Ventura et al. [99]
Aspergillus nidulans	MIC, MFC	Phosphonium	Alkyl	Cl	Petkovic et al. [98]
E. coli, *S. enteritidis*, *L. monocytogenes*, *S. aureus*	MIC, MBC	Cholinium	–	18 amino acids	Hou et al. [46]
12 bacteria	MIC, MBC	Ammonium, imidazolium, pyridinium	Alkyl, alkoxy	Cl	Feder-Kubis and Tomczuk [65]
B. subtilis, *E. coli*, *P. fluorescens*, *P. putida* (CP1), *P. putida* (KT2400)	IC_{50}	Imidazolium, pyridinium	Alkyl, methylenedioxy, dimethoxy-mandelic	[1COO], Br	Ventura et al. [67]
3 Gram-positive bacteria, 3 Gram-negative bacteria, 4 fungi	MIC	Phosphonium	Sulfobuyl, sulfophenyl, alkyl, chloroalkyl, hydroxyl, ether	[HSO$_4$]	Banothu et al. [100]
10 bacteria	MIC	Imidazolium, triazolium	Alkyl	Br, I	Borowiecki et al. [64]
3 cocci, 3 rods, 1 fungus, 1 bacillus	MIC	Imidazolium, pyridinium	Ester, alkyl	Br	Garcia et al. [72]
9 bacteria	MIC, MBC	Imidazolium	Alkyl	[AgBr$_2$], [CuCl$_2$]	Gilmore et al. [73]
B. subtilis, *E. coli*, *Klebsiella* sp., *P. aeruginosa*, *C. albicans*	MIC	Imidazolium	Alkyl, ether	Br, Cl, [BF$_4$], [PF$_6$]	Messali et al. [75]

S. aureus, *S. aureus* (MRSA), *E. faecalis*, *E. coli*, *P. aeruginosa*	MIC	Imidazolium	Alkyl	Br	Postleb et al. [76]
Listeria monocytogenes, *S. aureus* S244, *E. coli* E149, *A. hydrophila* A97	Growth	Imidazolium	Alkyl, hydroxyethyl	[N(CN)$_2$], [SCN], [BF$_4$], Cl, Br, [Gly], [Ala], [Ser], [Pro], [Asn]	Ben Ghanem et al. [101]
E. coli, *L. monocytogenes*	MIC, MBC	Imidazolium, pyridinium	Alkyl	[1OSO$_3$], Cl, [N(CN)$_2$], [SCN], [C(CN)$_3$]	Mester et al. [102]
B. subtilis, *E. coli*	Growth	Imidazolium, pyridinium, cholinium	Alkyl	Br, [Ala], [Arg], [Cys], [Gls], [Glu], [Gly], [Met], [Phe]	Gouveia et al. [69]
B. subtilis, *E. coli*	Growth, respiratory rate, dehydrogenase activity, viability	Ammonium	Dibenzyl, oleytrimethyl, alkylbenzyldimethyl	Theophyllinate	Borkowski et al. [103]
S. cerevisiae	Growth	Imidazolium	Alkyl	Cl	Dickinson et al. [19]
S. aureus	MIC	Piperazinium, guanidinium	Alkyl	[BF$_4$], [Lac], I, [2OSO$_3$]	Yu et al. [104]
8 bacteria, 12 fungi	MIC	Imidazolium, ammonium, pyridinium, pyrrolidinium	L-Phenylalanine, ether, ester, alkyl, L-tyrosine	Br	Jordan et al. [70]
S. aureus, *E. coli*, *P. aeruginosa*, *C. albicans*	Growth (agar disk diffusion method)	Imidazolium	Alkyl	[BF4]	Hodyna et al. [74]
15 *Candida* spp.	MIC	Ranitidinium, diphenylhydramin, glycinium, ethylglycinium	–	[Ibu], [Doc]	Frizzo et al. [71]
3 Gram-positive bacteria, 3 Gram-negative bacteria, 3 yeasts	MIC	Imidazolium/benzimidazolium	Alkylamine + HBr	Br	Kunduracıoğlu et al. [105]

(continued)

Table 2 (continued)

Test systems	End points	Head groups	Side chains	Anions	Reference
15 bacteria	MIC, bacterial growth	Cholinium	Alkyl, alkenol, alkanol	Br	Siopa et al. [106]
4 ATCC strains, 8 clinical isolates of fungi, 5 yeasts, 3 fungi	MIC	Imidazolium, pyridinium	Ester, alkyl, amide, heterocyclic	Br	Prydderch et al. [68]
Pseudomonas putida	Growth	Ammonium	Alkyl	Br, [dicamba], [MCPP]	Piotrowska et al. [66]
V. fischeri	Luminescence	Imidazolium	Alkyl	$[BF_4]$, $[PF_6]$, $[4MePhSO_3]$, Br, Cl	Ranke et al. [77]
V. fischeri	Luminescence	Imidazolium	Alkyl	Br, Cl, $[BF_4]$, $[PF_6]$	Garcia et al. [107]
V. fischeri	Luminescence	Imidazolium	Alkyl	$[1OSO_3]$, $[2OSO_3]$, Cl, $[PF_6]$	Romero et al. [108]
V. fischeri	Luminescence	Pyridinium, imidazolium, ammonium, phosphonium, cholinium	Alkyl	Br, $[(CF_3SO_2)_2N]$, $[(CH_3)_2PO_4]$	Couling et al. [109]
V. fischeri	Luminescence	Pyridinium, imidazolium	Alkyl	Br, $[N(CN)_2]$, Br	Docherty and Kulpa [110]
V. fischeri	Luminescence	Morpholinium, pyrrolidinium, ammonium, imidazolium, pyridinium, piperidinium	Dimethylamino, alkyl, ether, nitrile, hydroxyl	Cl, Br, $[(CF_3SO_2)_2N]$	Stolte et al. [111]
V. fischeri	Luminescence	Imidazolium	Ether, alkyl	$[BF_4]$, $[N(CN)_2]$	Samori et al. [78]
V. fischeri	Luminescence	Imidazolium, phosphonium, pyrrolidinium	Alkyl	$[BF_4]$, Br, $[CF_3SO_3]$, $[2OSO_3]$	Pinto et al. [112]
V. fischeri	Luminescence	Guanidinium, imidazolium, phosphonium	Alkyl, ether, ester	I, Cl, Br, [1SO3], [1OSO3], $[4MePhSO_3]$	Ventura et al. [79]

V. fischeri	Luminescence	Pyridinium, piperidinium, morpholinium, tropinium, imidazolium, quinuclidinium, cholinium	Alkyl, nitrile, hydroxyl, ester	Br, I, [SCN], [$1SO_3$], [N$(CN)_2$], [$(CF_3SO_2)_2N$], [$(C_2F_5)_3PF_3$], [$B(CN)_4$], [BPh_4]	Viboud et al. [80]
V. fischeri	Luminescence	Cholinium	Alkyl, hydroxyl, double bond	Br	Silva et al. [113]
V. fischeri	Luminescence	Cholinium	Benzyl	[DHCit], Cl, [But], [Prop], [DHPhosp], [Bic], [1COO], [Sal]	Ventura et al. [82]
V. fischeri	Luminescence	Imidazolium	Hydroxyethyl	[Ala], [Gly], [Ser], [Pro]	Ben Ghanem et al. [83]
V. fischeri	Luminescence	Imidazolium, pyridinium, ammonium	Alkyl	[PF_6], [BF_4], [$(CF_3SO_2)_2N$], [TFo], [$2OSO_3$], [1COO], [$1OSO_3$], [$MDEGSO_4$], Cl, [NO_3]	Montalban et al. [81]
Vibrio quinghaiensis sp. -Q67	Luminescence	Imidazolium	Alkyl	Br, Cl	Fan et al. [114]
V. fischeri	Luminescence	Guanidinium, cholinium	Alkyl, hydroxyl	[$FeCl_4$], [$MnCl_4$], [$CoCl_4$], [$GdCl_6$]	Sintra et al. [84]
V. fischeri	Luminescence	Guanidinium, tetramethylguanidinium, cholinium	Alkyl	Isostearate, decanoate, neodecanoate	Rantamaeki et al. [115]

ATCC American Type Culture Collection, *IC_{50}* half-maximal inhibitory concentration, *MBC* minimal biocidal concentration, *MBEC* minimal biofilm eradication concentration, *MFC* minimal fungicidal concentration, *MIC* minimal inhibitory concentration, *MRSA* methicillin-resistant *Staphylococcus aureus* found that 10 of these 27 anions (bis-[1,2-benzendiolato(2-)]-borate, bis-[oxalate (2-)]-borate, tris(trifluoromethylsulfonyl)methide, bis(trifluoromethylsulfonyl) imide, hexafluoroantimonate, bis(trifluromethyl)imide, tris(pentafluoroethyl) trifluorophosphate, tris(heptafluoropropyl)trifluorophosphate, bis(pentafluoroethyl) phosphinate, and cobalt tetracarbonyl) showed significant toxic effects on IPC-81 viability. The toxicity mechanism of anions may be due to their lipophilicity; another cause could be their instability in an aqueous phase [118]. Typically, fluorine-containing anions such as [BF_4], [PF_6], and [SbF_6] can be hydrolyzed and produce highly toxic HF [9]. Although two other fluorine-containing anions [$(CF_3SO_2)_2N$] and [$(C_2F_5)_3PF_3$]) are stable, they are involved in the species of rather toxic compounds [45]. A detailed investigation on the latter cases was performed by Steudte et al. [45]. Since such anions – along with [$B(CN)_4$], [$N(CN)_2$], and [$C(CN)_3$] – have high hydrophobicity, high stability, and low viscosity, they have been of interest in producing an electrochemical window [45]. Anionic properties that can lead to a binary phase with aqueous phase and well mixing may be attractive to extract/separate some targeted chemicals in bioprocess engineering. Thus, Steudte et al. [45] studied the toxicity impact of anions with an [IM12] cation on animal cells, luminescent bacteria, algae, aquatic plants, and crustaceans. The tested anions had high toxicity, which could be strongly correlated with their hydrophobicity values. Therefore, when selecting IL anions, toxic effects and environmental stability should be considered. The relevant studies are listed in Table 3.

Toxic Effects of ILs on Microalgae

Microalgae living in surface water are key species that provide food sources to their upper trophic levels and produce oxygen, an essential element for environmental organisms. Because they are located at the level of an important position, understanding of their biological responses to toxicants is important. Therefore, the toxic impacts of ILs on microalgae have been investigated by several researchers. For examination of chemical toxicity, using algae, usually the algal growth rate in the presence of a toxicant has been assessed by measuring and comparing algal growth rates in control conditions and in the presence of the toxicant. In other methods, algal physiological activities (e.g., photosynthetic activity [7, 8, 129], CO_2 absorption [130], and the degree of oxidative stress [131–133]) have been assessed. The biological responses of microalgae differ according to the algal species; thus, several species have been assessed in these studies. Toxicological studies using microalgae are listed in Table 4.

Latala et al. [134] investigated the toxic effects of ILs on two different marine algae (*Oocystis submarina* and *Cyclotella meneghiniana*) taken from the southern Baltic Sea. When measuring the toxicities of the ILs, they mainly focused on the effect of the alkyl chain and the types of substitutes (alkyl and aryl groups) in imidazolium-based ILs; moreover, the influence of salinity in the toxicity testing system on toxicity was measured [134]. The results showed that the toxicity response of the microalgae was dependent on the species. The growth rates of

Table 3 Viability tests of ionic liquids, using several animal cell lines

Cell lines	Cations	Side chains	Anions	Reference
IPC-81	Imidazolium	Alkyl	Br, $[(CF_3SO_2)_2N]$	Jastorff et al. [36]
IPC-81, C6 rat	Imidazolium	Alkyl	$[BF_4]$, $[PF_6]$, Cl	Ranke et al. [77]
IPC-81	Imidazolium, quinolinium, pyrrolidinium, pyridinium, ammonium, phosphonium	Alkyl, hydroxyl, ether, aryl, carboxyl	Cl, Br, $[BF_4]$, $[PF_6]$	Ranke et al. [119]
IPC-81	Imidazolium	Alkyl	Cl, $[BF_4]$, $[8OSO_3]$, $[(CF_3SO_2)_2N]$, $[(CF_3)_2N]$, $[(2\text{-}OPhO)_2B]$	Matzke et al. [38]
IPC-81	Imidazolium	Alkyl	27 anions	Stolte et al. [118]
IPC-81	4-(Dimethylamino) pyridinium, pyridinium, imidazolium, morpholinium, pyrrolidinium, piperidinium, ammonium	Alkyl, alkoxy, ether	Br, Cl	Stolte et al. [120]
IPC-81	Cholinium, ammonium, pyrrolidinium	Alkyl, hydroxyl	$[1OSO_3]$, $[1SO_3]$, $[(CF_3SO_2)_2N]$	Stolte et al. [121]
IPC-81	Morpholinium	Aryl, alkyl	17 anions	Pernak et al. [122]
IPC-81	Imidazolium	Alkyl	$[CF_3SO_2]_2N$, $[(C_2F_5)_3PF_3]$, $[B(CN)_4]$, $[N(CN)_2]$, $[C(CN)_3]$	Steudte et al. [45]
IPC-81	Imidazolium	Aryl	Br, Cl, I	Stolte et al. [47]
IPC-81	Ammonium	Hydroxyalkyl	Alkanoate	Peric et al. [34]
HeLa	Imidazolium	Alkyl	$[BF_4]$, $[PF_6]$, Cl	Stepnowski et al. [116]
HT29, Caco-2	Imidazolium, cholinium, guanidinium, ammonium, phosphonium	Alkyl, ether, hydroxyl	$[BF_4]$, $[PF_6]$, $[(CF_3SO_2)_2N]$, $[N(CN)_2]$, acetasulfame, saccharinate	Frade et al. [117]
Caco-2	Imidazolium	Alkyl, aryl, alkylaryl	$[PF_6]$, Cl, $[1OSO_3]$, $[2OSO_3]$	Garcia-Lorenzo et al. [123]
IPC-81	Hydroxypyridinium	Alkoxymethyl	Acetasulfame, saccharinate, Cl	Stasiewicz et al. [32]

(continued)

Table 3 (continued)

Cell lines	Cations	Side chains	Anions	Reference
PC12	Imidazolium	Alkyl	Cl	Li et al. [124]
CCO	Imidazolium	Alkyl, aryl, dimethylaminoalkyl	$[(CF_3SO_2)_2N]$, $[BF_4]$, $[PF_6]$, Br	Bubalo et al. [125]
CCO of *Ictalurus punctatus*	Imidazolium	Alkyl	$[(CF_3SO_2)_2N]$, $[BF_4]$, $[PF_6]$	Cvjetko et al. [126]
CCO of *Ictalurus punctatus*	Imidazolium	Alkyl	$[(CF_3SO_2)_2N]$, $[BF_4]$, $[PF_6]$, Br	Radosevic et al. [127]
CCO of *Ictalurus punctatus*	Phosphonium	Alkyl	Alkanoate	Ruokonen et al. [128]

CCO channel catfish ovary, *HT29* human colon carcinoma, *PC12* rat pheochromocytoma

both species in the presence of ILs were reduced, but the growth rate of *O. submarina* in the presence of 50 μM of ILs recovered after approximately 10 days, whereas the growth rate of *C. meneghiniana* did not recover. The presence of salts reduced the toxic effects because it might have reduced the permeability of the ILs [134]. Similar effects were observed in subsequent studies [142].

Wells and Coombe [135] studied the toxic effects of ILs on *Selenastrum capricornutum* (also known as *Pseudokirchneriella subcapitata* and *Raphidocelis subcapitata*), which is a standard bioindicator species for toxicity assessment of substances. In their study, they selected phosphonium-, ammonium-, and imidazolium-based ILs with differently elongated alkyl chains, and one pyridinium IL. The estimated algal toxicity showed that when the alkyl chain of phosphonium was increased from C4 to C6, the toxicity was steeply increased. However, when the chains of imidazolium were increased from C12 to C18, the toxicity values were similar, which is known as a “cutoff” effect. Later, the alkyl chain effect of toxicity on the same algal species was clearly observed by Cho et al. [136]. In that study, with an increase in the alkyl side chain from propyl to octyl, the toxicity was considerably increased; moreover, with an additional increase in the incubation time from 46 h to 96 h, the toxicity was also increased [136]. A similar trend of increasing toxic effects with an increasing alkyl chain has been observed in several different species in several reports [111, 138, 144, 147, 150, 154]. However, the study by Latala et al. did not show the same trend, because of the high toxicity of a short-chained IL ($[IM12][BF_4]$) compared with long-chained ILs ([IM14], [IM16], and [IM18] with $[BF_4]$) [134].

The effects of IL head groups have been investigated. Stolte et al. [120] studied the effects of a head group on toxicity and stated that it was of minor relevance and was related to lipophilicity. Nevertheless, among the tested cations, a morpholinium head group was found to be recommendable [111]; later, it was again proved that morpholinium-based ILs with alkoxy groups were low-toxicity structures when

Table 4 Studies on toxic effects of ionic liquids on microalgae

Species	End points	Head groups of cations	Side chains	Anions	References
Oocystis submarina, *Cyclotella meneghiniana*	Growth	Imidazolium	Alkyl, aryl	$[BF_4]$	Latala et al. [134]
P. subcapitata	Growth for 72 h; EC_{50}	Imidazolium, ammonium, phosphonium, pyridinium	Alkyl	$[PF_6]$, Cl, $[(EtO)_2PO_2]$, $[(CF_3SO_2)_2N]$, $[1OSO_3]$	Wells and Coombe [135]
P. subcapitata	Growth for 48, 72, and 96 h; EC_{50}	Imidazolium	Alkyl	Br	Cho et al. [136]
P. subcapitata	Growth for 96 h; EC_{50}	Imidazolium	Alkyl	$[SbF_6]$, $[PF_6]$, $[BF_4]$, $[CF_3SO_3]$, $[8OSO_3]$, Br, Cl	Cho et al. [9]
Scenedesmus vacuolatus	Reproducibility for 72 h; EC_{50}	Imidazolium	Alkyl	Cl, $[BF_4]$, $[8OSO_3]$, $[(CF_3SO_2)_2N]$, $[(CF_3)_2N]$, $[(2\text{-}OPhO)_2B]$	Matzke et al. [38]
S. vacuolatus	Reproducibility for 72 h; EC_{50}	Imidazolium	Alkyl	$[BF_4]$, $[(CF_3SO_2)_2N]$	Matzke et al. [137]
S. vacuolatus	Reproducibility for 72 h; EC_{50}	Pyridinium, morpholidinium, piperidinium, imidazolium, ammonium	Dimethylamino, alkyl	Br, Cl, $[(CF_3SO_2)_2N]$	Stolte et al. [111]
Scenedesmus quadricauda, *Chlamydomonas reinhardtii*	Growth for 96 h	Imidazolium	Alkyl	Br	Kulacki and Lamberti [138]
P. subcapitata	Photosynthetic activity; growth; EC_{50}	Imidazolium	Alkyl	Br, $[BF_4]$	Pham et al. [7, 8]

(continued)

Table 4 (continued)

Species	End points	Head groups of cations	Side chains	Anions	References
O. submarina, *Chlorella vulgaris*, *C. meneghiniana*, *Geitlerinema amphibium*	Growth for 72 h	Imidazolium	Alkyl	Cl, $[BF_4]$, $[N(CN)_2]$, $[CF_3SO_3]$, $[1OSO_3]$, $[MPEGSO_4]$	Latala et al. [139, 140]
Bacillaria paxillifer, *G. amphibium*	Growth for 72 h	Imidazolium	Alkyl	Cl, $[BF_4]$, $[N(CN)_2]$, $[CF_3SO_3]$, $[1OSO_3]$, $[MPEGSO_4]$	Latala et al. [139, 140]
P. subcapitata	Growth for 72 h	Imidazolium, cholinium, morpholodinium, sulfonium	Ethylchloride, alkyl, hydroxyl	Br, Cl, $[(CF_3SO_2)_2N]$	Pretti et al. [141]
O. submarina, *C. vulgaris*, *C. meneghiniana*, *G. amphibium*	Growth for 72 h	Imidazolium	Alkyl	Cl	Latala et al. [142]
Chlamydomonas reinhardtii	Growth for 96 h	Imidazolium, pyridinium, ammonium	Alkyl	Br	Sena et al. [143]
S. obliquus, *Chlorella ellipsoidea*	Reproducibility for 96 h; change in chlorophyll content	Imidazolium	Alkyl	Br	Ma et al. [144]
P. subcapitata	Growth for 96 h	Imidazolium	Alkyl	$[(CF_3SO_2)_2N]$	Ventura et al. [145]
S. vacuolatus	Reproducibility for 72 h; EC_{50}	Imidazolium	Alkyl, phenyl, carboxyl, ethoxy	I, Br, Cl	Stolte et al. [47]
C. vulgaris, *P. subcapitata*	–	Piperidinium, pyridinium, imidazolium	Alkyl, methylenedioxy, dimethoxy-mandelic	Br, Cl	Ventura et al. [67]
S. obliquus, *Euglena gracilis*	Growth for 24 h; oxidative stress tests	Imidazolium	Alkyl	L-(+)-Lactate, D-(−)-lactate	Chen et al. [146]
Propionibacterium freudenreichii subsp.	Growth for 48 h	Imidazolium	Alkyl	$[BF_4]$, [PTS], $[NO_3]$, $[1OSO_3]$, [SCN]	Hajfarajollah et al. [147]

S. vacuolatus	Growth	Morpholinium	Alkoxy, alkyl	$[2OSO_3]$, $[1OSO_3]$, $[FeCl_4]$	Salchner et al. [148]
S. quadricauda	Growth for 96 h; activity of antioxidative enzyme; membrane disruption; degree of oxidative stress	Imidazolium	Alkyl	Cl	Deng et al. [149]
S. quadricauda	Growth for 96 h; EC_{20}, EC_{50}, and EC_{80}	Imidazolium	Hydroxyl, alkyl	4 amino acids: [Gly], [Ser], [Ala], [Pro]	Ben Ghanem et al. [83]
S. obliquus	Growth for 24, 48, 72, and 96 h; chlorophyll concentration; chlorophyll fluorescence	Imidazolium	Alkyl	D-(−)-Tartrate, L-(+)-tartrate	Liu et al. [133]
S. obliquus	Growth for 24, 48, 72, and 96 h; IC_{50}; cell membrane permeability; morphology analysis	Imidazolium	Alkyl	Cl	Liu et al. [150]
Raphidocelis subcapitata	Ammonium consumption; growth for 72 h	Pyrrolidinium	Alkyl, ether	$[BF_4]$, $[(CF_3SO_2)_2N]$	Samori et al. [151]
Scenedesmus rubescens	Growth for 24, 48, 72, and 96 h; IC_{50}	Imidazolium	Alkyl	$[BF_4]$	Tsarpali and Dailianis [152]
Dunaliella tertiolecta	Growth; analysis of carotenoid and chlorophyll *a* content	Imidazolium	Alkyl	$[BF_4]$	Tsarpali et al. [153]
P. subcapitata	Growth for 48, 72, and 96 h	Imidazolium, pyridinium	Alkyl	Br	Pham et al. [154]
S. obliquus	Degree of oxidative stress; morphology analysis	Imidazolium	Alkyl	L-(+)-Tartrate, D-(−)-tartrate	Liu et al. [133]
Phaeodactylum tricornutum	Growth for 24, 48, 72, and 96 h; degree of oxidative stress	Imidazolium	Alkyl	$[BF_4]$	Deng et al. [131]
P. tricornutum	Growth for 24, 48, 72, and 96 h; degree of oxidative stress	Imidazolium	Alkyl	$[BF_4]$	Deng et al. [132]

(continued)

Table 4 (continued)

Species	End points	Head groups of cations	Side chains	Anions	References
Navicula sp.	Growth; adhesion and biofilm formation; MIC	Imidazolium, ammonium	Alkyl	Cl, Br	Reddy et al. [155]
S. quadricauda, *C. vulgaris*, *Botryococcus braunii*	CO_2 absorption	Imidazolium, pyridinium, piperidinium, pyrrolidinium	Alkyl	$[(CF_3SO_2)_2N]$	Quraishi et al. [130]

EC_{20} 20% of maximal effective concentration, EC_{50} half-maximal effective concentration, EC_{80} 80% of maximal effective concentration, IC_{50} half-maximal inhibitory concentration, *MIC* minimal inhibitory concentration tested with *S. vacuolatus*, with an EC_{50} of >1,000 μM. Pretti et al. [141] tested the ecotoxicity of several IL head groups with *P. subcapitata* and revealed that aromatic ring moieties (e.g., pyridinium and imidazolium) were more toxic than nonaromatic ones (e.g., pyrrolidinium, ammonium, and morpholinium) [141]; in particular, it was suggested that sulfonium and thiphenium moieties could be considered harmless structures [141]. Another comparison study between aromatic and nonaromatic structures was performed by Ventura et al. [145], who also found that nonaromatic structures had less toxicity than aromatic ILs. Samori et al. [151] also reported that nonaromatic pyrrolidinium had low toxicity (EC_{50} >100 mg/L) but that ILs could alter the cellular morphology of *R. subcapitata* [151].

However, Latala et al. [142] insisted that although cations have strong relevance to IL toxicity, they might not pose a strong influence on microorganisms in real systems, since the cations will undergo adsorption by natural materials [139]. Therefore, understanding of IL anions is important. Several researchers have sought to identify their toxic effects. Matzke et al. [38] tested the toxic effects of six different anions (Cl, [BF_4], [$(CF_3SO_2)_2N$], [$(CF_3)_2N$], [$8OSO_3$], and [$(2\text{-}OPhO)_2B$]) on *Scenedesmus vacuolatus* and identified that two anions ([$(CF_3SO_2)_2N$], and [$(2\text{-}OPhO)_2B$]) had remarkable toxicities [38]. Cho et al. [9] selected seven anions with [IM14] and sodium or potassium, and revealed that the toxicity values of the ILs were 135–2,884 μM; the order of the toxicity was as follows: [SbF_6] > [PF_6] > [BF_4] > [CF_3SO_3] > [$8OSO_3$] > Br ~ Cl. Additionally, this study showed that a hydrolyzed anion ([BF_4]) could have much higher toxicity depending on the hydrolysis rate [9]. The hydrolysis effect was also shown in the work of Pham et al. [7, 8], in which the inhibition of algal photosynthetic activity by hydrolyzed [IM14][BF_4] was greater than that of a less hydrolyzed form [9, 140]. To avoid highly toxic effects of IL anions, some researchers have employed less toxic anionic structures (e.g., mandelic acid) with EC_{50} values 10^3–10^7 times than those of other ILs with butadecyl or octadecyl side chains [67]. Ben Ghanem et al. [83] synthesized amino acid–based ILs that were nontoxic to algae. Additionally, some mechanisms of the toxicity of ILs to algae were suggested. Deng et al. reported that ILs interrupted the algal membrane structure [156] and could cause oxidative stress in algae [131, 132].

Toxic Effects of ILs on Invertebrates

The initial toxicity testing of ILs with invertebrates was focused on *D. magna* and compared IL toxicity with that of traditional organic solvents. Bernot et al. [157] selected [IM14]-based ILs, tested their toxic impact (i.e., lethality) on *Daphnia*, and compared it with those of several organic solvents (e.g., benzene, methanol, acetonitrile, and tetrachloromethane). The tested ILs showed lethal effects similar to those of phenol, trichloromethane, and tetrachloromethane, and had more toxic effects than benzene, methanol, and acetonitrile [157]. Similar results were reported by Wells and Coombe [135]. Garcia et al. [107] showed that alkylimidazolium-based ILs were more toxic than dichloromethane and chloroform, and less toxic than cationic surfactants; the IC_{50} values of the [IM14]-based ILs were 300 times those of the cationic surfactants.

From the structural viewpoint of ILs, as expected, the phenomenon of an increase in toxicity with an increase in alkyl chains was observed in toxicity studies using *D. magna* [107, 141]. In the case of the head group effect (mentioned in Sect. 3.1.1.4), in general, aromatic moieties showed higher toxicity than nonaromatic ones [145, 158, 159]. However, comparisons between aromatic and nonaromatic moieties should be done using those with similar alkyl chain lengths and anions. Previously, Bado-Nilles et al. [160] compared [IM12]-based ILs and phosphonium-based ILs with long alkyl chains and reported that phosphonium-based ILs were more toxic. In the case of a long-alkyl-chained IL such as [IM18]Br, a doubling of the incubation time in the presence of the IL enhanced its toxic effect on *D. magna*. In a long incubation test, it was also observed that [IM18]Br remarkably inhibited body growth and reproducibility; 21 days later, the damaged physiological properties were recovered [161]. In a detailed experiment, it was further observed that the tested IL could affect the antioxidant defense system of *D. magna* [162]. Meanwhile, a short-chained IL, [IM13][$(CF_3SO_2)_2N$], was found to be moderately toxic to *D. magna* and *D. longispina* [163]. For less environmental risk, oxygen functionalization on the alkyl chain (e.g., using an ether group) could lead to less toxicity than nonfunctionalization [78, 151, 164]. For example, the EC_{50} values of [IM14][BF_4] and [IM12O2][BF_4] were found to be 5.18 mg/L and 209 mg/L, respectively [78]. A similar study of chlorine and oxygen functionalization was reported by Pretti et al. [141]. Toxicological studies on crustaceans are listed in Table 5.

Toxic Effects of ILs on Vertebrates

Pretti et al. [167] estimated the toxic impacts of 15 ILs on the fish *Danio rerio* (zebrafish). The experiments, with observation of fish mortality, were performed at a concentration of 100 mg/L for 96 h; additionally, the authors performed histopathological examinations. Among the tested ILs, two compounds (AMMOENG 100 and AMMOENG 130) (Fig. 1) had LC_{50} values within the tested concentration of 100 mg/L; their LC_{50} values were in the range of 5.9 mg/L and 5.2 mg/L. The LC_{50} values of the ILs were much lower than those of traditional organic solvents (dichloromethane, methanol, acetonitrile, aniline, and triethylamine) [167]. In the histopathological examination of the fish after exposure to ILs at 10 parts per million (ppm), abnormal skin alterations (e.g., epithelial hyperplasia, formation of keratinocyte vesiculation, and erosion) were observed [167]. Similar results were observed in a toxicity test of tetrabutylammonium Br by Dumitrescu et al. [168]. Exposure to an IL concentration of 3,000–5,000 mg/L induced histological alterations by degrading fish (Danio rerio) cell structures i.e., breathing area, interrupting external barrier integrity and periglandular edema formation, etc. Similar abnormal histological alterations in the gill filaments and intestine were observed by Li et al. [169]. Perez et al. [170] visually proved, by an electrospray

Table 5 Studies on toxic effects of ionic liquids on crustaceans

Species	End points	Head groups	Side chains	Anions	Reference
D. magna	Lethality for 48 h, LC_{50}, reproductive ability, brood size for 21 days	Imidazolium	Alkyl	Cl, Br, $[PF_6]$, $[BF_4]$	Bernot et al. [157]
D. magna	Immobilization for 24 h, IC_{50}	Ammonium	Alkyl	Br, Cl, $[BF_4]$, $[PF_6]$	Garcia et al. [107]
D. magna	Lethal concentration for 24 and 48 h, LC_{50}	Ammonium, pyridinium, imidazolium, phosphonium	Alkyl	Br, $[PF_6]$	Couling et al. [109]
D. magna	Lethality for 48 h, EC_{50}	Imidazolium	Alkyl	$[PF_6]$, Cl, $[(EtO)_2PO_2]$, $[1OSO_3]$, $(CF_3SO_2)_2N]$	Wells and Coombe [135]
D. magna	Immobilization for 48 h, EC_{50}	Imidazolium	Alkyl, ether	$[BF_4]$, $[N(CN)_2]$	Samori et al. [78]
D. magna	Lethality for 24 and 48 h, LC_{50}, body development and reproducibility for 21 days	Imidazolium	Octyl, methyl	Br	Luo et al. [161]
D. magna	Lethality for 48 h, inhibition of antioxidative enzymes	Imidazolium	Alkyl	Br	Yu et al. [162]
D. magna	Immobilization for 24 and 48 h, EC_{50}	Imidazolium, pyrrolidinium, ammonium, sulfonium, pyridinium, morpholidinium, cholinium	Alkyl, chloroalkyl, alkylsilyl, hydroxyalkyl	$[(CF_3SO_2)_2N]$, Cl, Br	Pretti et al. [141]
D. magna	Reproducibility for 24 h	Imidazolium, pyridinium (biodegradation products)	Alkyl	Br	Docherty et al. [165]

(continued)

Table 5 (continued)

Species	End points	Head groups	Side chains	Anions	Reference
D. magna, Daphnia longispina	Immobilization for 48 h, EC_{50}, reproducibility and body development for 21 days	Imidazolium	Alkyl	$[(CF_3SO_2)_2N]$	Ventura et al. [163]
D. magna	Immobilization test for 48 h, EC_{50}	Imidazolium	Alkyl, alkoxyl	$[BF_4]$, $[N(CN)_2]$	Samori et al. [164]
D. magna	Immobilization test for 48 h, EC_{50}	Ammonium, pyrrolidinium	Alkyl, hydroxyl	$[1SO_3]$, $[1OSO_3]$, $[(CF_3SO_2)_2N]$	Stolte et al. [121]
D. magna	Immobilization test for 24 and 48 h, EC_{50}, reproducibility	Imidazolium, pyrrolidinium, piperidinium	Alkyl	$[(CF_3SO_2)_2N]$	Ventura et al. [145]
D. magna	Immobilization test for 24 and 48 h, EC_{50}	Imidazolium, phosphonium	Alkyl	$[N(CN)_2]$, $[1OSO_3]$, [1SO3], Cl, $[(iC8)2PO_2]$, $[4MePhSO_3]$	Bado-Nilles et al. [160]
D. magna	Immobilization test for 24 and 48 h, EC_{50}	Ammonium, imidazolium, pyrrolidinium, phosphonium	Alkyl	$[PF_6]$, $[BF_4]$, $[(CF_3SO_2)_2N]$, Cl, [N1,1,1(O),1(OH)]	Costa et al. [158]
D. magna	Immobilization test for 48 h, EC_{50}	Pyrrolidinium	Ether, alkyl	$[(CF_3SO_2)_2N]$, $[BF_4]$	Samori et al. [151]
D. magna	Immobilization test for 48 h, EC_{50}	Cholinium	–	[Prop], [But], [Bit], [DHCit], Cl	Santos et al. [166]
D. magna	Lethal concentration for 48 h, LC_{50}	Imidazolium, pyridinium, pyrrolidinium, piperidinium	Alkyl	Br	Wang et al. [159]

EC_{50} half-maximal effective concentration, IC_{50} half-maximal inhibitory concentration, LC_{50} half-maximal lethal concentration

Fig. 1 Ionic liquid structures of AMMOENG 110 and AMMOENG 130

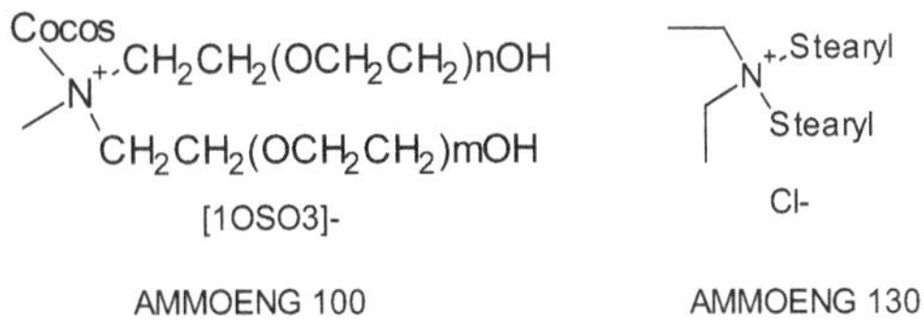

ionization mass spectrometry imaging method, that AMMOENG 130 and its metabolites can be accumulated in the zebrafish's gills, nervous system, and respiratory system; also, the metabolites could penetrate the blood–brain barrier. Pretti et al. [141] performed a toxicity test of 18 ILs comprising several head groups (ammonium, pyridinium, pyrrolidinium, imidazolium, morpholinium, sulfonium, and theophenium) with the anions [$(CF_3SO_2)_2N$], [$1OSO_3$], Cl, and Br, using the same species and methods. As reported previously, the ammonium salts (AMMOENG 100 and AMMOENG 130) repeatedly showed high toxicity, while sulfonium- and morpholinium-based ILs had low toxicity [141].

Wang et al. [171] performed a toxicity study of [IM18] Br by recording physiological responses during the development of embryos of the goldfish *Carassius auratus* for 72 h. In that study, they estimated the lethal concentration and physiological activities (e.g., closure of the blastopore, heart beating, cleavage, and early gastrula) [171]. The results showed that exposure of the fish embryo to the IL reduced hatching rates in comparison with the control conditions; also, it caused increasing embryo malformation and a lethal ratio. Therefore, Wang et al. [171] concluded that [IM18] Br has a toxic effect on fish embryo development and aquatic ecosystems. In 2012, Du et al. [172] assessed the toxic effect of [IM18][PF_6] on the antioxidant enzyme system and DNA of *D. rerio*. For the tests, the antioxidative activities of enzymes (superoxide dismutase and catalase), the amount of reactive oxygen species (ROS), and DNA in liver cells were estimated. The results revealed that the tested ILs could damage fish DNA and increase ROS [172]. Dong et al. [173] tested another IL ([IM1-10] Br), using the same testing methods. They showed that antioxidant enzyme activities were remarkably decreased in its presence, and exposure of fish to it led to higher production of ROS and malondialdehyde content, as well as it damaged DNA as [IM18][PF_6]. Fish liver was exposed to similar ILs ([IM18]Cl and [IM18][BF_4]) at different concentrations of 0 to 40 ppm. The results showed that as the exposure dose increased, the toxicity increased because ROS production was increased and DNA was damaged. In 2016, Hafez and coworkers [174] estimated the toxic effects of 17 pyridinium-based ILs with different alkyl chains and different numbers of substitutes on guppy fish (*Poecilia reticulata*). The results showed that elongation of the alkyl chain was related to an increase in toxic effects. Among them, short-alkyl-chained ILs with C4 and C6 were considered structures that were harmless to fish (LC_{50} >1,000 ppm), while those with C8 and C10 were considered practically nontoxic structures. However, dodecyl-chained pyridinium had a slightly toxic effect on fish. In the same year, Baharuddin et al. [175] introduced environmentally friendly phosphonium-, ammonium-, and cholinium-based IL structures comprising amino acids as anionic parts. All of the

Table 6 Studies on toxic effects of ionic liquids on vertebrates

Species	End point	Cations	Side chains	Anions	Reference
D. rerio	96-h LC_{50}	Imidazolium, ammonium, pyrrolidinium, pyridinium	Alkyl	$[PF_6]$, $[(CF_3SO_2)_2N]$, $[BF_4]$, $[N(CN)_2]$, $[CF_3SO_3]$, $[4MePhSO_3]$, $[NO_3]$	Pretti et al. [167]
D. rerio	48-h LC_{50}	Imidazolium, pyrrolidinium, ammonium, sulfonium, pyridinium, morpholidinium, cholinium	Alkyl, chloroalkyl, alkylsilyl, hydroxyalkyl	$[PF_6]$, $[(CF_3SO_2)_2N]$, Br, Cl	Pretti et al. [141]
Growth stage of *Carassius auratus* larva	24-h LC_{50}	Imidazolium	Alkyl	Br	Li et al. [176]
Poecilia reticulata	96-h LC_{50}	Pyridinium	Alkyl	Br	Hafez et al. [174]
D. rerio	96-h LC_{50}	Phosphonium, ammonium, cholinium	Alkyl	Phenylalaninate, taurinate	Baharuddin et al. [175]
Embryo of *D. rerio*	120-h LC_{50}	Phosphonium, imidazolium, ammonium	Alkyl	6 alkanoates	Ruokonen et al. [128]

LC_{50} half-maximal lethal concentration

tested ILs had LC_{50} values >100 ppm, which could be considered practically nontoxic. Studies on the toxic effects of ILs on fish are listed in Table 6.

Phytotoxicity of ILs

It has also been reported that ILs can inhibit the growth rate of plants. Jastorff et al. [36] and Larson et al. [177] examined growth rate inhibition in duckweed (*Lemna minor*) caused by imidazolium-based ILs with different alkyl chain lengths. As expected, an increase in the side chain increased the growth rate inhibition in *L. minor* [177]. This trend was also observed in a different plant (watercress [178]) and with a different cationic moiety [179]. For examination of the effects of the anion and head group, several types of plants (e.g., *L. minor*, cress, spring barley, wheat, and radish) were exposed to several types of ILs. This showed that there was no clearly identified influence of the head group [180], but it was reported that the toxicity of the anion was dependent on the plant species that was tested. For example result, $[(CF_3SO_2)_2N]$ was more toxic to wheat than small anionic molecules (e.g., Cl, $[BF_4]$, and $[HSO_4]$), whereas there was no clear observation in *L. minor* [180]. Other research showed that the presence of $[NO_3]$ in soil reduced the germination rate of radish, but this toxic effect was not observed in barley [181]. Moreover, as was observed in toxicity studies using algae and fish, ILs were shown to affect the activities of antioxidative enzymes and the production of ROS [182]. Again, ILs increased the production of oxidative stress indicators (malondialdehyde (MDA), hydrogen peroxide (H_2O_2), superoxide dismutase (SOD), peroxidase (POD), and catalase (CAT)). The relevant studies are listed in Table 7.

3.2 QSAR Prediction of IL Toxicity

Since the number of ILs is very numerous, it is time consuming and material consuming to determine their toxic impacts experimentally. To solve this problem and allow efficient estimation, a computational approach based on the concept of the quantitative structure–activity relationship (QSAR) has been suggested. The QSAR method can be defined as predicting or correlating a specific response variable using physicochemical properties or theoretical molecular parameters of a chemical, and this can provide many advantages. For example, the QSAR method helps to minimize the need to perform experiments, allowing economical and safe research. Because of the benefits, QSAR modeling has been applied to predict the physicochemical and environmental properties of ILs. In this section, previous QSAR studies performed to predict IL toxicity are summarized. In particular, linear free energy relationship (LFER) modeling is stressed, since this concept provides good accuracy and helps us to interpret the toxic effects of ILs on a molecular basis.

3.2.1 Various QSAR Approaches

Various types of QSAR models for predicting the biological responses of ILs as toxicants have been presented. First, simple prediction models for predicting the antimicrobial activity of ILs were developed by Pernak et al. [55]. They were of a second-order model type ($a \log k^2 + b \log k + c$) and predictability could be achieved with an R^2 value of 0.733–0.791. Here, the k value stands for cationic lipophilicity, which was experimentally measured using a lipophilic stationary phase in a high-performance liquid chromatography (HPLC) system. Couling et al. [109] developed genetic function approximation (GFA) statistical models for predicting the toxicity values of ILs in *V. fischeri* and *D. magna*. The developed models showed good predictabilities of 0.78 and 0.86, respectively [109]. However, since the prediction models were developed using a small data set, their application domain for other ILs not involved in the modeling set may be limited. Other modeling tests were performed by the Zentrum für Umweltforschung und nachhaltige Technologien (UFT) group [37]. To predict and simplify the toxic effect of ILs on enzyme activity, Arning et al. [39] correlated the toxicity values with the cationic lipophilicity ($\log k_o$), the origin of which was a comparison between the octanol–water partitioning

Table 7 Studies on phytotoxic effects of ionic liquids

Species	End points	Head groups	Side chains	Anions	Reference
Duckweed (*Lemna minor*), *Lepidium sativum*	Growth (numbers of pods and seedlings)	Imidazolium	Alkyl	$[BF_4]$	Jastorff et al. [36]
L. minor	Growth (pond area)	Pyridinium, morpholinium, piperidinium, pyrrolidinium, imidazolium, ammonium	Dimethylamino, alkyl	Cl, Br, I, $[(CF_3SO_2)_2N]$	Stolte et al. [111]
Spring barley (*Hordeum vulgare*), common radish (*Raphanus sativus* L. subvar. *radicula* Pers.)	Germination, growth	Imidazolium	Alkyl	Cl, Br, $[4MePhSO_3]$, $[PF_6]$, $[NO_3]$, $[CF_3SO_3]$, [C6H5CH(OH)C(O)O]	Balczewski et al. [181]
Wheat (*Triticum aestivum*), cress (*Lepidium sativum*), *L. minor*	Growth; EC_{50}	Imidazolium	Butyl	Cl, $[BF_4]$, $[8OSO_3]$, $[(CF_3)_2N]$, $[(2\text{-}OPhO)_2B]$	Matzke et al. [38]
L. minor	Growth (frond, root), EC_{50}	Imidazolium	Alkyl	Br	Larson et al. [177]
T. aestivum	Growth	Imidazolium	Butyl	$[BF_4]$, Cl, $[HSO_4]$, $[(CF_3SO_2)_2N]$	Matzke et al. [183]
T. aestivum	Germination, growth (root, shoot)	Imidazolium	Butyl	$[BF_4]$	Wang et al. [171]
Watercress (*Lepidium sativum* L.)	Germination, seedling growth	Imidazolium	Alkyl	Cl	Studzińska and Buszewski [178]
T. aestivum	Pigment and proline content assay, enzyme activity assay, growth	Imidazolium	Octyl	Br	Liu et al. [184]

Rice (*Oryza sativa* L.)	Growth, Hill reaction activity	Imidazolium	Octyl	Cl	Liu et al. [185]
H. vulgare, *R. sativus* L. subvar. *radicula* Pers.	Growth (germination and weight (dry and fresh))	Imidazolium	Alkylthiomethylene	Cl	Biczak et al. [186]
H. vulgare, *R. sativus* L. subvar. *radicula* Pers.	Germination, weight-based growth (dry and fresh)	Phosphonium	Triphenylalkyl	I	Biczak et al. [187]
T. aestivum	Germination (number of seeds), shoot length, root length, dry weight	Imidazolium	Octyl	$[PF_6]$	Liu et al. [188]
H. vulgare	Germination (number of seeds), growth (shoot height and root length)	Imidazolium	Alkyl	Br, $[CH_3CO_2]$, $[BF_4]$	Bubalo et al. [182]
H. vulgare, *R. sativus* L. subvar. *radicula* Pers.	Weight-based growth	Imidazolium	Alkyl, menthoxymethyl	$[BF_4]$	Biczak et al. [189]
Vicia faba	Seedlings (shoot length, root length, dry weight)	Imidazolium	Butyl	Cl	Liu et al. [190]
Rice (*Oryza sativa* L.)	Growth (length)	Imidazolium	Alkyl	Cl	Liu et al. [179]
Spring barley (*Hordeum vulgare*), common radish (*Raphanus sativus* L. subvar. *radicula* Pers.)	Weight-based growth, chlorophyll content, POD activity	Ammonium	Alkyl	$[BF_4]$	Biczak [191]
H. vulgare, *R. sativus* L. subvar. *radicula* Pers.	Chlorophyll and carotenoid content, antioxidative enzyme inhibition	Ammonium	Alkyl	$[PF_6]$	Biczak et al. [192]
H. vulgare, *R. sativus* L. subvar. *radicula* Pers.	Weight-based growth, oxidative stress (measured as antioxidative enzyme inhibition)	Imidazolium	Alkyl, substituent	$[PF_6]$	Biczak et al. [193]
T. aestivum	Growth (weight and length of shoot and root)	Imidazolium	Alkyl	Cl, Br	Liu et al. [194]

(continued)

Table 7 (continued)

Species	End points	Head groups	Side chains	Anions	Reference
V. faba	Shoot length, root length, dry weight, pigment content, oxidative stress	Imidazolium	Alkyl	Cl, Br, [NO_3]	Liu et al. [195]
H. vulgare, *R. sativus* L. subvar. *radicula* Pers.	Growth; inhibition of plant length, root length and plant fresh weight; oxidative stress analysis	Ammonium	Ethyl	Br, Cl	Pawlowska and Biczak [196]
Gallant soldier (*Galinsoga parviflora* Cav.), white goosefoot (*Chenopodium album* L.), common sorrel (*Rumex acetosa* L.)	Pigment content	Ammonium	Alkyl	[BF_4], [PF_6]	Biczak et al. [197]
H. vulgare, *R. sativus* L. subvar. *radicula* Pers.	Growth (length and weight of plant shoots and roots)	Pyrrolidinium, piperidinium, pyridinium	Butyl	[PF_6]	Biczak et al. [198]
H. vulgare, *R. sativus* L. subvar. *radicula* Pers.	Growth (length and weight of plant shoots and roots), antioxidative enzyme inhibition	Ammonium	Alkyl	I	Biczak et al. [199]

POD peroxidase

coefficient and the retention characteristics of the IL cation in an ether-functionalized C18 column in the HPLC system [200]. In the correlation between toxicity values and log k_o, it was shown that the toxic effect was strongly related to the cationic lipophilicity. Using the same parameters, Ranke et al. [119] correlated them with toxicity values of ILs in a leukemia rat cell line (IPC-81) and observed a similar correlation factor ($R^2 = 0.78$). After compiling a larger toxicity data set for ILs, several research groups developed prediction models for *V. fischeri*, enzyme activity, animal cell lines, *D. magna*, and bacteria. Modeling studies are summarized in Tables 8, 9, 10, 11, and 12.

3.2.2 LFER Model

Among the prediction studies, modeling using the LFER originated by Abraham and Acree [241] is one of the most advanced approaches because the model comprises well-defined solute descriptors and can help to achieve good predictabilities. Therefore, it has been applied to predict several physicochemical, biological, and environmental solute properties (e.g., partitioning coefficients in solvent–solvent and solvent–sorbent systems, solubility in solvents, and toxic effects). The model is as follows:

$$\text{Log toxicity} = eE + sS + aA + bB + vV + j^{-}J^{-} + j^{+}J^{+} + c \quad (1)$$

where an experimentally determined toxicity value (log 1/toxicity) is used as an independent variable solute property. The capital letters stand for solute-dependent descriptors, as follows: E is the excess molar refraction [cm^3 mol^{-1}/100]; S is the dipolarity/polarizability [dimensionless]; A and B are the hydrogen bonding acidity [dimensionless] and hydrogen bonding basicity [dimensionless], respectively; V is the McGowan volume [cm^3 mol^{-1}/10]; and J^+ and J^- are the ionic interactions of the cations and anions. The lowercase letters (e, s, a, b, v, j^-, and j^+) are system-dependent coefficients.

For application of the model to ILs, Eq. (1) needs to be extended to address the two parts of the cation and the anion, since most ILs are dissociated in an aqueous phase.

$$\begin{aligned}\text{Log toxicity} = {} & e_c E_c + s_c S_c + a_c A_c + b_c B_c + v_c V_c + e_a E_a + s_a S_a \\ & + a_a A_a + b_a B_a + v_a V_a + c \end{aligned} \quad (2)$$

where the subscripted c and a denote the cation and anion, respectively. For use of Eq. (2), solute descriptors of IL ions are required. However, experimentally determined descriptors were not available. Therefore, Cho et al. [242] developed calculation models for the descriptors on the basis of ionic and neutral compounds whose descriptors had previously been determined in experimental ways. For the modeling, Cho et al. [242] employed several subparameters calculated from density functional theory [243], a conductor-like screening model [244], and Obprop [245] internet freeware in the Turbomole [246] environment. Consequently, the solute descriptors

Table 8 Studies on ecotoxic effects of ionic liquids on *V. fischeri*

NO_{IL}	Methods	Types of descriptors or programs	NO_{Des}	Accuracy (R^2)	Reference
25	GFA statistical method	Electrotopological state, solvent-accessible surface areas, van der Waals radii	4	0.78	Couling et al. [109]
43	MLR	Group contribution method	9	0.925	Luis et al. [201]
22	Linear regression	Hansch factors from HyperChem	7	–	Lacrămăet al. [202]
96	MLR	Group contribution method using Polymath version 5.0	15	0.924	Luis et al. [203]
148	PLS-DA	PLS-Toolbox version 5.2 for MatLab	–	0.978 (training set), 0.93 (test set)	Alvarez-Guerra and Irabien [204]
33	HM	Electronic, geometric, energetic, and quantum chemical parameters from Gaussian 03 and CODESSA	4	0.9012 (gas phase), 0.921 (water phase)	Bruzzone et al. [205]
54	Multiple correspondence analysis	Total number of carbons in cation	1	0.861	Viboud et al. [80]
57	PCA	WHIM descriptors from Molden, Mopac, and Dragon calculations	–	–	Sosnowska et al. [206]
157	MLR	Topological index	28	0.908	Yan et al. [207]
69	GFA, LSSVM	HyperChem (version 8.0), Dragon (version 6.0)	5	0.893 (GFA training set), 0.903 (GFA test set), 0.910 (LSSVM training set), 0.933 (LSSVM test set)	Ma et al. [208]
103	MLR	LFER descriptors from DFT and COSMO calculations	10	0.762 (training set), 0.812 (test set)	Cho et al. [209]
146	LDA, MLR, PLS, GFA	Dragon (version 6)	5	0.694 (training set), 0.739 (test set)	Das and Roy [210]
24	MLR	Energy of the lowest unoccupied molecular orbital and molecular	2	0.954	Wang et al. [159]

(continued)

Table 8 (continued)

NO_{IL}	Methods	Types of descriptors or programs	NO_{Des}	Accuracy (R^2)	Reference
		volume by DFT and Gaussian 09W programs			
108	MLR	LFER descriptors from DFT and COSMO calculations	15	0.69	Cho et al. [211]
56	MLR	WHIM descriptors from Mopac and Dragon (version 6.0) calculations	3	0.78 (training set), 0.72 (validation set), 0.75 (external validation)	Grzonkowska et al. [212]
60	PLS	Structural descriptors from VolSurf+	3	0.789	Paterno et al. [213]
305	MLR, PLS, GFA statistical method	Two-dimensional variables from Dragon (version 6) and PaDEL-Descriptor (version 2.11), Morphy	3–7	0.697–0.748	Das et al. [214]
110	MLR, MLP neural network	Charge distribution density from COSMO calculation	5	0.906 (MLR training set), 0.978 (MLP training set), 0.961 (MLP validation set), 0.979 (MLP test set)	Ben Ghanem et al. [215]

CODESSA Comprehensive Descriptors for Structural and Statistical Analysis, *COSMO* Conductor-Like Screening Model, *DA* discriminant analysis, *DFT* density functional theory, *GFA* genetic function approximation, *HM* heuristic method, *LDA* linear discriminant analysis, *LFER* linear free energy relationship, *LSSVM* least squares support vector machine, *MLP* multilayer perceptron, *MLR* multiple linear regression, NO_{Des} number of descriptors, NO_{IL} number of ionic liquids, *PCA* principal component analysis, *PLS* partial least squares, R^2 coefficient of determination, *WHIM* weighted holistic invariant molecular

could be calculated for an R^2 of 0.711–0.99, as given in Eq. (3)–Eq. (9). The calculated LFER descriptors are provided in reference [211].

$$V\left[\left(\mathrm{cm}^3\ \mathrm{mol}^{-1}\right)/100\right] = 0.639\ (0.002)\ V_{\mathrm{cosmo}}\left[\mathrm{nm}^3\right] - 0.0046\ (0.002) \tag{3}$$

$$\begin{aligned} E\left[\mathrm{cm}^3\ \mathrm{mol}^{-1}/10\right] = {} & 0.341\ N_{\mathrm{Ring}} + 0.007\ \mathrm{PSA} + 0.057\ \mathrm{MR} - 1.762\ V_{\mathrm{cosmo}}/100 \\ & - 0.113\ E_{\mathrm{vdw}} + 0.275\ \mathrm{MW} + 0.135\sigma_1 + 0.015\ \sigma_4 \\ & - 0.037 \\ R^2 = {} & 0.949, \mathrm{SE} = 0.136, N = 992, F = 2274.8 \end{aligned} \tag{4}$$

Table 9 Studies on ecotoxic effects of ionic liquids on acetylcholinesterase

NO_{IL}	Methods	Types of descriptors or programs	NO_{Des}	Accuracy (R^2)	Reference
23	Linear regression	Log k_o	1	0.790	Arning et al. [31]
153	PCA, MLR, nonlinear regression (RB, MLP)	MatLab (nonlinear model), Statgraphics Plus (MLR)	12	0.814–0.973	Torrecilla et al. [216]
221	MLR	Topological index	17	0.884 (training set), 0.823 (test set)	Yan et al. [217]
292	LDA, MLR, PLS, GFA	PaDEL-Descriptor version 2.11, Cerius2, Statistica, SPSS, Minitab, Simca-P, Euclideon, Discovery Studio 2.5	11	0.918 (training set), 0.861 (test set)	Das and Roy [218]
–	PCR, PLS, decision tree model	R, Statistica	–	0.62 (PCR), 0.64 (PLS), 0.992 (decision tree)	Kurtanjek [219]
236	PCA	WHIM descriptors from Molden, Mopac, and Dragon calculations	82 (cation) + 82 (anion)	–	Sosnowska et al. [206]
61	Group contribution method	IBM SPSS Statistics 20	10	0.992	Peric et al. [220]
232	Four-category classification, CCN, SVM	Moses.Descriptors Community Edition	7	0.934–0.972 (training set), 0.907–0.938 (test set)	Basant et al. [221]
233	MLR	LFER descriptors from COSMO and DFT calculations	4–6	0.711–0.879	Cho and Yun [222]
230	PCA, PLS	Structural descriptors from VolSurf+	9	0.732	Paterno et al. [223]

CCN cascade correlation network, *COSMO* Conductor-Like Screening Model, *DFT* density functional theory, *GFA* genetic function approximation, *LDA* linear discriminant analysis, *LFER* linear free energy relationship, *log* k_o cationic lipophilicity, *MLP* multilayer perceptron, *MLR* multiple linear regression, NO_{Des} number of descriptors, NO_{IL} number of ionic liquids, *PCA* principal component analysis, *PCR* principal component regression, *PLS* partial least squares, R^2 coefficient of determination, *RB* radial basis, *SVM* support vector machine, *WHIM* weighted holistic invariant molecular

Table 10 Studies on ecotoxic effects of ionic liquids on IPC-81

NO_{IL}	Methods	Types of descriptors or programs	NO_{Des}	Accuracy (R^2)	Reference
74	Linear regression model	R 2.12.1		0.78	Ranke et al. [119]
153	MLR, RB, MLP, NN	PCR	12	0.982	Torrecilla et al. [224]
96	Nonlinear NN analysis, QSAM	COSMO-RS descriptors	–	0.90	Torrescilla et al. [224]
227	Genetic algorithm, NN, MLP, LM algorithm	Molecular weight, H-bond acceptor, TPSA, heavy atom count, and formal charge from HyperChem (version 7.0) and Dragon calculations	5	0.91 (MLR), 0.98 (MLP)	Fatemi and Izadiyan [225]
173	MLR	Topological index (e.g., using atom radius, atom electronegativity, atom position)	27	0.938	Yan et al. [226]
97	MLR	LFER descriptors from COSMO and DFT calculations	10	0.778	Cho et al. [209]
100	MLR, nonlinear SVM	CODESSA and Libsvm	4	MLR (0.918), SVM (0.959)	Zhao et al. [227]
243	LDA, PLS	2D descriptors including ETA indexes and topological non-ETA parameters	5–10	0.784–0.918 (training set), 0.869 (test set)	Das et al. [228]
198	CCN, PNN, GRNN, QSAR models	Moses.Descriptors Community Edition (Xlog P, TPSA, Polariz, Dipole, NAtoms, Span, InteriaZ)	5–6	0.927 (CCN–QSAR training set), 0.907 (CCN–QSAR test set), 0.915 (GRNN–QSAR training set), 0.952 (GRNN–QSAR test set)	Gupta et al. [229]
243	PCA, PLS	Structural descriptors from VolSurf+	9	0.725	Paterno et al. [223]
228	MLR	WHIM descriptors from Mopac and Dragon	4	0.77	Sosnowska et al. [230]

(continued)

Table 10 (continued)

NO_{IL}	Methods	Types of descriptors or programs	NO_{Des}	Accuracy (R^2)	Reference
		(version 6.0) calculations			
242	PCA	WHIM descriptors from Molden, Mopac, and Dragon calculations	82 (cation) + 82 (anion)	–	Sosnowska et al. [206]
17	MLR	Electrophilic indexes (ω), E_{HOMO} and E_{LUMO}, energy gap (ΔE) from DFT code of Dmol	4	0.999 (training set)	Salam et al. [231]

2D two-dimensional, *CCN* cascade correlation network, *CODESSA* Comprehensive Descriptors for Structural and Statistical Analysis, *COSMO* Conductor-Like Screening Model, *COSMO-RS* Conductor-Like Screening Model for Real Solvents, *DFT* density functional theory, E_{HOMO} energy of highest occupied molecular orbital, E_{LUMO} energy of lowest unoccupied molecular orbital, *ETA* extended topochemical atom, *GRNN* generalized regression neural network, *LDA* linear discriminant analysis, *LFER* linear free energy relationship, *LM* Levenberg–Marquardt, *MLP* multilayer perceptron, *MLR* multiple linear regression, *NN* neural network, NO_{Des} number of descriptors, NO_{IL} number of ionic liquids, *PCA* principal component analysis, *PCR* principal component regression, *PLS* partial least squares, *PNN* probabilistic neural network, *QSAM* quantitative structure–activity map, *QSAR* quantitative structure–activity relationship, R^2 coefficient of determination, *RB* radial basis, *SVM* support vector machine, *TPSA* topological polar surface area, *WHIM* weighted holistic invariant molecular

$$\begin{aligned} A\ [\text{dimensionless}] &= 171\ \mathrm{HBD_1}^2 - 0.047\ \mathrm{HBD_3}^2 + 0.032\ \mathrm{HBD_4}^2 + 73.511\ \mathrm{HBD_1} \\ &\quad + 0.654\ \mathrm{HBD_2} + 0.208\ \mathrm{HBD_3} + 0.203\ N_{\mathrm{OH}} + 0.080\ N_{\mathrm{HN}} \\ &\quad - 0.019\ E_{\mathrm{vdw}} - 0.132 \\ R^2 &= 0.936, \mathrm{SE} = 0.148, N = 976, F = 1626.2 \end{aligned} \tag{5}$$

$$\begin{aligned} B\ [\text{dimensionless}] &= 0.391\ \sigma_2 + 1.00\ \sigma_3 + 0.421\ \sigma_4 - 0.117\ \sigma_5 - 0.055\ \sigma_6 \\ &\quad + 0.112\ {\sigma_1}^2 - 0.149\ {\sigma_2}^2 - 0.070\ \mathrm{HBA_2}/V_{\mathrm{cosmo}} \\ &\quad + 0.074\ \mathrm{HBA_3}/V_{\mathrm{cosmo}} + 0.032 \\ R^2 &= 0.973, \mathrm{SE} = 0.160, N = 985, F = 3851 \end{aligned} \tag{6}$$

$$\begin{aligned} J^+\ [\text{dimensionless}] &= -0.124\ (0.234) - 0.106\ (0.016)\ E_{\mathrm{vdw}} + 0.421\ (0.181)\ \sigma_3 \\ &\quad + 0.292\ (0.125)\ \sigma_4 + 64.928\ (15.48)\ \mathrm{HBD_1} \\ &\quad + 0.661\ (0.163)\ \mathrm{HBD_2} - 0.049\ (0.017)\ \mathrm{HBA_2}/V_{\mathrm{C}} \\ &\quad - 0.092\ (0.033)\ {\sigma_6}^2/100 \\ R^2 &= 0.816, \mathrm{SE} = 0.351, N = 111, F = 65.4 \end{aligned} \tag{7}$$

Table 11 Studies on ecotoxic effects of ionic liquids (ILs) on *Daphia magna*

NO_{IL}	Methods	Types of descriptors or programs	NO_{Des}	Accuracy (R^2)	Reference
17	GFA	Mopac	3	0.86	Couling et al. [109]
64	MLR, polynomial method	MatLab	25	0.974	Hossain et al. [232]
62	MLR, PLS	Dragon and PaDEL-Descriptor	6	0.948 (training set), 0.802 (test set), 0.816 (total set)	Roy and Das [233]
49	MLR, PLS	GaussView and Gaussian 03	7	0.955 (training set), 0.848 (test set)	Roy et al. [234]
–	MLR, PLS	PaDEL-Descriptor, GaussView and Gaussian 03	3 including a measured IL toxicity value for algae (*S. vacuolatus*)	0.964	Das et al. [235]
24	MLR	Gaussian 09	2	0.954	Wang et al. [159]
53	Gradient-boosted tree, bagged decision tree	Moses.Descriptor Community Edition	3	0.953–0.983	Singh et al. [236]
15	PCA	WHIM descriptors from Molden, Mopac, and Dragon calculations	82 (cation) + 82 (anion)	–	Sosnowska et al. [206]
71	MLR	LFER descriptors from DFT and COSMO calculations	3	0.88 (training set), 0.848 (test set), 0.867 (total set)	Cho and Yun [237]
70	MLR	LFER descriptors from DFT and COSMO calculations	12	0.81	Cho et al. [211]

COSMO Conductor-Like Screening Model, *DFT* density functional theory, *GFA* genetic function approximation, *LFER* linear free energy relationship, *MLR* multiple linear regression, NO_{Des} number of descriptors, NO_{IL} number of ionic liquids, *PCA* principal component analysis, *PLS* partial least squares, R^2 coefficient of determination, *WHIM* weighted holistic invariant molecular

Table 12 Studies on toxic effects of ionic liquids on Caco-2 cells and bacteria

Testing species	NO_{IL}	Methods	Types of descriptors or programs	NO_{Des}	Accuracy (R^2)	Reference
Caco-2 cells	15	MLR	Topological sub-structural molecular design from ModesLab 1.5	3	0.98	Garcia-Lorenzo et al. [123]
E. coli, HeLa viability test, MCF-7 viability test, *Pseudokirchneriella subcapitata*, *Scenedemus vacuolatus*	*E. coli* (89), HeLa viability test (21), MCF-7 viability test (13), *Pseudokirchneriella subcapitata* (10), *Scenedemus vacuolatus* (38)	PCA	WHIM descriptors from Molden, Mopac, and Dragon calculations	82(cation)+82 (anion)	–	Sosnowska et al. [206]
E. coli, *S. aureus*, *C. albicans*	48–76	MLR	LFER descriptors from COSMO and DFT calculations	3–7	0.900 (*E. coli* MIC), 0.934 (*E. coli* MBC), 0.910 (*S. aureus* MIC), 0.947 (*S. aureus* MBC), 0.892 (*C. albicans* MIC), 0.803 (*C. albicans* MBC)	Cho et al. [238]
A. hydrophila, *E. coli*, *L. monocytogenes*, *S. aureus*	25	PCA, MLR	ETA indexes, interaction features, calculated log k_o	8 (MLR), 2 (PLS)	0.972 (*A. hydrophila* MLR), 0.966 (*E. coli* MLR), 0.963 (*L. monocytogenes* MLR), 0.966 (*S. aureus* MLR), 0.933 (*A. hydrophila* PLS), 0.925 (*E. coli* PLS), 0.925 (*L. monocytogenes* PLS), 0.923 (*S. aureus* PLS)	Das and Roy [239]

Listeria monocytogenes, *S. aureus*, *E. coli*, *A. hydrophila*	25	Linear regression	COSMOtherm (Version 6.2) and Turbomole package	8	0.951 (*L. monocytogenes*), 0.949 (*S. aureus*), 0.953 (*E. coli*), 0.958 (*A. hydrophila*)	Ben Ghanem et al. [101]
B. subtilis, *P. aeruginosa*	*B. subtilis* (47), *P. aeruginosa* (83)	ASNN, k-nearest neighbor method, Weka RF	OChem in Excel format	*B. subtilis* (191), *P. aeruginosa* (146)	0.83 (*B. subtilis*), 0.88 (*P. aeruginosa*)	Hodyna et al. [240]

ASNN associative neural network, *COSMO* Conductor-Like Screening Model, *DFT* density functional theory, *ETA* extended topochemical atom, *LFER* linear free energy relationship, $log\ k_o$ cationic lipophilicity, *MBC* minimal biocidal concentration, *MIC* minimal inhibitory concentration, *MLR* multiple linear regression, NO_{Des} number of descriptors, NO_{IL} number of ionic liquids, *OChem* Online Chemical Modeling Environment, *PCA* principal component analysis, *PLS* partial least squares, R^2 coefficient of determination, *RF* RandomForest, *WHIM* weighted holistic invariant molecular

$$J^- \text{ [dimensionless]} = 1.331\ (0.468) + 4.712\ (0.803)\ \sigma_2 - 2.770\ (0.262)\ \sigma_3 - 0.832\ (0.168)\ {\sigma_2}^2 + 0.300\ (0.040)\ {\sigma_3}^2 - 0.012\ (0.004)\ {\sigma_4}^2 - 0.155\ (0.021)\ \text{HBA}_2 + 0.238\ (0.023)\ \text{HBA}_3 - 0.292\ (0.141)\ N_{\text{OH}} + 0.183\ (0.065)\ N_{\text{Ring}}$$

$$R^2 = 0.711, \text{SE} = 0.291, N = 168, F = 43.2 \quad (8)$$

$$S \text{ [dimensionless]} = -1.441\ \sigma_1 + 0.206\ {\sigma_2}^2 - 0.009\ {\sigma_4}^2 - 0.122\ \text{HBA}_4 + 0.511\ \text{Calc.}E + 1.524\ \text{Calc.}B + 0.856\ \text{Calc.}J^+ + 3.308\ (\sigma_1{}^*\ \text{Calc.}J^-)/\text{Calc.}B - 0.099$$

$$R^2 = 0.940, \text{SE} = 0.378, N = 981, F = 1900.7 \quad (9)$$

The calculated descriptors based on Eqs. (3)–(9) were applied to predict the toxicity values of ILs for several microorganisms. They were first applied to predict the toxic effect of ILs on *D. magna*. For the modeling, 71 data values (which were experimentally measured after 48-h incubation) were collected. Next, the collected data set was randomly divided into a training set and a test set. The former was used for model development and the latter was used for model validation. In a statistical analysis, after inserting all calculated descriptors as independent variables and experimentally determined toxicity values as dependent variables in the training set, multiple linear regression (MLR) was performed with an option of stepwise analysis, which was for excluding the descriptors that made ignorable contributions to the modeling. In the results, it was observed that only three terms were selected, as below:

$$\text{Log } 1/\text{LC}_{50}\ [\text{M}] = -0.841 + 1.712\ E_c - 10.16\ B_c + 4.05\ J^+$$

$$R^2 = 0.880, {R^2}_{\text{adj}} = 0.872, \text{SE} = 0.359, N = 48, F = 107.5 \quad (10)$$

where the R^2 value, the SE (standard error), and the adjusted R^2 value (${R^2}_{adj}$), which was close to the R^2 value, proved that the descriptor selection was reasonable. Next, Eq. (10) was validated by correlating experimentally measured values with the calculated ones in the test set. The comparison for the model validation showed that they had a good correlation, with an R^2 value of 0.848 and an SE value of 0.479. The fitting is given in Fig. 2. When Eq. (10) was applied to the total data set, the R^2 value was 0.867 and the SE value was 0.391. For checking its clarification, Eq. (10) was internally validated using leave-one-out cross validation (${Q^2}_{\text{Loo}}$), which can be done by excluding one data point, by developing a model using the rest, and by predicting that one data point using the developed model. This process has to be repeated as many times as the number of data points in the data set. Finally, the experimental values need to be compared with the ones predicted by each of the models developed at each of the steps. The estimated ${Q^2}_{\text{Loo}}$ value was 0.832, which was much higher than the standard value of acceptance (0.5), indicating that the model was internally robust. For further validation, an external validation study was

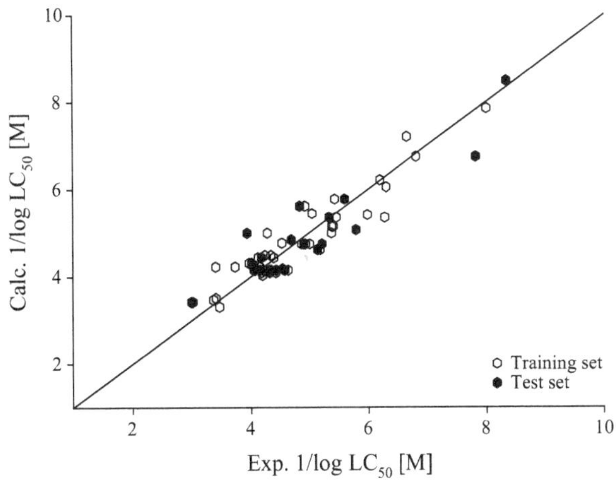

Fig. 2 Correlation between experimentally measured (*Exp.*) half-maximal lethal concentration (LC_{50}) values and those predicted (*Calc.*) by Eq. (10)

performed by estimating $Q^2_{\text{ext (test)}}$, $r_m^2{}_{\text{(test)}}$, and $\Delta r^2_{m\text{ (test)}}$, using software available online via the following links: http://dtclab.webs.com/softwaretools or http://teqip.jdvu.ac.in/QSAR_Tools/ [247]. The estimated $Q^2_{\text{ext (test)}}$ (100% data), $r_m^2{}_{\text{(test)}}$, and $\Delta r^2_{m\text{ (test)}}$ values were 0.847, 0.753, and 0.138, respectively [237]. The result of the Δr^2_m value of 0.138, which was higher than 0.2 Δr^2_m, implied the most stringent criterion of external validation [248], indicating the robustness of the prediction model. The developed model explained that with increases in the excess molar refraction and the columbic interactions of the cation, the toxic effect increases, while as H-bonding basicity rises, the toxicity of ILs decreases.

Using the same model (i.e., Eq. (10)) and procedure, the antimicrobial effects [238] of ILs on the Gram-negative bacteria *E. coli*, the Gram-positive bacteria *S. aureus*, and the diploid fungus *C. albicans*, as well as the enzyme (acetylcholinesterase) inhibition activities [222] of the ILs, were predicted. The end points for antimicrobial activities were the MIC and MBC, and EC_{50} values were used as the end point for enzyme inhibition activities. In the modeling, it was clearly observed that the toxic effects could be explained by the calculated descriptors, but the descriptor selection differed depending on the specific toxicity mechanism, as shown in Table 13. In the case of the QSAR modeling for the enzyme inhibition activities, the modeling was performed according to different head groups (imidazolium and pyridinium). As a result, the predictabilities for the MIC and MBC of the ILs for the three species were in the range of 0.788–0.969. All equations (with the exception of that for the MBC of *C. albicans*, because of the limited toxicity data set) were validated using the test sets, and the results showed that all of the models could predict the test set within an R^2 value of 0.921. In the case of the enzyme inhibition activity of the ILs, when the toxicities of imidazolium-based and pyridinium-based ILs were predicted separately, the predictabilities (R^2 values) were 0.828 and 0.879, respectively. The fittings are shown in Fig. 3.

Table 13 System coefficients for predicting antimicrobial effects of ionic liquids (ILs) on *E. coli*, *S. aureus*, and *C. albicans*, and acetylcholinesterase inhibition activity of ILs

Species	End points [μM]	c	e_c	s_c	a_c	b_c	v_c	j^+	e_a	s_a	a_a	b_a	v_a	j^-	R^2	R^2_{adj}	SE	N	F
E. coli	Log 1/MIC	−5.76	−1.39	–	1.35	−3.47	2.50	−1.91	–	–	–	–	–	–	0.900	0.893	0.43	76	126.4
	Log 1/MBC	−5.10	–	–	–	–	2.22	−2.42	–	–	–	0.34	–	0.40	0.968	0.965	0.26	41	274.3
S. aureus	Log 1/MIC	−5.01	−1.23	–	1.39	–	3.05	−3.30	–	–	–	–	–	–	0.910	0.905	0.47	70	164.9
	Log 1/MBC	−7.11	1.55	–	–	−3.90	1.34	–	–	–	–	–	–	0.49	0.972	0.969	0.27	39	298.4
C. albicans	Log 1/MIC	−20.87	–	10.72	−3.18	−13.59	2.27	–	–	–	9.36	–	−0.83	–	0.892	0.875	0.62	54	54.2
	Log 1/MBC	−14.89	–	3.81	–	−19.51	–	–	–	–	–	–	–	–	0.803	0.788	0.23	29	71.8
Acetylcholinesterase	Log 1/EC_{50} for imidazolium-based ILs	−2.53	–	–	−0.64	−1.81	0.77	–	–	0.03	–	–	–	–	0.828	0.820	0.27	98	111.7
	Log 1/EC_{50} for pyridinium-based ILs	−2.48	3.94	–	−1.64	–	–	–	−0.24	–	–	–	–	–	0.879	0.871	0.30	67	112.9

EC_{50} half-maximal effective concentration, *F* Fischer statistic, *LFER* linear free energy relationship, *MBC* minimal biocidal concentration, *MIC* minimal inhibitory concentration, *N* number of data points used, R^2 coefficient of determination, R^2_{adj} adjusted R^2 value, *SE* standard error

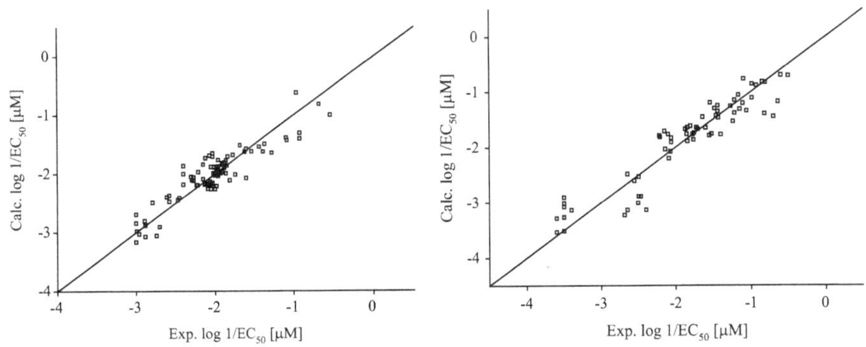

Fig. 3 Correlations between experimentally estimated (*Exp.*) and calculated (*Calc.*) enzyme (acetylcholinesterase) inhibition activity of imidazolium-based ionic liquids (*left*) and pyridinium-based ionic liquids (*right*). EC_{50} half-maximal effective concentration

3.2.3 Comprehensive Prediction Model

Since the toxic effects of ILs differ according to the types of ILs and the types of organisms exposed to them, the overall toxic effects of ILs cannot be explained. Therefore, using the calculated LFER descriptors, Cho et al. [211] developed one equation that could predict the results of 50 types of toxicity testing methods. To develop it, Eq. (1) had to be modified because the toxic effects differ according to the toxicity testing methods that are used. The modified model is shown in Eq. (11):

$$\begin{aligned}\text{Log}\,[1/\text{EC}_{50}, 1/\text{LC}_{50}, 1/\text{MIC}, 1/\text{MBC}, \text{or } 1/\text{IC}_{50}] \text{ in log units of mM} \\ = e_c\,E_c + s_c\,S_c + a_c\,A_c + b_c\,B_c + v_c\,V_c + j^+\,J^+ + e_a\,E_a + s_a\,S_a + a_a\,A_a \\ + b_a\,B_a + v_a\,V_a + j^-\,J^- + c + z_x Z_x \qquad (11)\end{aligned}$$

where the subscripted x in the zZ term stands for an indicator of toxicity testing methods, and the magnitude of the Z value, as part of the sensitivity of the toxicity testing methods, means the sensitivity difference (y-axis) from that of the reference method at the averaged points of the data set for each method. In general, the reference method is determined according to the number of data sets. Here, the reference method is the leukemia rat cell line. When the z term is higher than zero, it means the sensitivity is higher than that of the reference method; if it is lower, it indicates a lower trend. For the solute property, several end points expressed in millimolar units and in silico–calculated LFER descriptors were used. In the statistical analysis, after including all data sets and calculated descriptors, MLR was performed. Here, less important descriptors, with higher p values of ≥ 0.05, were automatically excluded by the stepwise analysis. Through the step, the system parameters were determined as follows:

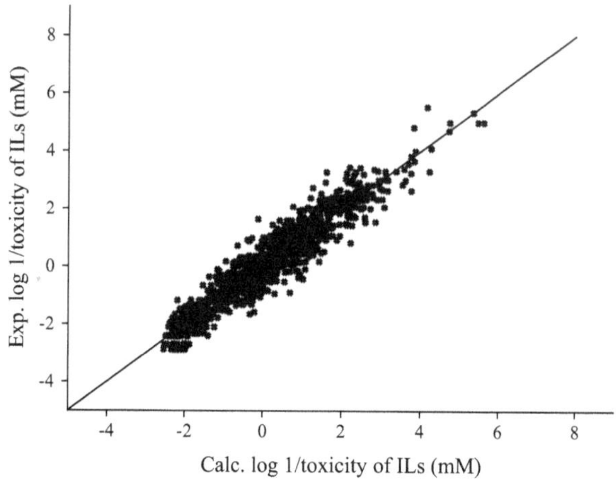

Fig. 4 Correlation between toxicity values of ionic liquids (*ILs*) observed (*Exp.*) using 50 testing methods and those calculated (*Calc.*) by Eq. (13)

$$\begin{aligned}&\text{Log}\left[1/\text{EC}_{50}, 1/\text{LC}_{50}, 1/\text{MIC}, 1/\text{MBC}, \text{or } 1/\text{IC}_{50}\right] \text{ in log units of mM}\\&\quad = 2.254\ E_c - 2.545\ S_c + 0.646\ A_c - 1.471\ B_c + 1.650\ V_c + 2.917\ J^+\\&\qquad - 0.201\ E_a\ 0.418\ V_a + 0.131\ J^- - 0.709 + z_x Z_x\\&R^2 = 0.880, \text{SE} = 0.465 \text{ log units}, N = 1633, \text{number of systems} = 44, F\\&\quad = 218.7\end{aligned} \quad (12)$$

However, the data plotted by the correlation between the experimentally determined values and the ones predicted by Eq. (12) were not distributed on an identical line (1:1); they had another fitting slope. Therefore, Cho et al. [211] addressed the differences by adding two more terms (α_x and β_x) as Eq. (13). The determined sensitivity-related terms were stated in their report [211].

$$\begin{aligned}&\text{Log}\left[1/\text{EC}_{50}, 1/\text{LC}_{50}, 1/\text{MIC}, 1/\text{MBC}, \text{or } 1/\text{IC}_{50}\right] \text{ in log units of mM}\\&\quad = \alpha_x{}^* \text{ Calc.value by Eq.}(4) + \beta_x\end{aligned} \quad (13)$$

From this step, the correlation accuracy improved to an R^2 value of 0.901 and an SE value of 0.426 log units, compared with those obtained using Eq. (12). Its fitting is shown in Fig. 4.

To check whether the step explained above can be applied to other toxicity testing methods, in addition, toxicity values from different methods (testing the growth rates of the microalgal species *B. paxillifer, G. amphibium, C. vulgaris, O. submarina, S. marinoi*, and *C. meneghiniana*) were collected. From the validation study, it was checked that the predicted values of each of the toxicity testing methods could be correlated with the measured R^2 values of 0.82–0.98.

Such a comprehensive model can provide several advantages in the field of IL toxicity research. First, on the basis of this method, 50 types of toxicity values can be predicted using solely computational methods. Second, the overall toxic interaction between ILs and microorganisms can be interpreted. In particular, the anion effects on toxicity can be clearly explained. This information will be helpful for designing low-toxicity IL structures. Third, the biological responses (z_x, α_x, and β_x) to ILs according to the species and end points can be explained numerically, which will be helpful for investigation of toxic effects for specific purposes – for example, high antimicrobial activity for medicinal use and low toxicity for sustainable chemistry. Fourth, a few toxicity data sets are needed for developing toxicity prediction models for other toxicity testing methods. Actually, for developing a prediction model, numerous toxicity data values are required because the prediction model has to be able to explain a wide spectrum of chemical structures. However, since the comprehensive model was developed on the basis of many IL structures, a few data points were needed for determination of the sensitivity terms (z_x, α_x, and β_x).

On the basis of the comprehensive prediction model, the toxic effects of ILs on several microorganisms could be explained at the molecular level. From Eq. (12), it could be interpreted that IL toxicity is mainly determined by the cationic part, rather than by the anion. Indeed, it was observed that more interactions by IL cations contributed to toxicity prediction modeling, and the magnitudes of the system parameters of the cation are larger than those of the anion. Nevertheless, the anion has a considerable effect on toxicity. To select less toxic IL structures, some structural concepts based on Eq. (12) can be suggested. To explain them, increases in the molecular volumes of the cation and anion increase the toxicity; also, the ionic forces of the cation and anion follow the trend of the molecular volume effect on toxicity. Also, increases in the H-bonding acidity and the excess molar refraction of the cation cause a rise in toxicity. On the other hand, increases in dipolarity/polarizability and the H-bonding basicity of the cation reduce IL toxicity. Through these findings, less toxic IL structures can be designed. Consequently, QSAR approaches can contribute to rapid and efficient estimation of the toxic impacts of ILs.

4 Concerns Regarding Biodegradation

Biodegradation testing is important when investigating the environmental fates of a chemical. Depending on their persistence, the chemical activities of chemicals in the environment can differ; therefore, their biodegradation properties can be categorized as follows:

- Primary biodegradation: loss of chemical properties of a chemical through alteration of the chemical structure by biological action
- Inherently biodegradable: 20–60% degradation of a compound
- Readily biodegradable: 60–100% degradation of a compound, meaning it will be degradable in aquatic environments
- Ultimately biodegradable (mineralization): complete degradation of a compound

Several accepted biodegradation testing methods [249] accepted by the Organization for Economic Co-operation and Development (OECD) are listed in Table 14.

Table 14 Biodegradation tests and analytical methods

Test	Analytical method
DOC die-away	Dissolved organic carbon
CO_2 evolution	Respiratory: CO_2 evolution
Ministry of International Trade and Industry	Respiratory: O_2 consumption
Closed bottle	Respiratory: dissolved oxygen
Modified OECD screening	Dissolved organic carbon
Manometric respirometry	O_2 consumption

DOC dissolved oxygen concentration, *OECD* Organization for Economic Co-operation and Development

Some of the methods estimate the concentrations of tested molecules, while some are based on the physiological activities of chemicals. Selection of an appropriate method will depend on the purpose of the experiment because each method has different applicability. Accordingly, the different analytical methods of the tests can generate different results.

Approximately 50 years ago, when the concept of ILs as green solvents had not yet been introduced, a biodegradation test of 4-carboxyl-1-methylpyridinium Cl was performed by Wright and Cain [250] using a biological extract from a pure bacterium, *Achromobacter* D; the metabolic fate of the compound was identified using a ^{14}C radioisotope study [251]. Fully fledged research on the biodegradation of various IL structures began in 2002. The main purposes of this research were to estimate the biodegradability of ILs and, on the basis of the results, to find more environmentally friendly chemical structures because IL structures are very numerous.

Many research teams have been dedicated to studying the biodegradability of various ILs and have attempted to understand the effects of the structures of ILs on their biodegradability.

4.1 *Biodegradation of Nonfunctionalized ILs*

Wells and Coombe [135] estimated the biodegradability of imidazolium-, pyridinium-, phosphonium-, and ammonium-based ILs with different anions by measuring oxygen uptake in closed bottles and reported that the tested ILs were extremely poorly biodegradable. In the same year, Kumar et al. [252] reported that soil or wastewater microorganisms could degrade [IM14][BF_4] and produced some biodegradation products, which were detected by gas chromatography–mass spectrometry (GC-MS) analysis. In 2007, Docherty et al. [253] estimated the biodegradability of imidazolium- and pyridinium-based ILs with propyl, hexyl, and octyl alkyl chains by the dissolved oxygen concentration (DOC) die-away testing method and found that imidazolium-based ILs with hexyl and octyl alkyl chains were 54% and 41% biodegradable, respectively, and pyridinium-based ILs with hexyl and octyl alkyl chains were almost mineralized. Moreover, by analyzing the nitrogen source from the samples that were taken, they showed that a pyridinium ring was opened by biological activity, while an imidazolium ring was not opened [253]. Later, biodegradation testing of imidazolium-based ILs with different alkyl chains was performed using a biochemical oxygen demand test (BOD5) for 10 days by Romero et al. [108]. During that time, the tested ILs were poorly degraded even in conditions with an additional carbon source (glucose), which was used to induce cometabolism. In 2008, to investigate more various IL structures, Stolte et al. [254] selected imidazolium-, pyridinium-, and 4-(dimethylamino)pyridinium-based ILs with Cl or Br. The results revealed that the imidazolium and pyridinium compounds with a long alkyl chain (e.g., an octyl, hydroxyl, or carboxyl group at the end of the alkyl chain) were fully degraded, while short-alkyl-chained ILs were recalcitrant. Next, the biodegradability of several ILs was tested in anaerobic conditions [255]. Here, with the exception of the [IM18OH] cation, the concentrations of other ILs were almost constant during the biodegradation period. The same groups further investigated the biodegradability of fluoro-organic and cyano-based IL anions, which can be used in some technical applications, under two conditions: aerobic and anaerobic. It was shown that none of the anions were biologically degraded. Similarly, Gotvajn et al. [256] investigated the biodegradability of common ILs ([Py4-3Me][$N(CN)_2$] and [IM14][BF_4]) in aerobic and anaerobic conditions. In the aerobic conditions, two compounds were poorly degraded, while the granular sludge in the anaerobic conditions degraded 34% of the IL concentration in the presence of glucose as a carbon source [256]. Moreover, Steudt et al. [35] studied dicationic ILs and showed that they were not biodegradable. The biodegradability was less than 5% of the degradation. More investigation on the biodegradability of ILs according to the type of structure was done by the same group [257]. That study evaluated a total of 27 ILs – including pyrrolidinium, pyridinium, imidazolium, morpholinium, and piperidinium with halide anions – and their cationic structures had different alkyl chain lengths and functional groups (i.e., ester, ether, hydroxyl, and cyan groups) [257]. The experimental results clearly showed that the biodegradability was dependent on the IL structure. In that study, pyrrolidinium with octyl and hydroxypropyl, piperidinium with hydroxypropyl, and pyridinium with hydroxyethyl were readily biodegraded. Pham et al. [154] investigated the effects of alkyl chain of ILs, including three different head groups (imidazolium, pyridinium, and pyrrolidinium). The alkyl chains were propyl, butyl, hexyl, and octyl. In the case of imidazolium- and pyrrolidinium-based ILs, hexyl- and octyl-chained ILs were 100% degradable by primary biodegradation, while the pyridinium-based ILs involved butyl, hexyl, and octyl chains. Several other researchers have also investigated the biodegradability of ILs for use in various applications [45, 258, 259].

4.2 *Biodegradation of Ether-Functionalized, Ester-Functionalized, and Amide-Functionalized ILs*

To design readily biodegradable and low-toxicity ILs, the effect of functionalization of ILs with ether, ester, or amide groups has often been investigated because hydrophilic functionalization can lead to less toxic impacts than those of nonfunctionalized ILs. Gathergood and Scammells [260] tested the biodegradability of [IM11COO1]Br, [IM11COO1][BF_4], and [IM14][PF_6] according to a modified Sturm testing method (ISO9439) for 20 days and showed that [IM11COO1]Br and [IM11COO1][BF_4] were 48% and 59% degraded, respectively, while [IM14][PF_6] was 60% degraded. The same group examined the biodegradability of the same cation with a different alkyl chain length (butyl to octyl) or amide group according to the guideline of the closed bottle test (OECD 301D) [261]. Here, as the alkyl chain length increased, the biodegradability also increased slightly, but this trend disappeared when the alkyl chain length was long [261]. This conclusion was later supported by findings reported by Liwarska-Bizukojc and Gendaszewska [262]. In a comparison of the biodegradability of ether- and amine-functionalized ILs, the former had higher biodegradation percentages ranging from 10% to 30%, while amine functionalization on the alkyl chain of the ILs led to poor biodegradation. Nevertheless, the functionalized ILs had greater biodegradability than the nonfunctionalized ILs ([IM14][BF_4] and [IM14][PF_6]), which had biodegradation values of nearly zero. Note that the estimated biodegradability of [IM14][PF_6] reported by Liwarska-Bizukojc and Gendaszewska [262] differed from that reported by Gathergood and Scammells [260]: 0% versus 60%. It is possible that the biodegradability could differ spatially and temporally. Detailed information is given in Sect. 4.3. Samori et al. [151] tested the biodegradability of methylpyrrolidinium-based ILs with alkyl chains or ether group(s) on fewer than six chains, and showed that none of the tested ILs were more than 5% degraded. Stasiewicz et al. [32] studied 1-alkoxymethyl-3-hydroxypyridinium, which has different alkyl chains from C3 to C18, combined with acesulfamates, saccharinates, and Cl. In their study, they found that as the alkyl chain length increased from C4 to C11, the biodegradability of the compounds increased. Nevertheless, only one compound (1-undecanoxymethyl-3-hydroxypyridinium saccharinate) was considered readily biodegradable, being 72% degraded [32].

Many cases have shown that ester functionalization of ILs with different counterions enhances degradation efficiency, since they can be enzymatically hydrolyzed (Fig. 5). Garcia et al. [107] performed comparative biodegradation tests of functionalization and nonfunctionalization of [IM14] and [IM11COO3] with different anions (Cl, Br, [BF_4], [PF_6], [$8OSO_3$], [$N(CN)_2$], and [$(CF_3SO_2)_2N$]) based on a closed bottle test (OECD 301D). In the case of [IM14]-based ILs, the combination of [$8OSO_3$] led to 25% biodegradation, while the other ILs remained almost intact and were less than 5% degraded [107] On the other hand, the ester-functionalized ILs showed higher biodegradability than the nonfunctionalized ones. Also, it was shown that the [$8OSO_3$] anion had the highest biodegradation rate of 49%, while the rest

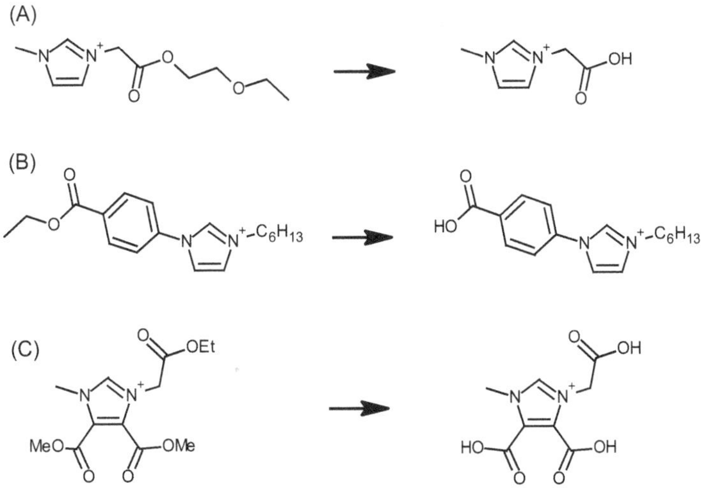

Fig. 5 Typical examples of enzymatic hydrolysis of ester-functionalized ionic liquids studied by (**a**) Deng et al. [92], (**b**) Stolte et al. [47], and (**c**) Prydderch et al. [68]

were around 10–30% degraded [107]. Later, the same group studied other ester-functionalized ILs with [8OSO$_3$] and showed that the combination of the cation and anion made them 49–56% degradable in a closed bottle test and 60–67% degradable in a CO_2 headspace test. The latter result meant that they could be considered readily biodegradable [263]. Harjani et al. [264, 265] demonstrated readily biodegradable IL structures designed on the basis of a pyridinium moiety with ester side chains, but a nicotinamide-based IL had low biodegradability of 30%. Then, using imidazolium-based ILs, they investigated structural effects (e.g., the alkyl chain lengths of the cation and anion, ester functionalization, and the type of anion) on biodegradability, and suggested that the introduction of lactate and saccharin as counteranions could lead to high mineralization abilities [266]. Morrissey et al. [267] studied biodegradation of 15 ester-functionalized ILs on the basis of a CO_2 headspace test. Some of them had degradation rates of 62–92%, meaning they could be considered readily biodegradable, and the others were 55–59% degradable, which nearly reached the level of being readily biodegradable. Similarly, it was observed by Deng et al. [92] that ester functionalization on the alkyl chain of imidazolium-based ILs helped to improve their biodegradability during incubation with each of three different microorganisms (*Pseudomonas viridiflava*, *Nocardia asteroides* strain 911, and *Candida parapsilcosis*). Moreover, Deng et al. [92] reported that although the ester-functionalized ILs were readily biodegradable, their common metabolite, [IM11COOH], could accumulate without further degradation. Later, aryl- and ester-functionalized ILs were prepared, and their biodegradability was tested [47]. Most of these prepared ILs showed no degradation, but addition of an ester group led to 100% primary degradation. Pernak et al. [122] showed that aryl functionalization did not help to enhance the biodegradability of 4-benzyl-4-

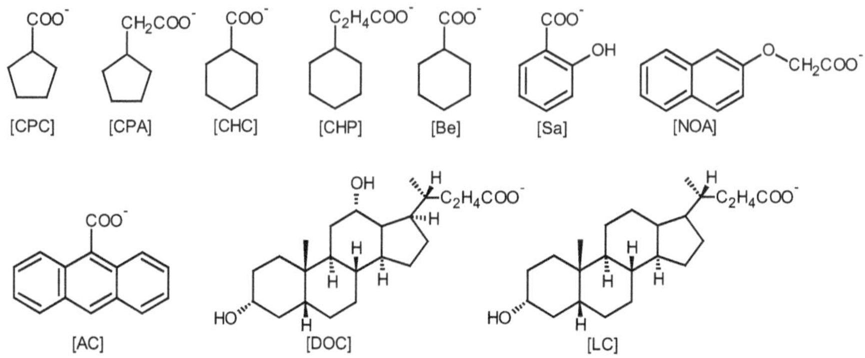

Fig. 6 Naphthenic acid–based anions studied by Yu et al. [271]

alkylmorpholinium-based ILs. Al-Mohammed et al. [268] designed tetrakis-imidazolium- and benzimidazolium-based ILs, which have different alkyl side chain and tetra-ester groups, and tested their biodegradability. The results showed that the ester group with long side chains and phenyl rings improved the biodegradability. In the case of long side chains (e.g., octyl, decyl, and dodecyl), the ILs reached the level of being readily biodegradable. However, ester functionalization does not always lead to high biodegradability. For example, Atefi et al. [269] prepared tetraalkylphosphonium- and trihexylphosphonium-based ILs with different anions and functions on the alkyl chain (ester, allyl, ether, and alcohol). The tested molecules were only modestly or poorly biodegradable. Similarly, imidazolium-based ILs with ester or amine groups could not be regarded as readily biodegradable [270].

To achieve more environmentally friendly (e.g., less toxic and more biodegradable) ILs, other functionalized ILs have been introduced. Yu et al. [271] prepared ILs with a naphthenic acid–based anion (NAILs) combined with [Ch] (Fig. 6), and performed biodegradation tests of these ILs using a closed bottle test method. With the exception of [Ch][NOA] and [Ch][1COO], the tested NAILs were more than 60% biodegradable. Moreover, Klein et al. [272] found that [Ch]-based ILs with different anions and different alkyl chain lengths from 8 to 16 (i.e., laurate, myristate, palmitate, and stearate) were easily degraded by bacteria in a wastewater treatment plant. With the same purpose, Ferlin et al. [273] studied natural organic acid–based anions (L-lactate, L-tartrate, malonate, succinate, L-malate, pyruvate, D-glucoronate, and D-galacturonate) combined with tetrabutylammonium. Although these compounds have less toxicity and higher biodegradability than general ILs (i.e., [N4444]Br, [N4444][OH], [N2222]Br, and [N11111]Br), they were less than 23% biodegradable. Thus, they could not be considered readily biodegradable compounds. Similarly, Boissou et al. [274] prepared levulinate-based ILs with [N1124] and showed that they were 54–79% biologically degraded according to the oxygen uptake method. Moreover, Hou et al. [46] produced several combinations between highly degradable [Ch] and amino acids. All of the prepared ILs passed the standard for being readily degradable. Peric et al. [34] prepared new types of ten protic ILs (PILs) comprising aliphatic amines and organic acids, and showed that these PILs had low toxicity and good biodegradability. Eight of the ten ILs could be considered readily biodegradable, and the remaining compounds were nearly 60% biodegraded.

Ford et al. [275] prepared pyridinium-based ILs with several substitutes and anions. Among them, hydroxyalkyl-substituted pyridinium had improved biodegradability, while in ILs with methyl or ethyl ether side chains the hydroxyalkyl substitution did not result in high biodegradability. Later, the same group [276] synthesized various types of ILs including pyridine, 1,4-dimethylpiperazine, 1,4-diazabicyclo[2,2,2]octane, and several alkyl side chains. The results, determined using a CO_2 headspace method, showed that only one compound, [Py41COOH] [$(CF_3SO_2)_2N$], was ultimately biodegradable, while the rest were poorly degradable.

Prydderch et al. [68] synthesized low-antimicrobial ILs based on mandelic acid, which is a biorenewable source; however, eight of ten ILs with ether groups had low biodegradability of 5–31%, tested in a closed bottle test, while the other two ILs with amide groups were only 0–1% biodegradable. Haiss et al. [277] studied fully mineralized phenylalanine-based ILs. These ILs were not readily biodegradable; 60% of the injected concentration was not degraded within 28 days. However, when the incubation time was increased to 42 days, microorganisms continued to break down three of ten IL structures to the level of ultimate biodegradation. Later, the same group with Prof. Kummerer introduced new L-phenylalanine-based, L-tyrosine-based, and tertiary amino–based ILs [70], but they were not readily degradable in a closed bottle test.

The effect of chiral structure and peralkylation on biodegradation efficiency has been tested [262, 278–280]. Ferlin et al. [278] investigated the biodegradability of chiral ILs, which had prolinate (*S*) or (*R*) as an anion combined with a tetrabutylammonium-based cation, and showed that both types were only 20% degradable. Liwarska-Bizukojc et al. [280] showed that peralkylated imidazolium-based ILs were also poorly degradable (less than 10%). The same group studied the metabolism of alkylimidazolium-based and peralkylated imidazolium-based ILs [279].

4.3 *Biodegradation of ILs by Microbiota or Axenic Culture*

The efficiency of IL biodegradation differs according to the type of microorganism inoculation; therefore, several trials have assessed the biodegradation efficiencies of ILs in several environments. As mentioned above, Docherty et al. [253] reported that the [Py4] cation was not degradable. However, in contrast to their previous results, they found that in a subsequent biodegradation test of the same IL [165], [Py4] was readily biodegradable, as was also reported by Pham et al. [281]. This finding was clearly confirmed by additional biodegradation testing using different microbial consortia taken from two different places [282, 283]. Additionally, the use of

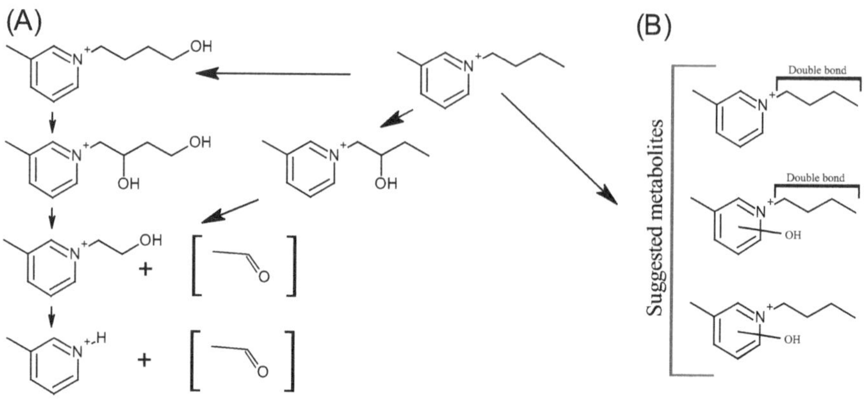

Fig. 7 (**a**) Biodegradation pathway of the [Py4] cation studied by Pham et al. [281]. (**b**) Metabolism of the [Py4] cation studied by Docherty et al. [165]

different microbial communities transformed the metabolism, as shown in Fig. 7. This trend was also observed in biodegradation testing of the [IM18] cation by Stolte et al. [254] and Cho et al. [284]. Stolte et al. [254] identified one degradation pathway (see pathway 1 in Fig. 8), while Cho et al. [284] observed more various degradation pathways (see pathways 2–4 in Fig. 8).

In another example, the biodegradation efficiency of herbicidal ILs, which were specified to enhance antimicrobial activity, could be differentiated according to microbiota taken from various areas [286]. Nevertheless, poor degradation of herbicidal and morpholinium-based ILs was observed in an ultimate biodegradation test. In a primary biodegradation test, cations had higher biodegradability than anions. Additionally, it was reported that high lipophilicity led to a high toxicity effect and low susceptibility, while hydrophilic compounds had less toxicity and slightly higher biodegradability. However, this is not always true. In a comparison of the susceptibility of [IM18] and [IM12], the first had higher biodegradability although it had high lipophilicity. On the other hand, Niemczak et al. [285] studied other herbicide-based ILs, using eight combinations between two ammonium-based cations (*N*-dodecylbetaine and *N*-(3-cocoamidopropyl)betaine) and four herbicidal anions (2,4-dichlorophenoxyacetic acid (2,4-D), 2-methyl-4-chlorophenoxyacetic acid (MCPA), methylchlorophenoxypropionic acid (MCPP), and dicamba) (Fig. 9). Among them, ILs with 2,4-D anions had high biodegradability of 72–73%, while the others were 42–62% biodegradable. The susceptibility of the anions in that study was in the order of 2,4-D > MCPA > MCPP > dicamba [285].

Moreover, the biodegradability of ILs has been checked in several environmental media. Modelli et al. [287] tested the biodegradability, based on the measurement of CO_2 evolution, of [IM14][BF_4], [IM14][$N(CN)_2$], [IM11O2][BF_4], and [IM11O2][$N(CN)_2$] in soil, but the tested ILs showed no significant change in CO_2 production. Also, Deive et al. [288] exposed ILs to microbial strains in soil from a region in Portugal to test the survival ability of bacteria and their ability to degrade the ILs. They showed that the bacterial community, exposed for a long time to a high-salt and

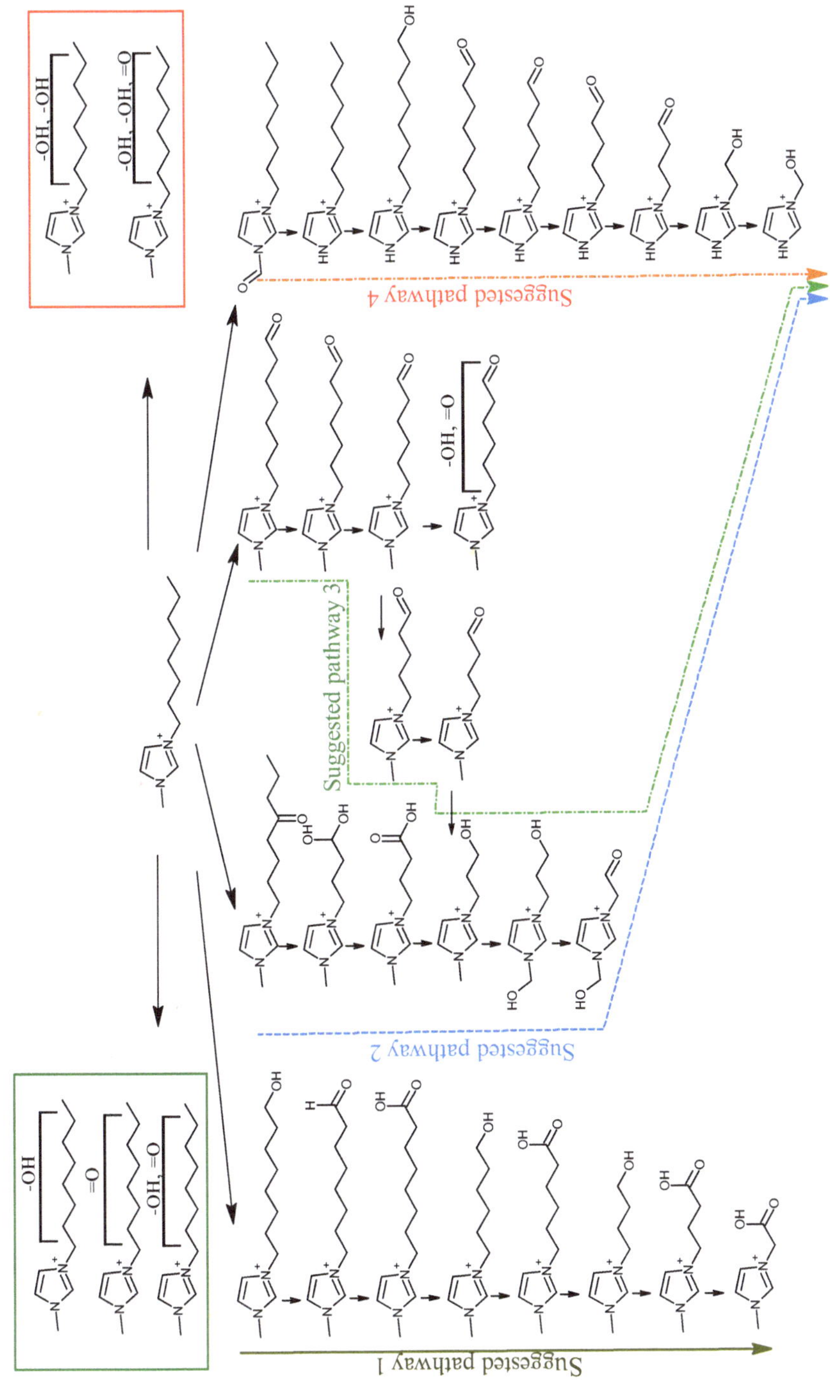

Fig. 8 Suggested biodegradation pathways: pathway 1 was studied by Stolte et al. [254], and pathways 2–4 were studied by Cho et al. [284]

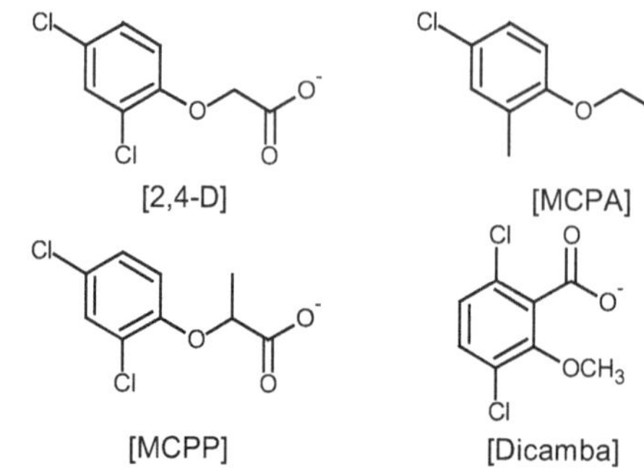

Fig. 9 Herbicidal acid–based anions studied by Niemczak et al. [285]. *2,4-D* 4-dichlorophenoxyacetic acid, *MCPA* 2-methyl-4-chlorophenoxyacetic acid, *MCPP* methylchlorophenoxypropionic acid

high-hydrocarbon environment, could survive in high concentrations of the tested ILs; however, they could not readily degrade them [288]. Garbaczewska and Hupka [289] estimated the biodegradability of [IM14][PF_6], [IM14][BF_4], [IM01][Lac], [IM04][Lac], [IM0-12][Lac], and [IM01O-4][Lac] in surface water for 77–98 days. The IL concentration was analyzed by HPLC. The results showed that the tested ILs were slightly degraded as time passed, and only [IM0-12][Lac] was fully degraded within 32 days. Sydow et al. [290] investigated the persistence of ammonium-based ILs (didecyldimethylammonium 3-amino-1,2,4-triazolate and benzalkonium 3-amino-1,2,4-triazolate) and phosphonium-based ILs ([P666-14]Cl and [P666-14] 1,2,4-triazolate) in urban park soil microcosms in Poznan, Poland, for 300 days. The biodegradability was assessed by measurement of CO_2 evolution. The results showed that [P666-14]Cl was 63% primarily degraded and 16% of it was mineralized. In case of the other ILs, the CO_2 evolution decreased below the control level. This might have been due to low respiration of the microorganisms in the soil.

For more specific study, an axenic culture of a pure microorganism was applied in a biodegradation study of ILs. In this research, several isolated microorganisms were applied. Esquivel-Viveros et al. [291] tested [IM14][PF_6], using *Fusarium*, and showed that the species used 80% of this compound as a carbon source. Abrusci et al. [292] isolated *S. paucimobilis* bacteria and incubated them in a batch with each of 36 ILs. Of these, 20 ILs, which had previously been considered recalcitrant compounds, were readily biodegraded. Zhang et al. [293] observed that *Corynebacterium* species could degrade [Py2] (also previously considered a recalcitrant compound) via ring opening, while *Pseudomonas fluorescens* metabolized the same cation (Fig. 10). However, the metabolite (i.e., the pyridinium core) was not further degraded [94]. In 2014, Markiewicz et al. [294] isolated nine bacteria from activated sludge and tested their ability to biodegrade [IM18]Cl. The results showed that a single bacterium had low degradation ability of around 30%, while in the same conditions, activated sludge achieved almost 60% degradation. To remove recalcitrant cyano-based anions, *Cupriavidus* spp. were applied, but no enhancement of biodegradability was observed. Deng et al. [295] applied *Rhodococcus rhodochrous*

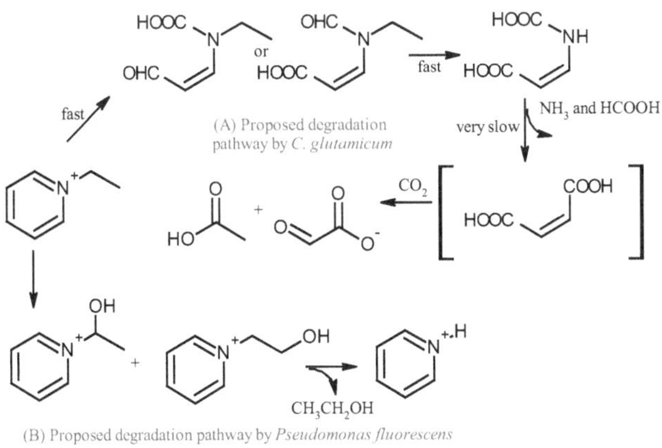

Fig. 10 Pathways of [Py2] biodegradation by (**a**) *Corynebacterium glutamicum* and (**b**) *Pseudonomas fluorescens*

ATCC 29672 to 10 ILs, including pyridinium, pyrrolidinium, and ammonium, which have relatively short chains. The results showed that nine of the ten ILs were readily biodegradable. Moreover, metabolites were also identified. In 2016, Thamke and Kodam [296] isolated a bacterium, *Rhodococcus hoagie VRT1*, and applied it to the biodegradation of [IM14]Br, which was considered a poorly degradable and toxic compound. The results showed that this pure bacterium could rapidly achieve almost the level of ready degradation within 8 days.

4.4 Bioaccumulation of ILs

Bioaccumulation of a chemical is an important environmental property, which is closely related to the octanol–water partitioning coefficient (log *P*). Crosthwaite et al. [297] measured the log *P* value for [IM14][$(CF_3SO_2)_2N$], which was extremely similar to the values for polar organic compounds, and forecasted that the bioaccumulation of this compound might be quite low. Similarly, Deng et al. [298] stated that hydrophilic ILs with low lipophilicity, which can be designed by oxygen functionalization (e.g., using ester and hydroxyl groups), may have low bioaccumulation, leading to less toxicity. Low-lipophilicity ILs have low passive diffusion across membranes; therefore, they have only a small possibility of being stored in the fatted area in organisms [298]. Estimated and predicted log *P* values have been provided in several reports [299–303]. In 2008, Cornmell et al. [304] experimentally estimated the bioaccumulation of the ILs [P666-14][$(CF_3SO_2)_2N$], [P666-14]Cl, [N1888]Cl, and [N1888][$(CF_3SO_2)_2N$] in *Escherichia coli*. The experiment was performed by incubation of the species with ILs and analysis by Fourier transform infrared (FTIR) spectroscopy. The results showed that among the tested ILs, the most hydrophobic ([P666-14][$(CF_3SO_2)_2N$]) specifically accumulated in the membranes of the cells. Nedzi et al. [305] experimentally measured the bioaccumulation of [IM14]Cl in *Chlorella vulgaris* algae, used as a food provider for *Balanus improvises* barnacles and *Mytilus trossulus* mussels. The measured bioconcentration factor (BCF) in the algae was 6.5, which was around six times the EC_{50} value. As the IL concentration in the water increased, the BCF in the barnacles increased [305]. With coexistence of the IL-contaminated algae and barnacles, the BCF in the barnacles increased by 80% [305]. The BCF in the mussels was similar to that in the barnacles. Dolzonek et al. [306] experimentally measured the membrane–water partitioning coefficients of ILs, using an artificial membrane test kit. On the basis of the estimated values, the bioconcentration potential of the ILs was predicted by the correlation between BCF values and the membrane–water partitioning coefficients of neutral compounds [306]. In that study, it was revealed that the bioconcentration potential of the ILs suggested by Dolzonek et al. [306] was much higher than the values expected from the log *P* values. The predicted BCF values of ILs may be relevant in terms of the "*B*" classification under the European Union's Registration, Evaluation, Authorisation and Restriction of Chemicals (REACH) regulation. Furthermore, the accumulation trend has been demonstrated by imaging. Perez et al. [170] reported that AMMOENG 130 and its metabolites accumulated in the nervous and respiratory systems of zebrafish, which was estimated by a desorption electrospray ionization mass spectrometry imaging method. Moreover, they reported that the ILs were capable of penetrating the blood–brain barrier of the fish, and thus could accumulate in that part of the body.

5 Adsorption, Mobility, and Adsorptive Removal

Information on the extent of sorption by environmental media is important for assessing the environmental mobility of ILs. Moreover, removal studies of ILs are important and necessary for sustainable development because most IL structures are not biodegradable or are hardly treatable. However, the extent of sorption of ILs by environmental media has rarely been studied.

Gorman-Lewis and Fein [307] experimentally investigated the adsorption of [IM14]Cl by Gram-positive soil bacteria (*Bacillus subtilis*) and minerals (gibbsite, quartz, and Na-montmorillonite), and showed that the IL did not have the ability to be adsorbed by natural media. Moreover, the authors implied that the mobility of ILs may be unimpeded since they have low adsorptive interactions with geological and biological surfaces [307]. Later, Stepnowski [308] evaluated adsorption of ILs by soils (agricultural soil, clayey soil, and peaty soil from northern and northeastern areas in Poland) and marine sediments from the southern Baltic Sea. The results showed that as the hydrophobicity increased, the adsorption of ILs by the soils decreased. The main mechanism of adsorption may be nonhydrophobic interaction (e.g., polar and electrostatic interactions). Moreover, the adsorption of imidazolium-based ILs by natural soils (i.e., clayey agricultural, fluvial meadow, forest, and fluvial agricultural soils) was further estimated by Stepnowski et al. [309]. The results showed that the maximum uptake of the adsorbents was above the cationic exchange capacity. It may be that adsorption involves the formation of a multilayer when a compound with a long alkyl chain is involved [309]. After the IL occupies the binding sites of the adsorbent surface by electrostatic interaction, further adsorption might occur by dispersive interaction. For more detailed understanding of the mobility of ILs, Mrozik et al. [310] tested the ability of ILs to be adsorbed by soil taken from a coastal region (Gdansk, Poland), using a column system, and confirmed that the degree of lipophilicity of the ILs was related to the degree of their retention in the soil column, while hydrophilic structures had weak adsorption. In 2012, Mrozik et al. [311] investigated nine ILs in 11 types of soils. Here, it was reported that structures with long alkyl chains were more adsorbed than hydrophilic structures. This indicates that short and hydrophilic structures may have more mobile properties in soil and sediments. In 2015, Sydow et al. [290] estimated the percentages of adsorption of ammonium-based ILs (didecyldimethylammonium 3-amino-1,2,4-triazolate and benzalkonium 3-amino-1,2,4-triazolate) and phosphonium-based ILs (trihexyl(tetradecyl)phosphonium chloride and trihexyl(tetradecyl)phosphonium 1,2,4-triazolate) by soils from an urban park in Poznan, Poland, and showed that 44% and 64% of the ammonium-based ILs were adsorbed, respectively, while only 13% and 29% of the phosphonium-based ILs were adsorbed, respectively. The tested ILs may exist persistently in the urban park. Reinert et al. [312] characterized the adsorptive interaction of imidazolium and pyridinium ILs with Na-montmorillonite, and determined that the order of the adsorption capacity was [Py4]Br > [Py8]Br ~ [IM12=1]Cl ~ [IM14]Cl > [IM18]Cl. Previously, Greenland and Quirk [313] studied the adsorptive interactions of 1-alkylpyridinium ILs with Na-montmorillonite and Ca-montrillonite, and reported that the main adsorption mechanisms were ionic and dispersion forces.

Enhancement of IL biodegradation in wastewater treatment plants using microbial adaptation to ILs has been discussed. Markiewicz et al. [314] tested the biodegradability of [IM18]Cl in activated sewage sludge on the basis of a die-away test and found that the threshold for the biodegradation of the compound was 0.2 mM. Moreover, through a 2-month adaptation of the sewage sludge, the biodegradation rate and treatable IL concentration could be increased, while additional carbon and nitrogen sources diminished the biodegradation rate [315]. Similarly, Alvarez et al. [316] and Gendaszewska and Liwarska-Bizukojc [317] observed similar adaptation effects in the microorganism *Pseudomonas stutzeri* and activated sludge, respectively. This adaptation process may lead to efficient biological treatment of ILs; however, further studies are needed.

Since ILs are hazardous and poorly biodegradable, removal methods should be developed. Therefore, several researchers have performed adsorptive removal studies using activated carbon, ion-exchange resin, biochar, biosorbents, etc. Anthony et al. [318] performed a removal study of [IM14][PF_6] using activated carbon, which is a commercially available sorbent, and reported that activated carbon was promising but inefficient for IL recovery. Using the same sorbent, Palomar et al. [319] found that activated carbon could be used for removal of hydrophilic ILs, and after the sorption, the sorbent could be regenerated using acetone. In 2012, Farooq et al.

[320] tested the capacities of several types of activated carbon, which were in granular and fabric forms, for adsorption of [IM18]Cl, [IM14]Cl, and [Py8]Cl. They reported that an increase in the number of oxygen-functionalized groups on the activated carbon enhanced the electrostatic interaction, while the original carbon was governed by dispersive interactions. In 2014, the same group [321] investigated the effects of IL structure on IL adsorption by activated carbon. They reported that the process was exothermic and the efficiency of activated carbon for IL removal was strongly dependent on the alkyl chain length of the IL; as the chain length increased, the adsorption by the adsorbent increased. Similarly, Lemus et al. [322] studied modified activated carbons with a variety of functional groups and chemical properties. The results showed that small numbers of polar functional groups on the activated carbon led to high adsorption capacity for hydrophilic ILs. Later, the same group estimated the removal efficacy of activated carbon for more ILs and reported that activated carbon could be used for removal and recovery of ILs, although it had a slower adsorption rate for ILs than phenol as a reference compound [323, 324]. With regard to carbon-based materials, hydrothermal carbonization of bioinspired materials (e.g., biochar) was shown to be effective for removal of ILs because it provided a large surface area with a high adsorption capacity, and was an environmentally friendly process. In 2013, Qi et al. [325] produced functional carbonaceous material based on carbonization of cellulose and applied it to adsorption of [IM14]Cl. The results showed that its adsorption capacity was similar to that of commercially available activated carbon, and it could be regenerated at least three times. In their next study they trebled the adsorption of the same IL [326] by modifying the cellulose using a preirradiation process to induce emulsion grafting of glycidyl methancylate and sulfonation. The modified cellulose microspheres could highly adsorb 1-alkyl-3-methylimidazolium Cl, with maximum uptake of 1.08 mmol/g, and an equilibrium state was reached within 40 min [327]. The used sorbent could be regenerated using 0.1 M HCl or NaCl [327]. In 2016, biochar of bamboo origin was prepared and applied for removal of [IM14]Cl [328]. Its adsorption capacity was 0.625 mmol/g [328]. After the sorption, it could be regenerated using 0.1 M HCl [104, 328]. Shi et al. [329] developed biochar based on straw and wood, and successfully applied it to remove water-soluble imidazolium-based ILs (e.g., [IM12][BF_4]).

With regard to other possible adsorbents to remove ILs, ionic exchange resins can be considered. Choi et al. [330] used 12 cationic exchange resins with different functional groups (e.g., thiourea, sulfonic acid, carboxylic acid, amonophephonic acid, and iminodiacetic acid) and reported that among the tested resins, a sulfonic acid–based one had the highest adsorption capacity (578.2–616.2 mg/g) for the cation of [IM12], which was not easily treatable in a wastewater treatment plant. A similar possibility was suggested by Li et al. [331], who estimated the maximum uptake of imidazolium-based ILs with different alkyl chain lengths and reported the following trend in the adsorption: sulfonic acid > carboxylic acid > nonfunctionalized IL. In a study similar to that performed by Choi et al. [330], Won et al. [332] used five chemically modified bacterial biosorbents for removal of [IM12]. Among them, succinated *E. coli* biomass had the highest uptake (72.6 mg/g) and a fast adsorption rate. Moreover, the researchers showed that the sorbent could be easily regenerated using acetic acid.

6 Concluding Remarks and Suggestions for Selecting ILs for Biotechnological Applications

Although ionic liquids (ILs) have negative impacts on the environment, as was described earlier, they are still attractive solvents because they provide numerous advantages in chemical and biological processes. Nevertheless, to apply ILs as solvents for achieving greener processes, more work is needed. Fortunately, because IL structures are tunable, their physicochemical properties and environmental properties can be designed. For selection of appropriate structures, it is mandatory to understand the relationships between IL structures and several of their properties. Therefore, it was decided to compile information on the properties of ILs as solvents and solutes, and on the basis of these data sets, several researchers have developed prediction models based on the concepts of the quantitative structure–activity relationship (QSAR) or the quantitative structure–property relationship (QSPR).

From many previous studies, several physicochemical properties and environmental properties of ILs have been added to databases and can be used for prediction modeling. Thus, some properties can be calculated without the need to perform experiments. However, many prediction models of the properties of ILs have not yet been developed because the experimental data sets have not yet been compiled or computational approaches have some limitations. Moreover, some prediction models have been developed on the basis of insufficient data sets and a narrow spectrum of IL structures, which may cause serious errors when making predictions about external IL structures that were not involved in the modeling set. Therefore, further validation and development of modeling studies must be performed.

On the basis of experimental, theoretical, and computational analyses of the toxic properties of ILs, low-toxicity IL structures can be suggested, as shown in Scheme 1. Moreover, to investigate the overall environmental impacts of ILs, their biodegradation properties, which determine the persistency of chemicals, should be considered. From the experimental results, some readily biodegradable ILs can be suggested. Firstly, cholinium-based ILs and protic-based ILs with short alkyl chains are considered readily biodegradable cations. In the cases of pyridinium and imidazolium, when they have longer alkyl chains of six and eight, they may be primarily degradable. However, it has been reported that in general application of activated sludge, the imidazolium core was not opened, though the pyridinium core was opened. Therefore, pyridinium-based ILs can be considered less persistent compounds than imidazolium. Biodegradation can also be considered by functionalizing IL structures. In particular, ether, carboxyl, and hydroxyl functionalization on the alkyl chain can enhance the process of breakdown of original IL structures, as well as leading to low toxicity. Nevertheless, the advantage of hydrophilic functionalization cannot be made available for all IL structures.

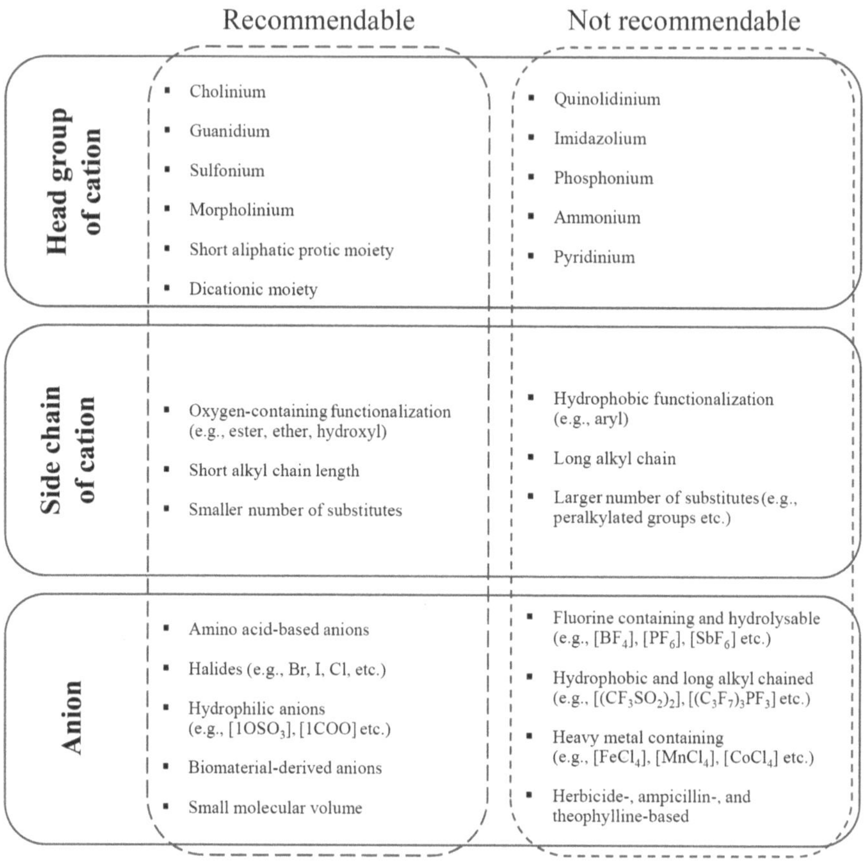

Scheme 1 Guideline for selection of low-toxicity ionic liquids

To date, a wealth of information on the environmental fate of ILs has been presented; there are still many tasks to be elucidated. These are summarized below:

- In comparison with the diversity of ILs, a broad spectrum of ILs has not been investigated. Therefore, along with further experimental estimation, theoretical analysis and prediction should be continued. In particular, the environmental fates (e.g., motility) of ILs are still rarely studied.
- Applications of ILs with other materials have been increasing. However, the synergistic and antagonistic toxic interactions between them should be investigated further.
- For sustainable industrial applications of ILs, their environmental aspects, as well as their functionalities, have to be considered. However, in general, not enough simultaneous consideration of these two aspects has been done. For efficient screening or design of IL structures, computational calculation is desirable, since experimental determination is limited to the study of particular IL structures and

is time and material consuming. Therefore, it is strongly recommended to develop prediction models of both aspects to help achieve the final aims of obtaining highly functionalized and environmentally friendly ILs. Prediction models should be used for better understanding of functionalities and environmental properties according to IL structures, and then greener and more functional ILs should be selected.

- Although ILs can be considered green solvents, the experimentally determined results make these chemicals seem less than green because most IL structures are not readily biodegradable and are more toxic than traditional organic solvents. However, ILs are still worth using in industrial fields because of their numerous benefits. Therefore, along with environmental and chemical research on ILs, economical, easily applicable, and efficient methods for removal of ILs must be developed.

References

1. Earle MJ, Seddon KR (2000) Ionic liquids. Green solvents for the future. Pure Appl Chem 72:1391–1398. https://doi.org/10.1351/pac200072071391
2. Marsh KN, Boxall JA, Lichtenthaler R (2004) Room temperature ionic liquids and their mixtures—a review. Fluid Phase Equilib 219:93–98. https://doi.org/10.1016/j.fluid.2004.02.003
3. McFarlane J, Ridenour WB, Luo H, Hunt RD, DePaoli DW, Ren RX (2005) Room temperature ionic liquids for separating organics from produced water. Sep Sci Technol 40:1245–1265. https://doi.org/10.1081/ss-200052807
4. Sheldon RA (2005) Green solvents for sustainable organic synthesis: state of the art. Green Chem 7:267–278. https://doi.org/10.1039/b418069k
5. Pham TPT, Cho C-W, Yun Y-S (2010) Environmental fate and toxicity of ionic liquids: a review. Water Res 44:352–372. https://doi.org/10.1016/j.watres.2009.09.030
6. Quijano G, Couvert A, Amrane A (2010) Ionic liquids: applications and future trends in bioreactor technology. Bioresour Technol 101:8923–8930. https://doi.org/10.1016/j.biortech.2010.06.161
7. Pham TPT, Cho C-W, Min J, Yun Y-S (2008) Alkyl-chain length effects of imidazolium and pyridinium ionic liquids on photosynthetic response of Pseudokirchneriella subcapitata. J Biosci Bioeng 105:425–428. https://doi.org/10.1263/jbb.105.425
8. Pham TPT, Cho CW, Vijayaraghavan K, Min JH, Yun YS (2008) Effect of imidazolium-based ionic liquids on the photosynthetic activity and growth rate of Selenastrum capricornutum. Environ Toxicol Chem 27:1583–1589
9. Cho CW, Pham TPT, Jeon YC, Yun YS (2008) Influence of anions on the toxic effects of ionic liquids to a phytoplankton Selenastrum capricornutum. Green Chem 10:67–72. https://doi.org/10.1039/b705520j
10. Kragl U, Eckstein M, Kaftzik N (2002) Enzyme catalysis in ionic liquids. Curr Opin Biotechnol 13:565–571. https://doi.org/10.1016/s0958-1669(02)00353-1
11. Sheldon RA, Lau RM, Sorgedrager MJ, van Rantwijk F, Seddon KR (2002) Biocatalysis in ionic liquids. Green Chem 4:147–151. https://doi.org/10.1039/b110008b
12. Matsumoto M, Mochiduki K, Kondo K (2004) Toxicity of ionic liquids and organic solvents to lactic acid–producing bacteria. J Biosci Bioeng 98:344–347. https://doi.org/10.1016/S1389-1723(04)00293-2

13. Matsumoto M, Mochiduki K, Fukunishi K, Kondo K (2004) Extraction of organic acids using imidazolium-based ionic liquids and their toxicity to Lactobacillus rhamnosus. Sep Purif Technol 40:97–101. https://doi.org/10.1016/j.seppur.2004.01.009
14. Santos AG, Ribeiro BD, Alviano DS, Coelho MAZ (2014) Toxicity of ionic liquids toward microorganisms interesting to the food industry. RSC Adv 4:37157–37163. https://doi.org/10.1039/C4RA05295A
15. Ouellet M et al (2011) Impact of ionic liquid pretreated plant biomass on Saccharomyces cerevisiae growth and biofuel production. Green Chem 13:2743–2749
16. Venkata Nancharaiah Y, Francis AJ (2011) Alkyl-methylimidazolium ionic liquids affect the growth and fermentative metabolism of Clostridium sp. Bioresour Technol 102:6573–6578. https://doi.org/10.1016/j.biortech.2011.03.042
17. Lovejoy KS et al (2013) Evaluation of ionic liquids on phototrophic microbes and their use in biofuel extraction and isolation. J Appl Phycol 25:973–981. https://doi.org/10.1007/s10811-012-9907-0
18. Bar R (1987) Phase toxicity in a water–solvent two-liquid phase microbial system. Stud Org Chem 29:147–153
19. Dickinson Q et al (2016) Mechanism of imidazolium ionic liquids toxicity in Saccharomyces cerevisiae and rational engineering of a tolerant, xylose-fermenting strain. Microb Cell Factories 15:17. https://doi.org/10.1186/s12934-016-0417-7
20. Xu J et al (2015) A novel ionic liquid–tolerant Fusarium oxysporum BN secreting ionic liquid–stable cellulase: consolidated bioprocessing of pretreated lignocellulose containing residual ionic liquid. Bioresour Technol 181:18–25. https://doi.org/10.1016/j.biortech.2014.12.080
21. Ruegg TL, Kim E-M, Simmons BA, Keasling JD, Singer SW, Lee TS, Thelen MP (2014) An auto-inducible mechanism for ionic liquid resistance in microbial biofuel production. Nat Commun 5:3490
22. Bubalo MC, Radosevic K, Redovnikovic IR, Halambek J, Srcek VG (2014) A brief overview of the potential environmental hazards of ionic liquids. Ecotoxicol Environ Saf 99:1–12. https://doi.org/10.1016/j.ecoenv.2013.10.019
23. Costa SPF, Azevedo AMO, Pinto P, Saraiva M (2017) Environmental impact of ionic liquids: recent advances in (eco)toxicology and (bio)degradability. ChemSusChem 10:2321–2347. https://doi.org/10.1002/cssc.201700261
24. Egorova KS, Gordeev EG, Ananikov VP (2017) Biological activity of ionic liquids and their application in pharmaceutics and medicine. Chem Rev 117:7132–7189. https://doi.org/10.1021/acs.chemrev.6b00562
25. Matzke M, Arning J, Ranke J, Jastorff B, Stolte S (2010) Design of inherently safer ionic liquids: toxicology and biodegradation. In: Anastas PT (ed) Handbook of green chemistry. Wiley, Weinheim. https://doi.org/10.1002/9783527628698.hgc069
26. Petkovic M, Seddon KR, Rebelo LPN, Pereira CS (2011) Ionic liquids: a pathway to environmental acceptability. Chem Soc Rev 40:1383–1403. https://doi.org/10.1039/c004968a
27. Ranke J, Stolte S, Stormann R, Arning J, Jastorff B (2007) Design of sustainable chemical products—the example of ionic liquids. Chem Rev 107:2183–2206. https://doi.org/10.1021/cr050942s
28. Samori C (2011) Ionic liquids and their biological effects towards microorganisms. Curr Org Chem 15:1888–1904
29. Zhao D, Liao Y, Zhang Z (2007) Toxicity of ionic liquids. Clean Soil Air Water 35:42–48. https://doi.org/10.1002/clen.200600015
30. Stock F, Hoffmann J, Ranke J, Stormann R, Ondruschka B, Jastorff B (2004) Effects of ionic liquids on the acetylcholinesterase—a structure–activity relationship consideration. Green Chem 6:286–290. https://doi.org/10.1039/b402348j
31. Arning J et al (2008) Structure–activity relationships for the impact of selected isothiazol-3-one biocides on glutathione metabolism and glutathione reductase of the human liver cell line Hep G2. Toxicology 246:203–212. https://doi.org/10.1016/j.tox.2008.01.011

32. Stasiewicz M et al (2008) Assessing toxicity and biodegradation of novel, environmentally benign ionic liquids (1-alkoxymethyl-3-hydroxypyridinium chloride, saccharinate and acesulfamates) on cellular and molecular level. Ecotoxicol Environ Saf 71:157–165. https://doi.org/10.1016/j.ecoenv.2007.08.011
33. Lai JQ, Li Z, Lu YH, Yang Z (2011) Specific ion effects of ionic liquids on enzyme activity and stability. Green Chem 13:1860–1868. https://doi.org/10.1039/c1gc15140a
34. Peric B et al (2013) (Eco)toxicity and biodegradability of selected protic and aprotic ionic liquids. J Hazard Mater 261:99–105. https://doi.org/10.1016/j.jhazmat.2013.06.070
35. Steudte S, Bemowsky S, Mahrova M, Bottin-Weber U, Tojo-Suarez E, Stepnowski P, Stolte S (2014) Toxicity and biodegradability of dicationic ionic liquids. RSC Adv 4:5198–5205. https://doi.org/10.1039/c3ra45675g
36. Jastorff B et al (2005) Progress in evaluation of risk potential of ionic liquids-basis for an eco-design of sustainable products. Green Chem 7:362–372. https://doi.org/10.1039/b418518h
37. UFT (Zentrum für Umweltforschung und nachhaltige Technologien). The UFT Ionic Liquids Biological Effects Database. University of Bremen. http://www.il-eco.uft.uni-bremen.de/index.php. Accessed 08-01-2017
38. Matzke M et al (2007) The influence of anion species on the toxicity of 1-alkyl-3-methylimidazolium ionic liquids observed in an (eco) toxicological test battery. Green Chem 9:1198–1207. https://doi.org/10.1039/b705795d
39. Arning J et al (2008) Qualitative and quantitative structure activity relationships for the inhibitory effects of cationic head groups, functionalised side chains and anions of ionic liquids on acetylcholinesterase. Green Chem 10:47–58. https://doi.org/10.1039/b712109a
40. Yu M, Li SM, Li XY, Zhang BJ, Wang JJ (2008) Acute effects of 1-octyl-3-methylimidazolium bromide ionic liquid on the antioxidant enzyme system of mouse liver. Ecotoxicol Environ Saf 71:903–908. https://doi.org/10.1016/j.ecoenv.2008.02.022
41. Boskin A, Tran CD, Franko M (2009) Oxidation of organophosphorus pesticides with chloroperoxidase enzyme in the presence of an ionic liquid as co-solvent. Environ Chem Lett 7:267–270. https://doi.org/10.1007/s10311-008-0161-2
42. Luo YR, Wang SH, Yun MX, Li XY, Wang JJ, Sun ZJ (2009) The toxic effects of ionic liquids on the activities of acetylcholinesterase and cellulase in earthworms. Chemosphere 77:313–318. https://doi.org/10.1016/j.chemosphere.2009.07.026
43. Schaffran T, Justus E, Elfert M, Chen T, Gabel D (2009) Toxicity of N,N,N-trialkylammoniododecaborates as new anions of ionic liquids in cellular, liposomal and enzymatic test systems. Green Chem 11:1458–1464. https://doi.org/10.1039/b906165g
44. Pinto P, Costa ADF, Lima J, Saraiva M (2011) Automated evaluation of the effect of ionic liquids on catalase activity. Chemosphere 82:1620–1628. https://doi.org/10.1016/j.chemosphere.2010.11.046
45. Steudte S, Stepnowski P, Cho C-W, Thoeming J, Stolte S (2012) (Eco)toxicity of fluoro-organic and cyano-based ionic liquid anions. Chem Commun 48:9382–9384. https://doi.org/10.1039/c2cc34955h
46. Hou XD, Liu QP, Smith TJ, Li N, Zong MH (2013) Evaluation of toxicity and biodegradability of cholinium amino acids ionic liquids. PLoS One 8:e59145. https://doi.org/10.1371/journal.pone.0059145
47. Stolte S, Schulz T, Cho CW, Arning J, Strassner T (2013) Synthesis, toxicity, and biodegradation of tunable aryl alkyl ionic liquids (TAAILs). ACS Sustain Chem Eng 1:410–418. https://doi.org/10.1021/sc300146t
48. Costa SPF, Justina VD, Bica K, Vasiloiu M, Pinto P, Saraiva M (2014) Automated evaluation of pharmaceutically active ionic liquids' (eco)toxicity through the inhibition of human carboxylesterase and Vibrio fischeri. J Hazard Mater 265:133–141. https://doi.org/10.1016/j.jhazmat.2013.11.052

49. Ge HL, Liu SS, Su BX, Zhu XW (2014) Two-stage prediction of the effects of imidazolium and pyridinium ionic liquid mixtures on luciferase. Molecules 19:6877–6890. https://doi.org/10.3390/molecules19056877
50. Cunha E, Passos MLC, Pinto P, Saraiva M (2015) Automated evaluation of the inhibition of glutathione reductase activity: application to the prediction of ionic liquids' toxicity. RSC Adv 5:78971–78978. https://doi.org/10.1039/c5ra04029a
51. Fan YC, Dong X, Yan LL, Li DD, Hua SF, Hu CB, Pan CC (2016) Evaluation of the toxicity of ionic liquids on trypsin: a mechanism study. Chemosphere 148:241–247. https://doi.org/10.1016/j.chemosphere.2016.01.033
52. Dong X, Fan YC, Zhang H, Zhong YY, Yang Y, Miao J, Hua SF (2016) Inhibitory effects of ionic liquids on the lactic dehydrogenase activity. Int J Biol Macromol 86:155–161. https://doi.org/10.1016/j.ijbiomac.2016.01.059
53. Wu YW et al (2016) Ionic liquids impact the bioenergy feedstock-degrading microbiome and transcription of enzymes relevant to polysaccharide hydrolysis. Msystems 1:e00120. https://doi.org/10.1128/mSystems.00120-16
54. Pernak J, Szymanowski J, Pujanek M, Kucharski S, Broniarz J (1979) Synthesis and bactericidal properties of some pyridinium chlorides with alkylthiomethyl hydrophobic groups. J Am Oil Chem Soc 56:830–833. https://doi.org/10.1007/bf02909528
55. Pernak J, Kalewska J, Ksycinska H, Cybulski J (2001) Synthesis and anti-microbial activities of some pyridinium salts with alkoxymethyl hydrophobic group. Eur J Med Chem 36:899–907. https://doi.org/10.1016/s0223-5234(01)01280-6
56. Pernak J, Goc I, Mirska I (2004) Anti-microbial activities of protic ionic liquids with lactate anion. Green Chem 6:323–329. https://doi.org/10.1039/b404625k
57. Cornellas A, Perez L, Comelles F, Ribosa I, Manresa A, Garcia MT (2011) Self-aggregation and antimicrobial activity of imidazolium and pyridinium based ionic liquids in aqueous solution. J Colloid Interface Sci 355:164–171. https://doi.org/10.1016/j.jcis.2010.11.063
58. Luczak J, Jungnickel C, Lacka I, Stolle S, Hupka J (2010) Antimicrobial and surface activity of 1-alkyl-3-methylimidazolium derivatives. Green Chem 12:593–601. https://doi.org/10.1039/b921805j
59. Pernak J et al (2007) Choline-derivative-based ionic liquids. Chem Eur J 13:6817–6827. https://doi.org/10.1002/chem.200700285
60. Petkovic M et al (2010) Novel biocompatible cholinium-based ionic liquids—toxicity and biodegradability. Green Chem 12:643–649. https://doi.org/10.1039/b922247b
61. Busetti A et al (2010) Antimicrobial and antibiofilm activities of 1-alkylquinolinium bromide ionic liquids. Green Chem 12:420–425. https://doi.org/10.1039/b919872e
62. Cole MR, Li M, El-Zahab B, Janes ME, Hayes D, Warner IM (2011) Design, synthesis, and biological evaluation of β-lactam antibiotic–based imidazolium- and pyridinium-type ionic liquids. Chem Biol Drug Des 78:33–41. https://doi.org/10.1111/j.1747-0285.2011.01114.x
63. Coleman D, Spulak M, Garcia MT, Gathergood N (2012) Antimicrobial toxicity studies of ionic liquids leading to a 'hit' MRSA selective antibacterial imidazolium salt. Green Chem 14:1350–1356. https://doi.org/10.1039/c2gc16090k
64. Borowiecki P, Milner-Krawczyk M, Brzezinska D, Wielechowska M, Plenkiewicz J (2013) Synthesis and antimicrobial activity of imidazolium and triazolium chiral ionic liquids. Eur J Org Chem 2013:712–720. https://doi.org/10.1002/ejoc.201201245
65. Feder-Kubis J, Tomczuk K (2013) The effect of the cationic structures of chiral ionic liquids on their antimicrobial activities. Tetrahedron 69:4190–4198. https://doi.org/10.1016/j.tet.2013.03.107
66. Piotrowska A, Syguda A, Wyrwas B, Chrzanowski L, Heipieper HJ (2017) Toxicity evaluation of selected ammonium-based ionic liquid forms with MCPP and dicamba moieties on Pseudomonas putida. Chemosphere 167:114–119. https://doi.org/10.1016/j.chemosphere.2016.09.140

67. Ventura SPM et al (2013) Imidazolium and pyridinium ionic liquids from mandelic acid derivatives: synthesis and bacteria and algae toxicity evaluation. ACS Sustain Chem Eng 1:393–402. https://doi.org/10.1021/sc3001299
68. Prydderch H, Haiss A, Spulak M, Quilty B, Kummerer K, Heise A, Gathergood N (2017) Mandelic acid derived ionic liquids: synthesis, toxicity and biodegradability. RSC Adv 7:2115–2126. https://doi.org/10.1039/c6ra25562k
69. Gouveia W et al (2014) Toxicity of ionic liquids prepared from biomaterials. Chemosphere 104:51–56. https://doi.org/10.1016/j.chemosphere.2013.10.055
70. Jordan A, Haiss A, Spulak M, Karpichev Y, Kummerer K, Gathergood N (2016) Synthesis of a series of amino acid derived ionic liquids and tertiary amines: green chemistry metrics including microbial toxicity and preliminary biodegradation data analysis. Green Chem 18:4374–4392. https://doi.org/10.1039/c6gc00415f
71. Frizzo CP et al (2016) Novel ibuprofenate- and docusate-based ionic liquids: emergence of antimicrobial activity. RSC Adv 6:100476–100486. https://doi.org/10.1039/c6ra22237d
72. Garcia MT, Ribosa I, Perez L, Manresa A, Comelles F (2013) Aggregation behavior and antimicrobial activity of ester-functionalized imidazolium- and pyridinium-based ionic liquids in aqueous solution. Langmuir 29:2536–2545. https://doi.org/10.1021/la304752e
73. Gilmore BF et al (2013) Enhanced antimicrobial activities of 1-alkyl-3-methyl imidazolium ionic liquids based on silver or copper containing anions. New J Chem 37:873–876. https://doi.org/10.1039/c3nj40759d
74. Hodyna D et al (2016) Efficient antimicrobial activity and reduced toxicity of 1-dodecy1-3-methylimidazolium tetrafluoroborate ionic liquid/beta-cyclodextrin complex. Chem Eng J 284:1136–1145. https://doi.org/10.1016/j.cej.2015.09.041
75. Messali M, Moussa Z, Alzahrani AY, El-Naggar MY, ElDouhaibi AS, Judeh ZMA, Hammouti B (2013) Synthesis, characterization and the antimicrobial activity of new eco-friendly ionic liquids. Chemosphere 91:1627–1634. https://doi.org/10.1016/j.chemosphere.2012.12.062
76. Postleb F, Stefanik D, Seifert H, Giernoth R (2013) Bionic liquids: imidazolium-based ionic liquids with antimicrobial activity. J Chem Sci 68:1123–1128. https://doi.org/10.5560/znb.2013-3150
77. Ranke J et al (2004) Biological effects of imidazolium ionic liquids with varying chain lengths in acute Vibrio fischeri and WST-1 cell viability assays. Ecotoxicol Environ Saf 58:396–404. https://doi.org/10.1016/s0147-6513(03)00105-2
78. Samori C, Pasteris A, Galletti P, Tagliavini E (2007) Acute toxicity of oxygenated and nonoxygenated imidazolium-based ionic liquids to Daphnia magna and Vibrio fischeri. Environ Toxicol Chem 26:2379–2382. https://doi.org/10.1897/07-066r2.1
79. Ventura SPM, Marques CS, Rosatella AA, Afonso CAM, Goncalves F, Coutinho JAP (2012) Toxicity assessment of various ionic liquid families towards Vibrio fischeri marine bacteria. Ecotoxicol Environ Saf 76:162–168. https://doi.org/10.1016/j.ecoenv.2011.10.006
80. Viboud S, Papaiconomou N, Cortesi A, Chatel G, Draye M, Fontvieille D (2012) Correlating the structure and composition of ionic liquids with their toxicity on Vibrio fischeri: a systematic study. J Hazard Mater 215:40–48. https://doi.org/10.1016/j.jhazmat.2012.02.019
81. Montalban MG, Hidalgo JM, Collado-Gonzalez M, Banos FGD, Villora G (2016) Assessing chemical toxicity of ionic liquids on Vibrio fischeri: correlation with structure and composition. Chemosphere 155:405–414. https://doi.org/10.1016/j.chemosphere.2016.04.042
82. Ventura SPM, Silva FAE, Goncalves AMM, Pereira JL, Goncalves F, Coutinho JAP (2014) Ecotoxicity analysis of cholinium-based ionic liquids to Vibrio fischeri marine bacteria. Ecotoxicol Environ Saf 102:48–54. https://doi.org/10.1016/j.ecoenv.2014.01.003
83. Ben Ghanem O et al (2015) Thermophysical properties and acute toxicity towards green algae and Vibrio fischeri of amino acid–based ionic liquids. J Mol Liq 212:352–359. https://doi.org/10.1016/j.molliq.2015.09.017
84. Sintra TE et al (2017) Ecotoxicological evaluation of magnetic ionic liquids. Ecotoxicol Environ Saf 143:315–321. https://doi.org/10.1016/j.ecoenv.2017.05.034

85. Ahlstrom B, Chelminskabertilsson M, Thompson RA, Edebo L (1995) Long-chain alkanoylcholines, a new category of soft antimicrobial agents that are enzymatically degradable. Antimicrob Agents Chemother 39:50–55
86. Pernak J, Rogoza J, Mirska I (2001) Synthesis and antimicrobial activities of new pyridinium and benzimidazolium chlorides. Eur J Med Chem 36:313–320. https://doi.org/10.1016/s0223-5234(01)01226-0
87. Cieniecka-Roslonkiewicz A, Pernak J, Kubis-Feder J, Ramani A, Robertson AJ, Seddon KR (2005) Synthesis, anti-microbial activities and anti-electrostatic properties of phosphonium-based ionic liquids. Green Chem 7:855–862. https://doi.org/10.1039/b508499g
88. Ganske F, Bornscheuer UT (2005) Lipase-catalyzed glucose fatty acid ester synthesis in ionic liquids. Org Lett 7:3097–3098. https://doi.org/10.1021/ol0511169
89. Petkovic M et al (2009) Exploring fungal activity in the presence of ionic liquids. Green Chem 11:889–894. https://doi.org/10.1039/b823225c
90. Carson L et al (2009) Antibiofilm activities of 1-alkyl-3-methylimidazolium chloride ionic liquids. Green Chem 11:492–497. https://doi.org/10.1039/b821842k
91. Singer SW, Reddy AP, Gladden JM, Guo H, Hazen TC, Simmons BA, VanderGheynst JS (2011) Enrichment, isolation and characterization of fungi tolerant to 1-ethyl-3-methylimidazolium acetate. J Appl Microbiol 110:1023–1031. https://doi.org/10.1111/j.1365-2672.2011.04959.x
92. Deng Y, Besse-Hoggan P, Sancelme M, Delort AM, Husson P, Gomes MFC (2011) Influence of oxygen functionalities on the environmental impact of imidazolium based ionic liquids. J Hazard Mater 198:165–174. https://doi.org/10.1016/j.jhazmat.2011.10.024
93. Wang H, Malhotra SV, Francis AJ (2011) Toxicity of various anions associated with methoxyethyl methyl imidazolium-based ionic liquids on Clostridium sp. Chemosphere 82:1597–1603. https://doi.org/10.1016/j.chemosphere.2010.11.049
94. Zhang C, Malhotra SV, Francis AJ (2011) Toxicity of imidazolium- and pyridinium-based ionic liquids and the co-metabolic degradation of N-ethylpyridinium tetrafluoroborate. Chemosphere 82:1690–1695. https://doi.org/10.1016/j.chemosphere.2010.10.085
95. Alberto EE, Rossato LL, Alves SH, Alves D, Braga AL (2011) Imidazolium ionic liquids containing selenium: synthesis and antimicrobial activity. Org Biomol Chem 9:1001–1003. https://doi.org/10.1039/c0ob01010c
96. Banothu J, Bavanthula R (2012) Bronsted acidic ionic liquid catalyzed highly efficient synthesis of chromeno pyrimidinone derivatives and their antimicrobial activity. Chin Chem Lett 23:1015–1018. https://doi.org/10.1016/j.cclet.2012.06.041
97. Li HQ, Yu CC, Chen R, Li J, Li JX (2012) Novel ionic liquid–type Gemini surfactants: synthesis, surface property and antimicrobial activity. Coll Surf 395:116–124. https://doi.org/10.1016/j.colsurfa.2011.12.014
98. Petkovic M, Hartmann DO, Adamova G, Seddon KR, Rebelo LPN, Pereira CS (2012) Unravelling the mechanism of toxicity of alkyltributylphosphonium chlorides in Aspergillus nidulans conidia. New J Chem 36:56–63. https://doi.org/10.1039/c1nj20470j
99. Ventura SPM, de Barros RLF, Sintra T, Soares CMF, Lima AS, Coutinho JAP (2012) Simple screening method to identify toxic/non-toxic ionic liquids: agar diffusion test adaptation. Ecotoxicol Environ Saf 83:55–62. https://doi.org/10.1016/j.ecoenv.2012.06.002
100. Banothu J, Gali R, Velpula R, Bavantula R (2013) Bronsted acidic ionic liquid catalysis: an efficient and eco-friendly synthesis of novel fused pyrano pyrimidinones and their antimicrobial activity. J Chem Sci 125:843–849
101. Ben Ghanem O, Mutalib MIA, El-Harbawi M, Gonfa G, Kait CF, Alitheen NBM, Leveque JM (2015) Effect of imidazolium-based ionic liquids on bacterial growth inhibition investigated via experimental and QSAR modelling studies. J Hazard Mater 297:198–206. https://doi.org/10.1016/j.jhazmat.2015.04.082
102. Mester P, Wagner M, Rossmanith P (2015) Antimicrobial effects of short chained imidazolium-based ionic liquids—influence of anion chaotropicity. Ecotoxicol Environ Saf 111:96–101. https://doi.org/10.1016/j.ecoenv.2014.08.032

103. Borkowski A et al (2016) Different antibacterial activity of novel theophylline-based ionic liquids—growth kinetic and cytotoxicity studies. Ecotoxicol Environ Saf 130:54–64. https://doi.org/10.1016/j.ecoenv.2016.04.004
104. Yu J, Zhang SS, Dai YT, Lu XX, Lei QF, Fang WJ (2016) Antimicrobial activity and cytotoxicity of piperazinium- and guanidinium-based ionic liquids. J Hazard Mater 307:73–81. https://doi.org/10.1016/j.jhazmat.2015.12.028
105. Kunduracioglu A, Gube O, Hames-Kocabas EE, Eyupoglu V, Sonmez F (2016) Synthesis, structural analysis and antimicrobial activities of novel water soluble ionic liquids derived from N-heterocyclic carbene salts. Croat Chem Acta 89:105–110. https://doi.org/10.5562/cca2599
106. Siopa F et al (2016) Choline-based ionic liquids: improvement of antimicrobial activity. ChemistrySelect 1:5909–5916. https://doi.org/10.1002/slct.201600864
107. Garcia MT, Gathergood N, Scammells PJ (2005) Biodegradable ionic liquids—part II. Effect of the anion and toxicology. Green Chem 7:9–14. https://doi.org/10.1039/b411922c
108. Romero A, Santos A, Tojo J, Rodriguez A (2008) Toxicity and biodegradability of imidazolium ionic liquids. J Hazard Mater 151:268–273. https://doi.org/10.1016/j.jhazmat.2007.10.079
109. Couling DJ, Bernot RJ, Docherty KM, Dixon JK, Maginn EJ (2006) Assessing the factors responsible for ionic liquid toxicity to aquatic organisms via quantitative structure–property relationship modeling. Green Chem 8:82–90. https://doi.org/10.1039/b511333d
110. Docherty KM, Kulpa CF (2005) Toxicity and antimicrobial activity of imidazolium and pyridinium ionic liquids. Green Chem 7:185–189. https://doi.org/10.1039/b419172b
111. Stolte S et al (2007) Effects of different head groups and functionalised side chains on the aquatic toxicity of ionic liquids. Green Chem 9:1170–1179. https://doi.org/10.1039/b711119c
112. Pinto P, Costa SPF, Lima J, Saraiva M (2012) Automated high-throughput Vibrio fischeri assay for (eco)toxicity screening: application to ionic liquids. Ecotoxicol Environ Saf 80:97–102. https://doi.org/10.1016/j.ecoenv.2012.02.013
113. Silva FAE et al (2014) Sustainable design for environment-friendly mono and dicationic cholinium-based ionic liquids. Ecotoxicol Environ Saf 108:302–310. https://doi.org/10.1016/j.ecoenv.2014.07.003
114. Fan Y, Liu SS, Qu R, Li K, Liu HL (2017) Polymyxin B sulfate inducing time-dependent antagonism of the mixtures of pesticide, ionic liquids, and antibiotics to Vibrio qinghaiensis sp.-Q67. RSC Adv 7:6080–6088. https://doi.org/10.1039/c6ra25843c
115. Rantamaki AH, Ruokonen SK, Sklavounos E, Kyllonen L, King AWT, Wiedmer SK (2017) Impact of surface-active guanidinium-, tetramethylguanidinium-, and cholinium-based ionic liquids on Vibrio fischeri cells and dipalmitoylphosphatidylcholine liposomes. Sci Rep 7:46673. https://doi.org/10.1038/srep46673
116. Stepnowski P, Skladanowski AC, Ludwiczak A, Laczynska E (2004) Evaluating the cytotoxicity of ionic liquids using human cell line HeLa. Hum Exp Toxicol 23:513–517. https://doi.org/10.1191/0960327104ht480oa
117. Frade RFM, Matias A, Branco LC, Afonso CAM, Duarte CMM (2007) Effect of ionic liquids on human colon carcinoma HT-29 and CaCo-2 cell lines. Green Chem 9:873–877. https://doi.org/10.1039/b617526k
118. Stolte S et al (2006) Anion effects on the cytotoxicity of ionic liquids. Green Chem 8:621–629. https://doi.org/10.1039/b602161a
119. Ranke J et al (2007) Lipophilicity parameters for ionic liquid cations and their correlation to in vitro cytotoxicity. Ecotoxicol Environ Saf 67:430–438. https://doi.org/10.1016/j.ecoenv.2006.08.008
120. Stolte S et al (2007) Effects of different head groups and functionalised side chains on the cytotoxicity of ionic liquids. Green Chem 9:760–767. https://doi.org/10.1039/b615326g
121. Stolte S, Steudte S, Areitioaurtena O, Pagano F, Thoming J, Stepnowski P, Igartua A (2012) Ionic liquids as lubricants or lubrication additives: an ecotoxicity and biodegradability assessment. Chemosphere 89:1135–1141. https://doi.org/10.1016/j.chemosphere.2012.05.102

122. Pernak J et al (2011) Synthesis, toxicity, biodegradability and physicochemical properties of 4-benzyl-4-methylmorpholinium-based ionic liquids. Green Chem 13:2901–2910. https://doi.org/10.1039/c1gc15468k
123. Garcia-Lorenzo A, Tojo E, Tojo J, Teijeira M, Rodriguez-Berrocal FJ, Gonzalez MP, Martinez-Zorzano VS (2008) Cytotoxicity of selected imidazolium-derived ionic liquids in the human Caco-2 cell line. Sub-structural toxicological interpretation through a QSAR study. Green Chem 10:508–516. https://doi.org/10.1039/b718860a
124. Li XY, Jing CQ, Zang XY, Yang SA, Wang JJ (2012) Toxic cytological alteration and mitochondrial dysfunction in PC12 cells induced by 1-octyl-3-methylimidazolium chloride. Toxicol in Vitro 26:1087–1092. https://doi.org/10.1016/j.tiv.2012.07.006
125. Bubalo MC, Radosevic K, Srcek VG, Das RN, Popelier P, Roy K (2015) Cytotoxicity towards CCO cells of imidazolium ionic liquids with functionalized side chains: preliminary QSTR modeling using regression and classification based approaches. Ecotoxicol Environ Saf 112:22–28. https://doi.org/10.1016/j.ecoenv.2014.10.029
126. Cvjetko M, Radosevic K, Tomica A, Slivac I, Vorkapic-Furac J, Srcek VG (2012) Cytotoxic effects of imidazolium ionic liquids on fish and human cell lines. Arh Hig Rada Toksikol 63:15–20. https://doi.org/10.2478/10004-1254-63-2012-2132
127. Radosevic K, Cvjetko M, Kopjar N, Novak R, Dumic J, Srcek VG (2013) In vitro cytotoxicity assessment of imidazolium ionic liquids: biological effects in fish channel catfish ovary (CCO) cell line. Ecotoxicol Environ Saf 92:112–118. https://doi.org/10.1016/j.ecoenv.2013.03.002
128. Ruokonen SK et al (2016) Effect of ionic liquids on zebrafish (Danio rerio) viability, behavior, and histology; correlation between toxicity and ionic liquid aggregation. Environ Sci Technol 50:7116–7125. https://doi.org/10.1021/acs.est.5b06107
129. Cho C-W, Pham TPT, Kim S, Kim Y-R, Jeon Y-C, Yun Y-S (2009) Toxicity assessment of common organic solvents using a biosensor based on algal photosynthetic activity measurement. J Appl Phycol 21:683–689. https://doi.org/10.1007/s10811-009-9401-5
130. Quraishi KS et al (2017) Ionic liquids toxicity on fresh water microalgae, Scenedesmus quadricauda, Chlorella vulgaris & Botryococcus braunii; selection criterion for use in a two-phase partitioning bioreactor (TPPBR). Chemosphere 184:642–651. https://doi.org/10.1016/j.chemosphere.2017.06.037
131. Deng XY, Chen B, Li D, Hu XL, Cheng J, Gao K, Wang CH (2017) Growth and physiological responses of a marine diatom (Phaeodactylum tricornutum) against two imidazolium-based ionic liquids (C(4)mim BF4 and C(8)mim BF4). Aquat Toxicol 189:115–122. https://doi.org/10.1016/j.aquatox.2017.05.016
132. Deng XY, Li D, Wang L, Hu XL, Cheng J, Gao K (2017) Potential toxicity of ionic liquid (C(12)mim BF4) on the growth and biochemical characteristics of a marine diatom Phaeodactylum tricornuturn. Sci Total Environ 586:675–684. https://doi.org/10.1016/j.scitotenv.2017.02.043
133. Liu HJ, Zhang XQ, Dong Y, Chen CD, Zhu SM, Ma XJ (2015) Enantioselective toxicities of chiral ionic liquids 1-alkyl-3-methyl imidazolium tartrate on Scenedesmus obliquus. Aquat Toxicol 169:179–187. https://doi.org/10.1016/j.aquatox.2015.10.024
134. Latala A, Stepnowski P, Nedzi M, Mrozik W (2005) Marine toxicity assessment of imidazolium ionic liquids: acute effects on the Baltic algae Oocystis submarina and Cyclotella meneghiniana. Aquat Toxicol 73:91–98. https://doi.org/10.1016/j.aquatox.2005.03.008
135. Wells AS, Coombe VT (2006) On the freshwater ecotoxicity and biodegradation properties of some common ionic liquids. Org Process Res Dev 10:794–798. https://doi.org/10.1021/op060048i
136. Cho CW, Pham TPT, Jeon YC, Vijayaraghavan K, Choe WS, Yun YS (2007) Toxicity of imidazolium salt with anion bromide to a phytoplankton Selenastrum capricornutum: effect of alkyl-chain length. Chemosphere 69:1003–1007. https://doi.org/10.1016/j.chemosphere.2007.06.023
137. Matzke M, Stolte S, Boschen A, Filser J (2008) Mixture effects and predictability of combination effects of imidazolium based ionic liquids as well as imidazolium based ionic liquids

and cadmium on terrestrial plants (Triticum aestivum) and limnic green algae (Scenedesmus vacuolatus). Green Chem 10:784–792. https://doi.org/10.1039/b802350f
138. Kulacki KJ, Lamberti GA (2008) Toxicity of imidazolium ionic liquids to freshwater algae. Green Chem 10:104–110. https://doi.org/10.1039/b709289j
139. Latala A, Nedzi M, Stepnowski P (2009) Toxicity of imidazolium and pyridinium based ionic liquids towards algae. Bacillaria paxillifer (a microphytobenthic diatom) and Geitlerinema amphibium (a microphytobenthic blue green alga). Green Chem 11:1371–1376. https://doi.org/10.1039/b901887e
140. Latala A, Nedzi M, Stepnowski P (2009) Toxicity of imidazolium and pyridinium based ionic liquids towards algae. Chlorella vulgaris, Oocystis submarina (green algae) and Cyclotella meneghiniana, Skeletonema marinoi (diatoms). Green Chem 11:580–588. https://doi.org/10.1039/b821140j
141. Pretti C, Chiappe C, Baldetti I, Brunini S, Monni G, Intorre L (2009) Acute toxicity of ionic liquids for three freshwater organisms: Pseudokirchneriella subcapitata, Daphnia magna and Danio rerio. Ecotoxicol Environ Saf 72:1170–1176. https://doi.org/10.1016/j.ecoenv.2008.09.010
142. Latala A, Nedzi M, Stepnowski P (2010) Toxicity of imidazolium ionic liquids towards algae. Influence of salinity variations. Green Chem 12:60–64. https://doi.org/10.1039/b918355h
143. Sena DW, Kulacki KJ, Chaloner DT, Lamberti GA (2010) The role of the cell wall in the toxicity of ionic liquids to the alga Chlamydomonas reinhardtii. Green Chem 12:1066–1071. https://doi.org/10.1039/c000899k
144. Ma JM, Cai LL, Zhang BJ, Hu LW, Li XY, Wang JJ (2010) Acute toxicity and effects of 1-alkyl-3-methylimidazolium bromide ionic liquids on green algae. Ecotoxicol Environ Saf 73:1465–1469. https://doi.org/10.1016/j.ecoenv.2009.10.004
145. Ventura SPM, Goncalves AMM, Sintra T, Pereira JL, Goncalves F, Coutinho JAP (2013) Designing ionic liquids: the chemical structure role in the toxicity. Ecotoxicology 22:1–12. https://doi.org/10.1007/s10646-012-0997-x
146. Chen H, Zou YQ, Zhang LJ, Wen YZ, Liu WP (2014) Enantioselective toxicities of chiral ionic liquids 1-alkyl-3-methylimidazolium lactate to aquatic algae. Aquat Toxicol 154:114–120. https://doi.org/10.1016/j.aquatox.2014.05.010
147. Hajfarajollah H, Mokhtarani B, Sharifi A, Mirzaei M, Afaghi A (2014) Toxicity of various kinds of ionic liquids towards the cell growth and end product formation of the probiotic strain, Propionibacterium freudenreichii. RSC Adv 4:13153–13160. https://doi.org/10.1039/c4ra00925h
148. Salchner R et al (2015) Structural and ecotoxicological profile of n-alkoxymorpholinhjm-based ionic liquids. Heterocycles 90:1018–1037. https://doi.org/10.3987/com-14-s(k)73
149. Deng Y, Beadham I, Wu J, Chen XD, Hu L, Gu J (2015) Chronic effects of the ionic liquid C(4)mim Cl towards the microalga Scenedesmus quadricauda. Environ Pollut 204:248–255. https://doi.org/10.1016/j.envpol.2015.05.011
150. Liu HJ, Zhang XQ, Chen CD, Du ST, Dong Y (2015) Effects of imidazolium chloride ionic liquids and their toxicity to Scenedesmus obliquus. Ecotoxicol Environ Saf 122:83–90. https://doi.org/10.1016/j.ecoenv.2015.07.010
151. Samori C et al (2015) Pyrrolidinium-based ionic liquids: aquatic ecotoxicity, biodegradability, and algal subinhibitory stimulation. ACS Sustain Chem Eng 3:1860–1865. https://doi.org/10.1021/acssuschemeng.5b00458
152. Tsarpali V, Dailianis S (2015) Toxicity of two imidazolium ionic liquids, bmim BF4 and omim BF4, to standard aquatic test organisms: role of acetone in the induced toxicity. Ecotoxicol Environ Saf 117:62–71. https://doi.org/10.1016/j.ecoenv.2015.03.026
153. Tsarpali V, Harbi K, Dailianis S (2016) Physiological response of the green microalgae Dunaliella tertiolecta against imidazolium ionic liquids bmim BF4 and/or omim BF4: the role of salinity on the observed effects. J Appl Phycol 28:979–990. https://doi.org/10.1007/s10811-015-0613-6

154. Pham TPT, Cho CW, Yun YS (2016) Structural effects of ionic liquids on microalgal growth inhibition and microbial degradation. Environ Sci Pollut Res 23:4294–4300. https://doi.org/10.1007/s11356-015-5287-8
155. Reddy GKK, Nancharaiah YV, Venugopalan VP (2017) Long alkyl-chain imidazolium ionic liquids: antibiofilm activity against phototrophic biofilms. Coll Surf 155:487–496. https://doi.org/10.1016/j.colsurfb.2017.04.040
156. Deng XY, Cheng J, Hu XL, Gao K, Wang CH (2015) Physiological and biochemical responses of a marine diatom Phaeodactylum tricornutum exposed to 1-octyl-3-methylimidazolium bromide. Aquat Biol 24:109–115. https://doi.org/10.3354/ab00643
157. Bernot RJ, Brueseke MA, Evans-White MA, Lamberti GA (2005) Acute and chronic toxicity of imidazolium-based ionic liquids on Daphnia magna. Environ Toxicol Chem 24:87–92. https://doi.org/10.1897/03-635.1
158. Costa SPF, Pinto P, Saraiva M, Rocha FRP, Santos JRP, Monteiro RTR (2015) The aquatic impact of ionic liquids on freshwater organisms. Chemosphere 139:288–294. https://doi.org/10.1016/j.chemosphere.2015.05.100
159. Wang C, Wei ZB, Wang LS, Sun P, Wang ZY (2015) Assessment of bromide-based ionic liquid toxicity toward aquatic organisms and QSAR analysis. Ecotoxicol Environ Saf 115:112–118. https://doi.org/10.1016/j.ecoenv.2015.02.012
160. Bado-Nilles A et al (2015) Coupling of OECD standardized test and immunomarkers to select the most environmentally benign ionic liquids option—towards an innovative "safety by design" approach. J Hazard Mater 283:202–210. https://doi.org/10.1016/j.jhazmat.2014.09.023
161. Luo YR, Li XY, Chen XX, Zhang BJ, Sun ZJ, Wang JJ (2008) The developmental toxicity of 1-methyl-3-octylimidazolium bromide on Daphnia magna. Environ Toxicol 23:736–744. https://doi.org/10.1002/tox.20382
162. Yu M, Wang SH, Luo YR, Han YW, Li XY, Zhang BJ, Wang JJ (2009) Effects of the 1-alkyl-3-methylimidazolium bromide ionic liquids on the antioxidant defense system of Daphnia magna. Ecotoxicol Environ Saf 72:1798–1804. https://doi.org/10.1016/j.ecoenv.2009.05.002
163. Ventura SPM, Goncalves AMM, Goncalves F, Coutinho JAP (2010) Assessing the toxicity on C(3)mim Tf2N to aquatic organisms of different trophic levels. Aquat Toxicol 96:290–297. https://doi.org/10.1016/j.aquatox.2009.11.008
164. Samori C et al (2010) Introduction of oxygenated side chain into imidazolium ionic liquids: evaluation of the effects at different biological organization levels. Ecotoxicol Environ Saf 73:1456–1464. https://doi.org/10.1016/j.ecoenv.2010.07.020
165. Docherty KM, Joyce MV, Kulacki KJ, Kulpa CF (2010) Microbial biodegradation and metabolite toxicity of three pyridinium-based cation ionic liquids. Green Chem 12:701–712. https://doi.org/10.1039/b919154b
166. Santos JI et al (2015) Environmental safety of cholinium-based ionic liquids: assessing structure-ecotoxicity relationships. Green Chem 17:4657–4668. https://doi.org/10.1039/c5gc01129a
167. Pretti C, Chiappe C, Pieraccini D, Gregori M, Abramo F, Monni G, Intorre L (2006) Acute toxicity of ionic liquids to the zebrafish (Danio rerio). Green Chem 8:238–240. https://doi.org/10.1039/b511554j
168. Dumitrescu G, Petculescu-Ciochină L, Bencsik I, Dronca D, Boca L (2010) Evaluation on acute toxicity of tetrabuthylammonium bromide ionic liquid at histological structure of some organs in zebrafish (Danio rerio). AACL Bioflux 3:404–414
169. Li XY, Zeng SH, Zhang WH, Liu L, Ma S, Wang JJ (2013) Acute toxicity and superficial damage to goldfish from the ionic liquid 1-methyl-3-octylimidazolium bromide. Environ Toxicol 28:207–214. https://doi.org/10.1002/tox.20712
170. Perez CJ, Tata A, de Campos ML, Peng C, Ifa DR (2017) Monitoring toxic ionic liquids in Zebrafish (Danio rerio) with desorption electrospray ionization mass spectrometry imaging (DESI-MSI). J Am Soc Mass Spectrom 28:1136–1148. https://doi.org/10.1007/s13361-016-1515-9

171. Wang LS, Wang L, Wang L, Wang G, Li ZH, Wang JJ (2009) Effect of 1-butyl-3-methylimidazolium tetrafluoroborate on the wheat (Triticum aestivum L.) seedlings. Environ Toxicol 24:296–303. https://doi.org/10.1002/tox.20435
172. Du ZK, Zhu LS, Dong MA, Wang JH, Wang J, Xie H, Zhu SY (2012) Effects of the ionic liquid Omim PF6 on antioxidant enzyme systems, ROS and DNA damage in zebrafish (Danio rerio). Aquat Toxicol 124:91–93. https://doi.org/10.1016/j.aquatox.2012.08.002
173. Dong M, Zhu LS, Zhu SY, Wang JH, Wang J, Xie H, Du ZK (2013) Toxic effects of 1-decyl-3-methylimidazolium bromide ionic liquid on the antioxidant enzyme system and DNA in zebrafish (Danio rerio) livers. Chemosphere 91:1107–1112. https://doi.org/10.1016/j.chemosphere.2013.01.013
174. Hafez NFM, Mutalib MIA, Bustam MA, El-Harbawi M, Leveque JM (2016) Ecotoxicity of pyridinium based ILs towards guppy fish and four bacterial strains. Procedia Eng 148:830–838
175. Baharuddin SH, Mustahil NA, Abdullah AA, Sivapragasam M, Moniruzzaman M (2016) Ecotoxicity study of amino acid ionic liquids towards danio rerio fish: effect of cations. Procedia Eng 148:401–408
176. Li XY, Zeng SH, Dong XY, Ma JG, Wang JJ (2012) Acute toxicity and responses of antioxidant systems to 1-methyl-3-octylimidazolium bromide at different developmental stages of goldfish. Ecotoxicology 21:253–259. https://doi.org/10.1007/s10646-011-0785-z
177. Larson JH, Frost PC, Lamberti GA (2008) Variable toxicity of ionic liquid–forming chemicals to Lemna minor and the influence of dissolved organic matter. Environ Toxicol Chem 27:676–681. https://doi.org/10.1897/06-540.1
178. Studzinska S, Buszewski B (2009) Study of toxicity of imidazolium ionic liquids to watercress (Lepidium sativum L.). Anal Bioanal Chem 393:983–990. https://doi.org/10.1007/s00216-008-2523-9
179. Liu HJ, Zhang SX, Zhang XQ, Chen CD (2015) Growth inhibition and effect on photosystem by three imidazolium chloride ionic liquids in rice seedlings. J Hazard Mater 286:440–448. https://doi.org/10.1016/j.jhazmat.2015.01.008
180. Matzke M, Stolte S, Arning U, Uebers U, Filser J (2008) Imidazolium based ionic liquids in soils: effects of the side chain length on wheat (Triticum aestivum) and cress (Lepidium sativum) as affected by different clays and organic matter. Green Chem 10:584–591. https://doi.org/10.1039/b717811e
181. Balczewski P, Bachowska B, Bialas T, Biczak R, Wieczorek WM, Balinska A (2007) Synthesis and phytotoxicity of new ionic liquids incorporating chiral cations and/or chiral anions. J Agric Food Chem 55:1881–1892. https://doi.org/10.1021/jf062849q
182. Bubalo MC, Hanousek K, Radosevic K, Srcek VG, Jakovljevic T, Redovnikovic IR (2014) Imidiazolium based ionic liquids: effects of different anions and alkyl chains lengths on the barley seedlings. Ecotoxicol Environ Saf 101:116–123. https://doi.org/10.1016/j.ecoenv.2013.12.022
183. Matzke M, Stolte S, Arning J, Uebers U, Filser J (2009) Ionic liquids in soils: effects of different anion species of imidazolium based ionic liquids on wheat (Triticum aestivum) as affected by different clay minerals and clay concentrations. Ecotoxicology 18:197–203. https://doi.org/10.1007/s10646-008-0272-3
184. Liu P, Ding Y, Liu H, Sun L, Li X, Wang J (2010) Toxic effects of 1-methyl-3-octylimidazolium bromide on the wheat seedling. J Environ Sci 22:1974–1979. https://doi.org/10.1016/S1001-0742(09)60348-X
185. Liu HJ, Zhang SX, Hu XN, Chen CD (2013) Phytotoxicity and oxidative stress effect of 1-octyl-3-methylimidazolium chloride ionic liquid on rice seedlings. Environ Pollut 181:242–249. https://doi.org/10.1016/j.envpol.2013.06.007
186. Biczak R, Balczewski P, Bachowska B, Pawlowska B, Kazmierczak-Baranska J, Cieslak M, Nawrot B (2013) Phytotoxicity and cytotoxicity of imidazolium ionic liquids containing sulfur atom. Phosphorus Sulfur Silicon Relat Elem 188:459–461. https://doi.org/10.1080/10426507.2012.737880

187. Biczak R, Balczewski P, Pawlowska B, Bachowska B, Rychter P (2014) Comparison of phytotoxicity of selected phosphonium ionic liquid. Ecol Chem Eng 21:281–295. https://doi.org/10.2478/eces-2014-0022
188. Liu T, Zhu LS, Xie H, Wang JH, Wang J, Sun FX, Wang FH (2014) Effects of the ionic liquid 1-octyl-3-methylimidazolium hexafluorophosphate on the growth of wheat seedlings. Environ Sci Pollut Res 21:3936–3945. https://doi.org/10.1007/s11356-013-2348-8
189. Biczak R, Pawlowska B, Feder-Kubis J (2015) The phytotoxicity of ionic liquids from natural pool of (-)-menthol with tetrafluoroborate anion. Environ Sci Pollut Res 22:11740–11754. https://doi.org/10.1007/s11356-015-4327-8
190. Liu T, Zhu LS, Wang JH, Wang J, Xie H (2015) The genotoxic and cytotoxic effects of 1-butyl-3-methylimidazolium chloride in soil on Vicia faba seedlings. J Hazard Mater 285:27–36. https://doi.org/10.1016/j.jhazmat.2014.11.028
191. Biczak R (2016) Quaternary ammonium salts with tetrafluoroborate anion: phytotoxicity and oxidative stress in terrestrial plants. J Hazard Mater 304:173–185. https://doi.org/10.1016/j.jhazmat.2015.10.055
192. Biczak R, Telesinski A, Pawlowska B (2016) Oxidative stress in spring barley and common radish exposed to quaternary ammonium salts with hexafluorophosphate anion. Plant Physiol Biochem 107:248–256. https://doi.org/10.1016/j.plaphy.2016.05.016
193. Biczak R, Pawlowska B, Telesinski A, Ciesielski W (2016) The effect of the number of alkyl substituents on imidazolium ionic liquids phytotoxicity and oxidative stress in spring barley and common radish seedlings. Chemosphere 165:519–528. https://doi.org/10.1016/j.chemosphere.2016.09.074
194. Liu T, Wang JH, Wang J, Zhu LS, Zhang J, Sun X (2016) Growth and physiological and biochemical responses of wheat seedlings to imidazolium-based ionic liquids 1-octyl-3-methylimidazolium chloride and 1-octyl-3-methylimidazolium bromide. Bull Environ Contam Toxicol 96:544–549. https://doi.org/10.1007/s00128-016-1747-1
195. Liu T, Zhu LS, Wang JH, Wang J, Tan MY (2016) Phytotoxicity of imidazolium-based ILs with different anions in soil on Vicia faba seedlings and the influence of anions on toxicity. Chemosphere 145:269–276. https://doi.org/10.1016/j.chemosphere.2015.11.055
196. Pawlowska B, Biczak R (2016) Evaluation of the effect of tetraethylammonium bromide and chloride on the growth and development of terrestrial plants. Chemosphere 149:24–33. https://doi.org/10.1016/j.chemosphere.2016.01.072
197. Biczak R, Pawlowska B, Platkowski M, Strek M, Telesinski A (2017) Effect of quaternary ammonium salts with fluorine atoms on selected weed species. Bull Environ Contam Toxicol 98:567–573. https://doi.org/10.1007/s00128-017-2033-6
198. Biczak R, Pawlowska B, Telesinski A, Kapusniak J (2017) Role of cation structure in the phytotoxicity of ionic liquids: growth inhibition and oxidative stress in spring barley and common radish. Environ Sci Pollut Res 24:18444–18457. https://doi.org/10.1007/s11356-017-9439-x
199. Biczak R, Snioszek M, Telesinski A, Pawlowska B (2017) Growth inhibition and efficiency of the antioxidant system in spring barley and common radish grown on soil polluted ionic liquids with iodide anions. Ecotoxicol Environ Saf 139:463–471. https://doi.org/10.1016/j.ecoenv.2017.02.016
200. Stepnowski P, Storoniak P (2005) Lipophilicity and metabolic route prediction of imidazollum ionic liquids. Environ Sci Pollut Res 12:199–204. https://doi.org/10.1065/espr2005.05.255
201. Luis P, Ortiz I, Aldaco R, Irabien A (2007) A novel group contribution method in the development of a QSAR for predicting the toxicity (Vibrio fischeri EC50) of ionic liquids. Ecotoxicol Environ Saf 67:423–429. https://doi.org/10.1016/j.ecoenv.2006.06.010
202. Lacrama AM, Putz MV, Ostafe V (2007) A spectral-SAR model for the anionic-cationic interaction in ionic liquids: application to Vibrio fischeri ecotoxicity. Int J Mol Sci 8:842–863. https://doi.org/10.3390/i8080842

203. Luis P, Garea A, Irabien A (2010) Quantitative structure–activity relationships (QSARs) to estimate ionic liquids ecotoxicity EC50 (Vibrio fischeri). J Mol Liq 152:28–33. https://doi.org/10.1016/j.molliq.2009.12.008
204. Alvarez-Guerra M, Irabien A (2011) Design of ionic liquids: an ecotoxicity (Vibrio fischeri) discrimination approach. Green Chem 13:1507–1516. https://doi.org/10.1039/c0gc00921k
205. Bruzzone S, Chiappe C, Focardi SE, Pretti C, Renzi M (2011) Theoretical descriptor for the correlation of aquatic toxicity of ionic liquids by quantitative structure–toxicity relationships. Chem Eng J 175:17–23. https://doi.org/10.1016/j.cej.2011.08.073
206. Sosnowska A, Barycki M, Zaborowska M, Rybinska A, Puzyn T (2014) Towards designing environmentally safe ionic liquids: the influence of the cation structure. Green Chem 16:4749–4757. https://doi.org/10.1039/c4gc00526k
207. Yan FY, Shang QY, Xia SQ, Wang Q, Ma PS (2015) Topological study on the toxicity of ionic liquids on Vibrio fischeri by the quantitative structure–activity relationship method. J Hazard Mater 286:410–415. https://doi.org/10.1016/j.jhazmat.2015.01.016
208. Ma SY, Lv M, Deng FF, Zhang XY, Zhai HL, Lv WJ (2015) Predicting the ecotoxicity of ionic liquids towards Vibrio fischeri using genetic function approximation and least squares support vector machine. J Hazard Mater 283:591–598. https://doi.org/10.1016/j.jhazmat.2014.10.011
209. Cho CW et al (2013) In silico modelling for predicting the cationic hydrophobicity and cytotoxicity of ionic liquids towards the leukemia rat cell line, Vibrio fischeri and Scenedesmus vacuolatus based on molecular interaction potentials of ions. SAR QSAR Environ Res 24:863–882. https://doi.org/10.1080/1062936x.2013.821092
210. Das RN, Roy K (2012) Development of classification and regression models for Vibrio fischeri toxicity of ionic liquids: green solvents for the future. Toxicol Res 1:186–195. https://doi.org/10.1039/c2tx20020a
211. Cho CW, Stolte S, Yun YS (2016) Comprehensive approach for predicting toxicological effects of ionic liquids on several biological systems using unified descriptors. Sci Rep 6:33403. https://doi.org/10.1038/srep33403
212. Grzonkowska M, Sosnowska A, Barycki M, Rybinska A, Puzyn T (2016) How the structure of ionic liquid affects its toxicity to Vibrio fischeri? Chemosphere 159:199–207. https://doi.org/10.1016/j.chemosphere.2016.06.004
213. Paterno A, Scire S, Musumarra G (2016) A QSPR approach to the ecotoxicity of ionic liquids (Vibrio fischeri) using VolSurf principal properties. Toxicol Res 5:1090–1096. https://doi.org/10.1039/c6tx00071a
214. Das RN, Sintra TE, Coutinho JAP, Ventura SPM, Roy K, Popelier PLA (2016) Development of predictive QSAR models for Vibrio fischeri toxicity of ionic liquids and their true external and experimental validation tests. Toxicol Res 5:1388–1399. https://doi.org/10.1039/c6tx00180g
215. Ben Ghanem O, Mutalib MIA, Leveque JM, El-Harbawi M (2017) Development of QSAR model to predict the ecotoxicity of Vibrio fischeri using COSMO-RS descriptors. Chemosphere 170:242–250. https://doi.org/10.1016/j.chemosphere.2016.12.003
216. Torrecilla JS, Garcia J, Rojo E, Rodriguez F (2009) Estimation of toxicity of ionic liquids in leukemia rat cell line and acetylcholinesterase enzyme by principal component analysis, neural networks and multiple lineal regressions. J Hazard Mater 164:182–194. https://doi.org/10.1016/j.jhazmat.2008.08.022
217. Yan F, Xia S, Wang Q, Ma P (2012) Predicting toxicity of ionic liquids in acetylcholinesterase enzyme by the quantitative structure–activity relationship method using topological indexes. J Chem Eng Data 57:2252–2257. https://doi.org/10.1021/je3002046
218. Das RN, Roy K (2014) Predictive in silico modeling of ionic liquids toward inhibition of the acetyl cholinesterase enzyme of electrophorus electricus: a predictive toxicology approach. Ind Eng Chem Res 53:1020–1032. https://doi.org/10.1021/ie403636q
219. Kurtanjek Z (2014) Chemometric versus random forest predictors of ionic liquid toxicity. Chem Biochem Eng Q 28:459–463

220. Peric B, Sierra J, Marti E, Cruanas R, Antonia Garau M (2015) Quantitative structure–activity relationship (QSAR) prediction of (eco)toxicity of short aliphatic protic ionic liquids. Ecotoxicol Environ Saf 115:257–262. https://doi.org/10.1016/j.ecoenv.2015.02.027
221. Basant N, Gupta S, Singh KP (2015) Predicting acetyl cholinesterase enzyme inhibition potential of ionic liquids using machine learning approaches: an aid to green chemicals designing. J Mol Liq 209:404–412. https://doi.org/10.1016/j.molliq.2015.06.001
222. Cho CW, Yun YS (2016) Interpretation of toxicological activity of ionic liquids to acetylcholinesterase inhibition via in silico modelling. Chemosphere 159:178–183. https://doi.org/10.1016/j.chemosphere.2016.06.005
223. Paterno A, Bocci G, Cruciani G, Fortuna CG, Goracci L, Scire S, Musumarra G (2016) Cyto- and enzyme toxicities of ionic liquids modelled on the basis of VolSurf+ descriptors and their principal properties. SAR QSAR Environ Res 27:221–244. https://doi.org/10.1080/1062936x.2016.1156571
224. Torrecilla JS, Palomar J, Lemus J, Rodriguez F (2010) A quantum-chemical-based guide to analyze/quantify the cytotoxicity of ionic liquids. Green Chem 12:123–134. https://doi.org/10.1039/b919806g
225. Fatemi MH, Izadiyan P (2011) Cytotoxicity estimation of ionic liquids based on their effective structural features. Chemosphere 84:553–563. https://doi.org/10.1016/j.chemosphere.2011.04.021
226. Yan F, Xia S, Wang Q, Ma P (2012) Predicting the toxicity of ionic liquids in leukemia rat cell line by the quantitative structure–activity relationship method using topological indexes. Ind Eng Chem Res 51:13897–13901. https://doi.org/10.1021/ie301764j
227. Zhao YS, Zhao JH, Huang Y, Zhou Q, Zhang XP, Zhang SJ (2014) Toxicity of ionic liquids: database and prediction via quantitative structure–activity relationship method. J Hazard Mater 278:320–329. https://doi.org/10.1016/j.jhazmat.2014.06.018
228. Das RN, Roy K, Popelier PLA (2015) Exploring simple, transparent, interpretable and predictive QSAR models for classification and quantitative prediction of rat toxicity of ionic liquids using OECD recommended guidelines. Chemosphere 139:163–173. https://doi.org/10.1016/j.chemosphere.2015.06.022
229. Gupta S, Basant N, Singh KP (2015) Nonlinear QSAR modeling for predicting cytotoxicity of ionic liquids in leukemia rat cell line: an aid to green chemicals designing. Environ Sci Pollut Res 22:12699–12710. https://doi.org/10.1007/s11356-015-4526-3
230. Sosnowska A, Grzonkowska M, Puzyn T (2017) Global versus local QSAR models for predicting ionic liquids toxicity against IPC-81 leukemia rat cell line: the predictive ability. J Mol Liq 231:333–340. https://doi.org/10.1016/j.molliq.2017.02.025
231. Salam MA, Abdullah B, Ramli A, Mujtaba IM (2016) Structural feature based computational approach of toxicity prediction of ionic liquids: cationic and anionic effects on ionic liquids toxicity. J Mol Liq 224:393–400. https://doi.org/10.1016/j.molliq.2016.09.120
232. Hossain MI, Samir BB, El-Harbawi M, Masri AN, Mutalib MIA, Hefter G, Yin CY (2011) Development of a novel mathematical model using a group contribution method for prediction of ionic liquid toxicities. Chemosphere 85:990–994. https://doi.org/10.1016/j.chemosphere.2011.06.088
233. Roy K, Das RN (2013) QSTR with extended topochemical atom (ETA) indices. 16. Development of predictive classification and regression models for toxicity of ionic liquids towards Daphnia magna. J Hazard Mater 254:166–178. https://doi.org/10.1016/j.jhazmat.2013.03.023
234. Roy K, Das RN, Popelier PLA (2014) Quantitative structure–activity relationship for toxicity of ionic liquids to Daphnia magna: aromaticity vs. lipophilicity. Chemosphere 112:120–127. https://doi.org/10.1016/j.chemosphere.2014.04.002
235. Das RN, Roy K, Popelier PLA (2015) Interspecies quantitative structure-toxicity-toxicity (QSTTR) relationship modeling of ionic liquids. Toxicity of ionic liquids to V. fischeri, D. magna and S. vacuolatus. Ecotoxicol Environ Saf 122:497–520. https://doi.org/10.1016/j.ecoenv.2015.09.014

236. Singh KP, Gupta S, Basant N (2014) Predicting toxicities of ionic liquids in multiple test species—an aid in designing green chemicals. RSC Adv 4:64443–64456. https://doi.org/10.1039/c4ra11252k
237. Cho CW, Yun YS (2016) Correlating toxicological effects of ionic liquids on Daphnia magna with in silico calculated linear free energy relationship descriptors. Chemosphere 152:207–213. https://doi.org/10.1016/j.chemosphere.2016.02.108
238. Cho CW, Park JS, Stolte S, Yun YS (2016) Modelling for antimicrobial activities of ionic liquids towards Escherichia coli, Staphylococcus aureus and Candida albicans using linear free energy relationship descriptors. J Hazard Mater 311:168–175. https://doi.org/10.1016/j.jhazmat.2016.03.006
239. Das RN, Roy K (2016) Computation of chromatographic lipophilicity parameter logk(0) of ionic liquid cations from "ETA" descriptors: application in modeling of toxicity of ionic liquids to pathogenic bacteria. J Mol Liq 216:754–763. https://doi.org/10.1016/j.molliq.2016.02.013
240. Hodyna D, Kovalishyn V, Rogalsky S, Blagodatnyi V, Petko K, Metelytsia L (2016) Antibacterial activity of imidazolium-based ionic liquids investigated by QSAR modeling and experimental studies. Chem Biol Drug Des 88:422–433. https://doi.org/10.1111/cbdd.12770
241. Abraham MH, Acree WE (2010) Solute descriptors for phenoxide anions and their use to establish correlations of rates of reaction of anions with iodomethane. J Org Chem 75:3021–3026. https://doi.org/10.1021/jo100292j
242. Cho CW, Stolte S, Yun YS, Krossing I, Thoming J (2015) In silico prediction of linear free energy relationship descriptors of neutral and ionic compounds. RSC Adv 5:80634–80642. https://doi.org/10.1039/c5ra13595h
243. Parr RG, Yang W (1989) Density-functional theory of atoms and molecules. Oxford University Press, New York
244. Klamt A, Schuurmann G (1993) COSMO—a new approach to dielectric screening in solvents with explicit expressions for the screening energy and its gradient. J Chem Soc 2:799–805. https://doi.org/10.1039/p29930000799
245. O'Boyle NM, Banck M, James CA, Morley C, Vandermeersch T, Hutchison GR (2011) Open Babel: an open chemical toolbox. J Cheminform 3:33. https://doi.org/10.1186/1758-2946-3-33
246. COSMOlogic (1989–2007) Turbomole V5.10. University of Karlsruhe, Karlsruhe
247. Roy K, Das RN, Ambure P, Aher RB (2016) Be aware of error measures. Further studies on validation of predictive QSAR models. Chemom Intell Lab Syst 152:18–33. https://doi.org/10.1016/j.chemolab.2016.01.008
248. Roy K, Mitra I, Kar S, Ojha PK, Das RN, Kabir H (2012) Comparative studies on some metrics for external validation of QSPR models. J Chem Inf Model 52:396–408. https://doi.org/10.1021/ci200520g
249. OECD (2006) Revised introduction to the OECD guidelines for testing of chemicals, Section 3. OECD Publishing, Paris
250. Wright KA, Cain RB (1972) Microbial metabolism of pyridinium compounds—radioisotope studies of metabolic fate of 4-carboxy-1-methylpyridinium chloride. Biochem J 128:561–568
251. Wright KA, Cain RB (1972) Microbial metabolism of pyridinium compounds—metabolism of 4-carboxy-1-methylpyridinium chloride, a photolytic product of paraquat. Biochem J 128:543–559
252. Kumar S, Ruth W, Sprenger B, Kragl U (2006) On the biodegradation of ionic liquids 1-butyl-3-methylimidazolium tetrafluoroborate. Chim Oggi 24:24–26
253. Docherty KM, Dixon JK, Kulpa CF (2007) Biodegradability of imidazolium and pyridinium ionic liquids by an activated sludge microbial community. Biodegradation 18:481–493. https://doi.org/10.1007/s10532-006-9081-7
254. Stolte S et al (2008) Primary biodegradation of ionic liquid cations, identification of degradation products of 1-methyl-3-octylimidazolium chloride and electrochemical wastewater

treatment of poorly biodegradable compounds. Green Chem 10:214–224. https://doi.org/10.1039/b713095c
255. Neumann J, Grundmann O, Thoming J, Schulte M, Stolte S (2010) Anaerobic biodegradability of ionic liquid cations under denitrifying conditions. Green Chem 12:620–627. https://doi.org/10.1039/b918453h
256. Gotvajn AZ, Tratar-Pirc E, Bukovec P, Plazl PZ (2014) Evaluation of biotreatability of ionic liquids in aerobic and anaerobic conditions. Water Sci Technol 70:698–704. https://doi.org/10.2166/wst.2014.283
257. Neumann J, Steudte S, Cho CW, Thoming J, Stolte S (2014) Biodegradability of 27 pyrrolidinium, morpholinium, piperidinium, imidazolium and pyridinium ionic liquid cations under aerobic conditions. Green Chem 16:2174–2184. https://doi.org/10.1039/c3gc41997e
258. Quijano G et al (2011) Toxicity and biodegradability of ionic liquids: new perspectives towards whole-cell biotechnological applications. Chem Eng J 174:27–32. https://doi.org/10.1016/j.cej.2011.07.055
259. Radosevic K, Bubalo MC, Srcek VG, Grgas D, Dragicevic TL, Redovnikovic IR (2015) Evaluation of toxicity and biodegradability of choline chloride based deep eutectic solvents. Ecotoxicol Environ Saf 112:46–53. https://doi.org/10.1016/j.ecoenv.2014.09.034
260. Gathergood N, Scammells PJ (2002) Design and preparation of room-temperature ionic liquids containing biodegradable side chains. Aust J Chem 55:557–560. https://doi.org/10.1071/ch02148
261. Gathergood N, Garcia MT, Scammells PJ (2004) Biodegradable ionic liquids: part I. Concept, preliminary targets and evaluation. Green Chem 6:166–175. https://doi.org/10.1039/b315270g
262. Liwarska-Bizukojc E, Gendaszewska D (2013) Removal of imidazolium ionic liquids by microbial associations: study of the biodegradability and kinetics. J Biosci Bioeng 115:71–75. https://doi.org/10.1016/j.jbiosc.2012.08.002
263. Gathergood N, Scammells PJ, Garcia MT (2006) Biodegradable ionic liquids—part III. The first readily biodegradable ionic liquids. Green Chem 8:156–160. https://doi.org/10.1039/b516206h
264. Harjani JR, Singer RD, Garcia MT, Scammells PJ (2008) The design and synthesis of biodegradable pyridinium ionic liquids. Green Chem 10:436–438. https://doi.org/10.1039/b800534f
265. Harjani JR, Singer RD, Garciac MT, Scammells PJ (2009) Biodegradable pyridinium ionic liquids: design, synthesis and evaluation. Green Chem 11:83–90. https://doi.org/10.1039/b811814k
266. Harjani JR, Farrell J, Garcia MT, Singer RD, Scammells PJ (2009) Further investigation of the biodegradability of imidazolium ionic liquids. Green Chem 11:821–829. https://doi.org/10.1039/b900787c
267. Morrissey S, Pegot B, Coleman D, Garcia MT, Ferguson D, Quilty B, Gathergood N (2009) Biodegradable, non-bactericidal oxygen-functionalised imidazolium esters: a step towards 'greener' ionic liquids. Green Chem 11:475–483. https://doi.org/10.1039/b812809j
268. Al-Mohammed NN, Hussen RSD, Ali TH, Alias Y, Abdullah Z (2015) Tetrakis-imidazolium and benzimidazolium ionic liquids: a new class of biodegradable surfactants. RSC Adv 5:21865–21876. https://doi.org/10.1039/c4ra16811a
269. Atefi F, Garcia MT, Singer RD, Scammells PJ (2009) Phosphonium ionic liquids: design, synthesis and evaluation of biodegradability. Green Chem 11:1595–1604. https://doi.org/10.1039/b913057h
270. Gore RG, Myles L, Spulak M, Beadham I, Garcia MT, Connon SJ, Gathergood N (2013) A new generation of aprotic yet Bronsted acidic imidazolium salts: effect of ester/amide groups in the C-2, C-4 and C-5 on antimicrobial toxicity and biodegradation. Green Chem 15:2747–2760. https://doi.org/10.1039/c3gc40992a

271. Yu YH, Lu XM, Zhou Q, Dong K, Yao HW, Zhang SJ (2008) Biodegradable naphthenic acid ionic liquids: synthesis, characterization, and quantitative structure–biodegradation relationship. Chem Eur J 14:11174–11182. https://doi.org/10.1002/chem.200800620
272. Klein R et al (2013) Biodegradability and cytotoxicity of choline soaps on human cell lines: effects of chain length and the cation. RSC Adv 3:23347–23354. https://doi.org/10.1039/c3ra42812e
273. Ferlin N et al (2013) Biomass derived ionic liquids: synthesis from natural organic acids, characterization, toxicity, biodegradation and use as solvents for catalytic hydrogenation processes. Tetrahedron 69:6150–6161. https://doi.org/10.1016/j.tet.2013.05.054
274. Boissou F et al (2014) Transition of cellulose crystalline structure in biodegradable mixtures of renewably-sourced levulinate alkyl ammonium ionic liquids, gamma-valerolactone and water. Green Chem 16:2463–2471. https://doi.org/10.1039/c3gc42396d
275. Ford L, Harjani JR, Atefi F, Garcia MT, Singer RD, Scammells PJ (2010) Further studies on the biodegradation of ionic liquids. Green Chem 12:1783–1789. https://doi.org/10.1039/c0gc00082e
276. Ford L, Ylijoki KEO, Garcia MT, Singer RD, Scammells PJ (2015) Nitrogen-containing ionic liquids: biodegradation studies and utility in base-mediated reactions. Aust J Chem 68:849–857. https://doi.org/10.1071/ch14499
277. Haiss A, Jordan A, Westphal J, Logunova E, Gathergood N, Kummerer K (2016) On the way to greener ionic liquids: identification of a fully mineralizable phenylalanine-based ionic liquid. Green Chem 18:4361–4373. https://doi.org/10.1039/c6gc00417b
278. Ferlin N et al (2013) Tetrabutylammonium prolinate-based ionic liquids: a combined asymmetric catalysis, antimicrobial toxicity and biodegradation assessment. RSC Adv 3:26241–26251. https://doi.org/10.1039/c3ra43785j
279. Liwarska-Bizukojc E, Maton C, Stevens C (2015) Biodegradation of imidazolium ionic liquids by activated sludge microorganisms. Biodegradation 26:453–463. https://doi.org/10.1007/s10532-015-9747-0
280. Liwarska-Bizukojc E, Maton C, Stevens CV, Gendaszewska D (2014) Biodegradability and kinetics of the removal of new peralkylated imidazolium ionic liquids. J Chem Technol Biotechnol 89:763–768. https://doi.org/10.1002/jctb.4187
281. Pham TPT, Cho CW, Jeon CO, Chung YJ, Lee MW, Yun YS (2009) Identification of metabolites involved in the biodegradation of the ionic liquid 1-butyl-3-methylpyridinium bromide by activated sludge microorganisms. Environ Sci Technol 43:516–521. https://doi.org/10.1021/es703004h
282. Docherty KM, Aiello SW, Buehler BK, Jones SE, Szymczyna BR, Walker KA (2015) Ionic liquid biodegradability depends on specific wastewater microbial consortia. Chemosphere 136:160–166. https://doi.org/10.1016/j.chemosphere.2015.05.016
283. Markiewicz M, Piszora M, Caicedo N, Jungnickel C, Stolte S (2013) Toxicity of ionic liquid cations and anions towards activated sewage sludge organisms from different sources—consequences for biodegradation testing and wastewater treatment plant operation. Water Res 47:2921–2928. https://doi.org/10.1016/j.watres.2013.02.055
284. Cho CW, Pham TPT, Kim S, Song MH, Chung YJ, Yun YS (2016) Three degradation pathways of 1-octyl-3-methylimidazolium cation by activated sludge from wastewater treatment process. Water Res 90:294–300. https://doi.org/10.1016/j.watres.2015.11.065
285. Niemczak M, Chrzanowski L, Praczyk T, Pernak J (2017) Biodegradable herbicidal ionic liquids based on synthetic auxins and analogues of betaine. New J Chem 41:8066–8077. https://doi.org/10.1039/c7nj01474k
286. Lawniczak L, Materna K, Framski G, Szulc A, Syguda A (2015) Comparative study on the biodegradability of morpholinium herbicidal ionic liquids. Biodegradation 26:327–340. https://doi.org/10.1007/s10532-015-9737-2
287. Modelli A, Sali A, Galletti P, Samori C (2008) Biodegradation of oxygenated and non-oxygenated imidazolium-based ionic liquid in soil. Chemosphere 73:1332–1327. https://doi.org/10.1016/j.chemosphere.2008.07.012

288. Deive FJ et al (2011) Impact of ionic liquids on extreme microbial biotypes from soil. Green Chem 13:687–696. https://doi.org/10.1039/c0gc00369g
289. Garbaczewska S, Hupko J (2007) Determination of ionic liquids by HPLC method. Involvement in biodegradation test. Pestycydy 3-4:61–66
290. Sydow M et al (2015) Persistence of selected ammonium- and phosphonium-based ionic liquids in urban park soil microcosms. Int Biodeterior Biodegrad 103:91–96. https://doi.org/10.1016/j.ibiod.2015.04.019
291. Esquivel-Viveros A, Ponce-Vargas F, Esponda-Aguilar P, Prado-Barragan LA, Gutierrez-Rojas M, Lye GJ, Huerta-Ochoa S (2009) Biodegradation of bmim PF6 using Fusarium sp. Rev Mex Ing Quim 8:163–168
292. Abrusci C, Palomar J, Pablos JL, Rodriguez F, Catalina F (2011) Efficient biodegradation of common ionic liquids by Sphingomonas paucimobilis bacterium. Green Chem 13:709–717. https://doi.org/10.1039/c0gc00766h
293. Zhang C, Wang H, Malhotra SV, Dodge CJ, Francis AJ (2010) Biodegradation of pyridinium-based ionic liquids by an axenic culture of soil Corynebacteria. Green Chem 12:851–858. https://doi.org/10.1039/b924264c
294. Markiewicz M, Henke J, Brillowska-Dabrowska A, Stolte S, Luczak J, Jungnickel C (2014) Bacterial consortium and axenic cultures isolated from activated sewage sludge for biodegradation of imidazolium-based ionic liquid. Int J Environ Sci Technol 11:1919–1926. https://doi.org/10.1007/s13762-013-0390-1
295. Deng Y et al (2015) When can ionic liquids be considered readily biodegradable? Biodegradation pathways of pyridinium, pyrrolidinium and ammonium-based ionic liquids. Green Chem 17:1479–1491. https://doi.org/10.1039/c4gc01904k
296. Thamke VR, Kodam KM (2016) Toxicity study of ionic liquid, 1-butyl-3-methylimidazolium bromide on guppy fish, Poecilia reticulata and its biodegradation by soil bacterium Rhodococcus hoagii VRT1. J Hazard Mater 320:408–416. https://doi.org/10.1016/j.jhazmat.2016.08.056
297. Crosthwaite JM, Ropel LJ, Anthony JL, Aki S, Maginn EJ, Brennecke JF (2005) Phase equilibria with gases and liquids of 1-n-butyl-3-methylimidazolium bis (trifluoromethylsulfonyl)imide. In: Rogers RD, Seddon KR (eds) Ionic liquids III A: fundamentals, progress, challenges, and opportunities, properties and structure. ACS Symposium Series, vol 901, pp 292–300
298. Deng Y et al (2012) Relevant parameters for assessing the environmental impact of some pyridinium, ammonium and pyrrolidinium based ionic liquids. Chemosphere 89:327–333. https://doi.org/10.1016/j.chemosphere.2012.04.050
299. Cho CW et al (2011) Ionic liquids: predictions of physicochemical properties with experimental and/or DFT-calculated LFER parameters to understand molecular interactions in solution. J Phys Chem B 115:6040–6050. https://doi.org/10.1021/jp200042f
300. Cho CW, Stolte S, Ranke J, Preiss U, Krossing I, Thoming J (2014) Quantitative analysis of molecular interaction potentials of ionic liquid anions using multi-functionalized stationary phases in HPLC. ChemPhysChem 15:2351–2358. https://doi.org/10.1002/cphc.201402092
301. Cho CW, Song MH, Yun YS (2017) Comment on "Filling environmental data gaps with QSPR for ionic liquids: modeling n-octanol/water coefficient". J Hazard Mater 329:348–350. https://doi.org/10.1016/j.jhazmat.2016.10.060
302. Cho CW, Stolte S, Yun YS (2018) Validation and updating of QSAR models for partitioning coeffcieints of ionic liquids in octanol–water and development of a new LFER model. Sci Total Environ 633:920–928. https://doi.org/10.1016/j.scitotenv.2018.03.225
303. Rybinska A, Sosnowska A, Grzonkowska M, Barycki M, Puzyn T (2016) Filling environmental data gaps with QSPR for ionic liquids: modeling n-octanol/water coefficient. J Hazard Mater 303:137–144. https://doi.org/10.1016/j.jhazmat.2015.10.023
304. Cornmell R, Winder CL, Tiddy GJ, Coodacre R, Stephens G (2008) Accumulation of ionic liquids in Escherchia coli. Green Chem 10:836–841. https://doi.org/10.1039/B807214K

305. Nedzi M, Latala A, Nichthauser J, Stepnowski P (2013) Bioaccumulation of 1-butyl-3-methylimidazolium chloride ionic liquid in a simple marine trophic chain. Oceanol Hydrobiol Stud 42:149–154. https://doi.org/10.2478/s13545-013-0068-9
306. Dolzonek J, Cho CW, Stepnowski P, Markiewicz M, Thoming J, Stolte S (2017) Membrane partitioning of ionic liquid cations, anions and ion pairs—estimating the bioconcentration potential of organic ions. Environ Pollut 228:378–389. https://doi.org/10.1016/j.envpol.2017.04.079
307. Gorman-Lewis DJ, Fein JB (2004) Experimental study of the adsorption of an ionic liquid onto bacterial and mineral surfaces. Environ Sci Technol 38:2491–2495. https://doi.org/10.1021/es0350841
308. Stepnowski P (2005) Preliminary assessment of the sorption of some alkyl imidazolium cations as used in ionic liquids to soils and sediments. Aust J Chem 58:170–173. https://doi.org/10.1071/ch05018
309. Stepnowski P, Mrozik W, Nichthauser J (2007) Adsorption of alkylimidazolium and alkylpyridinium ionic liquids onto natural soils. Environ Sci Technol 41:511–516. https://doi.org/10.1021/es062014w
310. Mrozik W, Jungnickel C, Ciborowski T, Pitner WR, Kumirska J, Kaczynski Z, Stepnowski P (2009) Predicting mobility of alkylimidazolium ionic liquids in soils. J Soils Sediments 9:237–245. https://doi.org/10.1007/s11368-009-0057-1
311. Mrozik W, Kotlowska A, Kamysz W, Stepnowski P (2012) Sorption of ionic liquids onto soils: experimental and chemometric studies. Chemosphere 88:1202–1207. https://doi.org/10.1016/j.chemosphere.2012.03.070
312. Reinert L, Batouche K, Leveque J-M, Muller F, Beny J-M, Kebabi B, Duclaux L (2012) Adsorption of imidazolium and pyridinium ionic liquids onto montmorillonite: characterisation and thermodynamic calculations. Chem Eng J 209:13–19. https://doi.org/10.1016/j.cej.2012.07.128
313. Greenland DJ, Quirk JP (1962) Adsorption on 1-n-alkylpyridinium bromides by montmorillonites. Proceedings of the Ninth National Conference on Clays and Clay Minerals. Pergamon, New York, pp 484–499
314. Markiewicz M, Jungnickel C, Markowska A, Szczepaniak U, Paszkiewicz M, Hupka J (2009) 1-methyl-3-octylimidazolium chloride-sorption and primary biodegradation analysis in activated sewage sludge. Molecules 14:4396–4405. https://doi.org/10.3390/molecules14114396
315. Markiewicz M, Stolte S, Lustig Z, Luczak J, Skup M, Hupka J, Jungnickel C (2011) Influence of microbial adaption and supplementation of nutrients on the biodegradation of ionic liquids in sewage sludge treatment processes. J Hazard Mater 195:378–382. https://doi.org/10.1016/j.jhazmat.2011.08.053
316. Alvarez MS, Rodriguez A, Sanroman MA, Deive FJ (2015) Microbial adaptation to ionic liquids. RSC Adv 5:17379–17382. https://doi.org/10.1039/c4ra10283e
317. Gendaszewska D, Liwarska-Bizukojc E (2016) Adaptation of microbial communities in activated sludge to 1-decyl-3-methylimidazolium bromide. Water Sci Technol 74:1227–1234. https://doi.org/10.2166/wst.2016.317
318. Anthony JL, Maginn EJ, Brennecke JF (2001) Solution thermodynamics of imidazolium-based ionic liquids and water. J Phys Chem B 105:10942–10949. https://doi.org/10.1021/jp0112368
319. Palomar J, Lemus J, Gilarranz MA, Rodriguez JJ (2009) Adsorption of ionic liquids from aqueous effluents by activated carbon. Carbon 47:1846–1856. https://doi.org/10.1016/j.carbon.2009.03.028
320. Farooq A, Reinert L, Leveque JM, Papaiconomou N, Irfan N, Duclaux L (2012) Adsorption of ionic liquids onto activated carbons: effect of pH and temperature. Microporous Mesoporous Mater 158:55–63. https://doi.org/10.1016/j.micromeso.2012.03.008
321. Hassan S, Duclaux L, Leveque JM, Reinert L, Farooq A, Yasin T (2014) Effect of cation type, alkyl chain length, adsorbate size on adsorption kinetics and isotherms of bromide ionic liquids

from aqueous solutions onto microporous fabric and granulated activated carbons. J Environ Manag 144:108–117. https://doi.org/10.1016/j.jenvman.2014.05.005
322. Lemus J, Palomar J, Heras F, Gilarranz MA, Rodriguez JJ (2012) Developing criteria for the recovery of ionic liquids from aqueous phase by adsorption with activated carbon. Sep Purif Technol 97:11–19. https://doi.org/10.1016/j.seppur.2012.02.027
323. Lemus J, Neves C, Marques CFC, Freire MG, Coutinho JAP, Palomar J (2013) Composition and structural effects on the adsorption of ionic liquids onto activated carbon. Environ Sci 15:1752–1759. https://doi.org/10.1039/c3em00230f
324. Lemus J, Palomar J, Gilarranz MA, Rodriguez JJ (2013) On the kinetics of ionic liquid adsorption onto activated carbons from aqueous solution. Ind Eng Chem Res 52:2969–2976. https://doi.org/10.1021/ie3028729
325. Qi XH, Li LY, Tan TF, Chen WT, Smith RL (2013) Adsorption of 1-butyl-3-methylimidazolium chloride ionic liquid by functional carbon microspheres from hydrothermal carbonization of cellulose. Environ Sci Technol 47:2792–2798. https://doi.org/10.1021/es304873t
326. Qi XH, Li LY, Wang Y, Liu N, Smith RL (2014) Removal of hydrophilic ionic liquids from aqueous solutions by adsorption onto high surface area oxygenated carbonaceous material. Chem Eng J 256:407–414. https://doi.org/10.1016/j.cej.2014.07.020
327. Xu M, Ao YY, Wang SJ, Peng J, Li JQ, Zhai ML (2015) Efficient adsorption of 1-alkyl-3-methylimidazolium chloride ionic liquids onto modified cellulose microspheres. Carbohydr Polym 128:171–178. https://doi.org/10.1016/j.carbpol.2015.04.018
328. Yu F et al (2016) Effective removal of ionic liquid using modified biochar and its biological effects. J Taiwan Inst Chem Eng 67:318–324. https://doi.org/10.1016/j.jtice.2016.07.038
329. Shi KS, Qiu YP, Li B, Stenstrom MK (2016) Effectiveness and potential of straw- and wood-based biochars for adsorption of imidazolium-type ionic liquids. Ecotoxicol Environ Saf 130:155–162. https://doi.org/10.1016/j.ecoenv.2016.04.017
330. Choi SB, Won SW, Yun YS (2013) Use of ion-exchange resins for the adsorption of the cationic part of ionic liquid, 1-ethyl-3-methylimidazolium. Chem Eng J 214:78–82. https://doi.org/10.1016/j.cej.2012.10.035
331. Li LY, Wang Y, Qi XH (2015) Adsorption of imidazolium-based ionic liquids with different chemical structures onto various resins from aqueous solutions. RSC Adv 5:41352–41358. https://doi.org/10.1039/c5ra04191k
332. Won SW, Choi SB, Mao J, Yun YS (2013) Removal of 1-ethyl-3-methylimidazolium cations with bacterial biosorbents from aqueous media. J Hazard Mater 244:130–134. https://doi.org/10.1016/j.jhazmat.2012.11.018

Index

X

FSC
www.fsc.org
MIX
Papier | Fördert gute Waldnutzung
FSC® C083411

Kolibri 360 GmbH
Ferdinand-Jühlke-Straße 7
99095 Erfurt, Deutschland
produktsicherheit@kolibri360.de